# Multivariate Analysis of Quality
# An Introduction

# Multivariate Analysis of Quality

## An Introduction

**Harald Martens**
*Norwegian University of Science and Technology, and*
*Technical University of Denmark*

**Magni Martens**
*The Royal Veterinary & Agricultural University, Denmark*

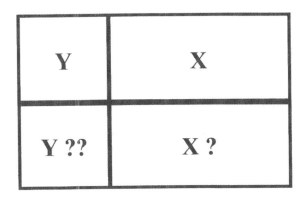

JOHN WILEY & SONS, LTD
Chichester • New York • Weinheim • Brisbane • Singapore • Toronto

*Other Wiley Editorial Offices*
New York, Weinheim, Brisbane, Singapore, Toronto

*Library of Congress Cataloging-in-Publication Data*
Martens, Harald.
   Introduction to multivariate data analysis for understanding quality / Harald Martens, Magni Martens.
      p. cm.
   Includes bibliographical references and index.
   ISBN 0-471-97428-5 (acid-free paper)
      1. Multivariate analysis. I. Martens, M. (Magni) II. Title.

QA278.M373 2000
519.5'35--dc21

00-043342

*British Library Cataloguing in Publication Data*

A catalogue record for this book is available from the British Library

ISBN 0 471 97428 5

Typeset in 10/12pt Times by Deerpark Publishing Services, Shannon, Ireland
Printed and bound in Great Britain Antony Rowe Ltd, Eastbourne
This book is printed on acid-free paper responsibly manufactured from sustainable forestry, in which at least two trees are planted for each one used for paper production.

# Preface

In 1972, while working as young biochemists at Makerere University in Uganda, both of us gained formative experiences, which resulted in this book.

People would rather die from malnutrition than eat what they do not *like*; our protein-enriched biscuits did not go down well. Perceived quality, complex as it is, must be considered by science. Since that time Magni Martens has worked in fields where hard and soft sciences meet and where the concept of quality is central. This is fertile ground, although sometimes muddy.

Researchers go selectively blind if they have to look at *too much data* at one time. To our frustration, the plant breeders then working in Uganda chose to ignore most of our painstakingly measured amino acid data of local, protein-rich millets. However, measured data reflect a latent reality – we knew that our amino acid measurements reflected the proteome – the protein patterns, which in turn reflect the DNA genome and the growth conditions. So why couldn't we use mathematics to extract this hidden, but decisive information from our mass of raw data? Without training in mathematics, Harald Martens initially learnt a lot about why not. But eventually, multivariate data analysis based on soft modelling in latent variables did the job. And with that, his career as a biochemist ended, and his role as full-time data analyst began.

Since then, each of us has worked in several different research fields. All along, we have shared our experiences with each other. Now, 28 years and several hundred publications later, we have seen that many basic problems are the same, across scientific barriers. One problem is how to ensure both human relevance and scientific rationale and innovation in research, as evidenced by the confusing uses of the word "Quality". Another problem concerns "Information". How the researchers can analyse real-world systems where the raw data required are too complex for the mind to grasp. Both topics call for multivariate data analysis. The present book is an attempt at sharing some of our many late Sunday morning discussions with other researchers.

Given the explosive increase in the amount of data available in research, there are simply not enough professional data analysts around. That leaves many data sets without proper analysis, and this wasteful situation is only getting worse. We have therefore chosen to stimulate researchers to analyse their *own* data.

In this introduction to the analysis of real-world data, we have selected just one single modelling approach, to minimise the mathematical or statistical burden. We use a narrative style of presentation, with many illustrations and application examples. More mathematically-oriented readers who consider this style too verbose may find some of the desired formalism in the more terse technical appendices, and the rest in the references.

Copenhagen, October 2001

Magni Martens                    Harald Martens

Photo (of Magni and Harald): Lars Munck.

# Acknowledgements

Over the years, a number of people have supported and contributed to our development of what is now collected in this book. We can only name some of them here. First and foremost we want to thank our colleague and friend Lars Munck (Royal Veterinary and Agricultural University, Copenhagen) for 25 years of cooperation and inspiration.

Also, we feel deeply indebted to the Norwegian Food Research Institute (Matforsk), which allowed both of us to play seriously for more than a decade. Here, we learnt to take inspiration in applied projects and yet to search for basic knowledge. Our many colleagues, in particular Einar Risvik and Tormod Næs, are thanked for cooperation, and our superiors Anton Skulberg, Hellmut Russwurm Jr. and Kjell Ivar Hildrum are thanked for their patience.

From our earliest years of data analysis we want to thank Terje Hertzberg at the Norwegian University of Science and Technology for his introduction to mathematical modelling. We thank Yasumoto Kyoden (then at Kyoto U.) for moral support when the project was in the morass, and Finn Tschudi at U. of Oslo for getting it on dry ground again. Forrest Young (U. of North Carolina), Joseph Kruskal (Bell Labs), Edmund Malinowski (Stevens Inst. Techn. Hoboken, NJ), Emil Spjøtvoll (Agricultural U. of Norway) and Rolf Volden (Norwegian Computing Center) provided further data analytical inspiration and insight.

Our data analysis development pace increased after having met chemometrician Svante Wold (Umeå U., Sweden) and statistician Herman Wold (†) in the beginning of the 1980s; we are grateful for their generous sharing of ideas with us. Chemometrics colleagues Bruce Kowalski (U. of Washington), Paul Geladi (Umeå U.), Kim Esbensen (Telemark U. College, Norway) and Frank Westad (Matforsk), as well as NIR experts Karl Norris (USDA) and Edward Stark (KES Analysis, NY), are also thanked for their cooperation and friendship over many years.

The sensory scientists Rose Marie Pangborn (†), Howard G. Schutz and Ann Noble (U. California Davis) are likewise thanked for two decades of support and inspiration, and so is Erik von Sydow (SIK, Gothenburg).

More recently, Klaus Diepold and his colleagues in Dynapel GmbH (Munich),

Rasmus Bro, Lars Nørgaard and Per B. Brockhoff (Royal Veterinary and Agricultural University), Martin Høy, Endre Anderssen and Kay Steen, (Norwegian U. of Science and Technology), and Jens Frisvad, Henrik Spliid and Agnar Höskuldsson (Technical U. of Denmark) have provided fruitful data analytical discussions. Wender Bredie, Garmt Dijksterhuis, Per Møller, Ep Köster and the other colleagues and students in the Sensory Science research group at the Royal Veterinary and Agricultural University provide constant inspiration concerning human perception and cognition.

We want to thank Peik Jenssen (Det Norske Veritas, Oslo), Rasmus Bro, Silje Martens and Johannes Martens for constructive comments on the book manuscript. Our present colleagues and students at the Royal Veterinary and Agricultural U., Copenhagen, the Norwegian U. of Science and Technology and the Technical U. of Denmark are thanked for accepting our prolonged absentmindedness while writing.

Nycomed Amersham Imaging AS and the Danish Government (FØTEK programme) are thanked for financial support.

We also wish to thank Jens Strøm for his drawing of the cartoons.

Dedicated to our beloved children, Silje and Johannes.

# Contents

# PART ONE: OVERVIEW
# (Chapters 1–3)

The book explains a practical approach to data analysis in complex systems: soft modelling. The method is illustrated by a wide range of research applications concerning the assessment of quality.

This first part of the book introduces basic concepts about quality, information and validity, and provides a layman's guide to multivariate data analysis.

Chapter 1 gives an overview of the content of the book, with an example of why multivariate data analysis should be chosen; addressing fundamental aspects of causality and selectivity.

Chapter 2 motivates the reader to work diligently with goal formulation, and criteria for reliable and relevant information. In particular, it discusses the concept of quality and the art of quantitative measurements.

Chapter 3 takes the reader through a cycle in the research process, from goal formulation via experiments to data analysis and presentation of conclusions. It outlines the bi-linear modelling (BLM) method, and shows a range of data analytical problems that may be solved by this method. It also demonstrates some useful ways to plot data.

The corresponding technical appendices in Part Four cover the following:

Appendix A1 puts the soft modelling approach to data analysis into a wider scientific modelling perspective.

Appendix A2 describes sensory science as an example of a field where quality is measured in a reproducible manner, yet with human relevance.

Appendix A3 gives an overview of the many different statistical methods that may be replaced by the single data analytical method in this book – the cross-validated BLM. It also outlines some common pitfalls in this data analytical approach.

# Chapter 1

# Why Multivariate Data Analysis?

*This book gives some basic concepts and tools for exploring, understanding and acting upon real-world complexity, particularly in relation to the extraction of information about quality. A powerful, versatile and reasonably safe way to analyse data, based on multivariate soft modelling, is presented. Data analysis is seen as a cognitive discipline, not as a question of difficult mathematics.*

*This introductory chapter focuses on motivation and guidelines for using the book, and gives a small example of multivariate data analysis for understanding quality.*

## 1.1. PURPOSE OF THE BOOK

'But the emperor has no clothes on!' said the little boy in HC Andersen's fairy tale (author Hans Christian Andersen, born 1805). Researchers, who take the little boy's keen, frank and courageous attitude may see the world a little more clearly. The present book provides a data analysis method that may help the reader in that direction.

Multivariate data analysis is only a tool, but a useful one. It allows us to move efficiently from problem to solution, from questions and intentions to reliable answers and relevant new understanding. It is important to work in the *context* of prior knowledge, but equally important to analyse data without being *locked* into our cage of prior conventions, theories and expectations.

There are many different data analytical traditions (Appendix A1), ranging from the traditional *hard* first-principles modelling in physics, via the statistical testing taught to all university students, to the *black box* neural nets modelling in informatics. They differ strongly in how they use *prior knowledge* (as a cage of accepted understanding, as a hypothesis, or not at all). They also differ in how the results are *validated*, and in how transparent this validation is for the user.

This book is an attempt at providing a graphically oriented tool for data analysis, which is compact, coherent and relatively transparent to the user. The tool makes use of prior knowledge for planning and interpretation, but lets the data speak for themselves during the actual mathematical modelling, although with strict validity requirements.

### 1.1.1. Multivariate Analysis of a Multivariate World

The physical world may be seen as self-organising structures emerging in complex, non-linear dynamic systems. Our cultures and our minds may be understood likewise. But to model all of this complexity is beyond our grasp.

Realising that the real world is multifacetted, and when addressing a new problem, we cannot know beforehand precisely what to measure. That is particularly so when dealing with the complex issue of measuring *quality*.

Fortunately, modern data-generating techniques provide a multitude of variables for a multitude of samples, with little effort. This is true for questionnaires in marketing, DNA fragment fingerprints in biotechnology, spectra in chemistry and physics, images in signal processing and medicine and process analysers in engineering.

All of these registrations and measurements may be collected into massive *data tables*. Computerised data analysis then allows us to study these data tables. But the human cognitive capacity is limited. Therefore, the data analysis must be able to bring out the reliable and relevant information for us, without loss of essential dimensionality, while not overwhelming us.

Multivariate data analysis is (a) the analysis of data with *many* observed variables (i.e. large data tables), as well as (b) the study of systems with *many* important types of variation.

In some texts, *multivariate* refers to data with many variables, while *multivariable* refers to a system that varies in many ways. Here, we use *different variation phenomena* or *several types of variation* to express the multivariable character of a system.

*Multivariate data analysis* has two distinct, but inter-related and equally important phases: *planning* and *modelling*. Disciplined experimental planning allows us to ask cost-effective and pertinent questions to the real world. Creative and critical modelling allows us to interpret the real world's rather confusing answers in more *objective* or intersubjective terms.

### 1.1.2. Why This Book?

The present book has been written to meet a need for an entry-level textbook on multivariate data analysis for the practical researcher in academia or industry. It is intended to explain and illustrate some ways in which the researcher can acquire improved understanding and make better decisions.

This may concern a scientific need for new fundamental knowledge, a practical need for solving an engineering or instrumentation problem, or it may concern gaining overview of the important types of quality variation in a set of products, samples, processes, services or organisations.

Being introductory, the methodology in the book is relatively simple. But it is better to use a simple method correctly and feel confident, than to misuse advanced methods and feel alienated. Or, as John Tukey said: 'It is better to be approximately right than precisely wrong' (Tukey, 1962).

### 1.1.3. One Approach to Multivariate Data Analysis

There is no single *best way* to analyse data; many different approaches will work, as long as they are used correctly. So the question for the researcher who does not want to spend too much time on data analysis, is to choose one versatile approach, and learn to use it well.

The present book will present one candidate for such a single approach to multivariate data analysis. This approach has, over the years, proven to be powerful and versatile in several fields. To make the approach complete and self-contained, it has been supplemented with several statistical details specifically developed for this book, concerning model validation and power assessment.

◻ CALIBRATION AND PREDICTION

The title page of this book symbolises its purpose and content. The upper half represents two multivariate data tables, **Y** and **X** for a *calibration set* of samples, obtained after conscious planning and diligent work. As the examples in the book will show, the input variables used as **X** and **Y** can be of widely different nature. We would like to reveal and interpret the reliable and relevant patterns within and between these X- and Y-data.

Frequently, we also want to use the available set of calibration samples to develop a *model* that predicts the variables in **Y** from the variables in **X**: $\mathbf{Y} \approx f(\mathbf{X})$. Then, as illustrated by the lower half of the title page, we would like to apply this model to a new set of *prediction* samples: we want to check that the data in the X-variables in the new samples resemble the previous ones, and if so, to use the model to predict the unknown values of their Y-variables from their X-variables.

As this book will show, the four data tables, **X** and **Y** for the calibration samples and for the new prediction samples, allow a broad range of data analytical problems to be solved, for an even broader range of input data types.

◻ SOFT MODELLING BY BLM

The present approach to multivariate data analysis is called *soft modelling*. This name alludes to the fact that the mathematical modelling works more or less automatically, which is simple for the user, and does not require *hard, a priori* assumptions. The soft modelling approach is based on conscious, but relaxed experimental planning, followed by interactive data analytical modelling by a particular method.

This method is the so-called *bi-linear modelling* (*BLM*). It is based on one single tool for mathematical extraction of relevant information from the input data, combined with extensive *graphics* and so-called *cross-validation*.

The BLM method is built on a combination of multivariate bi-linear series expansion and statistical regression theory. For simplicity, only relevant covariance eigenvalues are extracted. For robustness, only large eigenvalues with predictive ability are used in the final models, which are validated by graphical interpretation and cross-validation/jack-knifing.

With BLM, the soft modelling provides informative model *maps* that can be interpreted in light of background knowledge, with an open eye for the unexpected. The statistical validation guards against wishful thinking, and automatic outlier warnings alert the researcher to anomalies that could otherwise have been overlooked.

❑ QUALITY CRITERIA FOR BLM

Details on the BLM method will be given in Chapters 3–11. The main reasons for why the reader might want to use the BLM method are listed below. One particular property will be outlined in some more detail: the ability of BLM to turning the traditional *collinearity problem* into a stabilising advantage.

**The BLM method may be chosen because of its ability to:**
1. Work for both *soft* human data, *hard* technological data and design data
2. Work for both explorative and confirmative data analysis
3. Reveal the dominant structures in one data table **X**
4. Reveal the relevant structures within and between two data tables **X** and **Y**
5. Predict one set of variables **Y** from another set of variables **X**, even in new samples
6. Work for classification/discrimination issues
7. Handle cases with far more X-variables than samples, in contrast to many traditional methods
8. Allow collinear redundancies in the input data and actively utilise these in the modelling
9. Allow errors in both **X** and **Y**
10. Allow different error levels in different input variables in **X** and in **Y**
12. Handle a few missing values in a simple and robust way
13. Provide compact graphical overviews and statistical details, as chosen by the user
14. Give understandable statistical assessment of the validity of the model: predictive ability and model parameter stability
15. Be compatible with contemporary statistics, but incorporate experiences from other fields
16. Give automatic warnings for outliers and gross mistakes
17. Be available in several commercial program packages with documentation and support

❑ USING THE MULTIVARIATE CO-VARIATIONS

To have a certain excess of samples and variables in the input data should be an advantage, because it should counteract input noise and allow contextually meaningful interpretation. The BLM is particularly suited for utilising such redundancy in data tables.

When combining a number of potentially relevant input variables, it is only natural that *some of these variables will be more or less inter-correlated.* Contrary to some traditional statistical methods, the BLM method treats this *collinearity*, caused by the tendencies of co-variation between variables, as an advantage that stabilises the modelling and facilitates subsequent graphical model inspection.

Likewise, when observing a number of related samples, chosen by random sampling or by more strict experimental design, it is only natural that *some of these samples will be more or less similar to each other.* This is very valuable. In fact, one of the main purposes of the planning phase is to ensure that every important type of variation is spanned clearly by *two or more* samples. This stabilises the modelling. Furthermore, it enables the results to be *cross-validated* within the available data set, to guard against being fooled by noise and mistakes in the input data. The use of cross-validation more or less eliminates the need for alienating theoretical assumptions about degrees of freedom and error distributions. Finally, this redundancy also enables us to distinguish systematic variation phenomena from unexpected outliers (unique samples). Outliers can then be respected and handled in a proper way.

### 1.1.4. The Examples

In this book, the soft modelling approach will be explained, for several data analytical problems and for many different types of input data from a range of research situations (cf. overview of all the examples in Chapter 3.1). The reason for this wide presentation is two-fold:

1. Even within one single branch of research and within one laboratory, organisation or company, there are usually many different types of data analytical problems to be solved.
2. Although every data set is unique and should be thus treated, some fundamental data analytical problems are surprisingly common, across scientific barriers.

These common questions concern how to get informative input data, how to gain relevant overviews of the input data, how to minimise the effect of errors and identify mistakes, how to view and interpret the observed effects and phenomena, and how to assess the reliability of our conclusions.

The data analytical problems in this book thus range from *soft* exploration of new, interdisciplinary research areas, via the calibration of modern electronic instruments, to *hard* causal assessment of designed experiments. The types of input data range from linguistic, psychosocial and sensory studies, via economically important quality descriptors in microbiology and food technology, to quantum chemistry computations and to quality monitoring in the process industry. This broad range of examples is intended to demonstrate how different data analytical problems can be analysed by one and the same method.

The reader is recommended to focus on the examples that appear most relevant. But the reader is also invited to try to follow the examples that may appear more remote. Each example explains or demonstrates useful ways to look at empirical data. The most *remote* examples may in fact create new ideas for the reader's own research.

### 1.1.5. Data Analysis is a *Cognitive* Discipline, Not a *Mathematical* Exercise

In research it is important to simplify the problems to make them manageable. But it is also important not to over-simplify and thus come up with irrelevant or unreliable results. Real-world problems are usually multifacetted and cannot be described by one selected variable without serious misrepresentation. But multivariate data tables create their own problems. It is easy to get lost in large tables of raw data, and in panic look only at the few data that one already feels comfortable with!

The balancing act between the overwhelming and oversimplified is a challenge in research. Figure 1.1 illustrates how the researcher is blessed and cursed with the ability to take in a number of different sensory signals from the external world, some of which can be useful, and others not. He or she is also blessed and cursed with a variety of internal memories and prior expectations, some of which can be useful, and others not. After some mental information processing, in which the external and the internal stimuli are more or less consciously brought together, the researcher may come up with an understanding of the relationships in the world observed: sometimes this understanding may be correct, and at other times not.

It seems important to realise that the individual researcher's mind has no direct access to the real world itself; the world is under indirect observation (Jöreskog and Wold, 1982). This will be discussed in more detail in Chapter 2.1. Not all scientists and technologists seem to accept that their own brain is the central resource, but also the most important limitation, in their work. The cognitive aspects of the research process play an important role and will be addressed throughout the book. Three of the examples even concern input data of purely cognitive nature.

This book regards active data analysis as a cognitive discipline, not a mathematical exam or a computer game. The method and algorithm development is a technical question that has already been solved by specialists. With modern commercial software the difficult mathematical technicalities have already

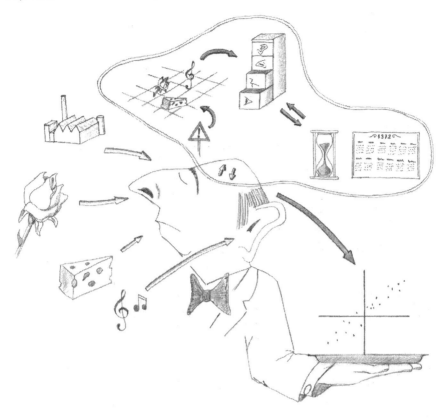

**Figure 1.1.** It all comes together in the brain whether we like it or not, our senses and minds are part of our research process. This allows flexibility and creativity, but calls for checks against wishful thinking. Illustration by Jens Strøm.

been implemented, and hidden inside software programs, like the motor is hidden under the hood of a car.

In practical data analysis, the user's ability to overview the data in light of his or her contextual knowledge is far more important than the last digits of statistical significance. The user should understand the basic principles of the modelling, and learn to use the method and software correctly. But the user has no need to be humiliated by mathematical derivations and burdened by statistical proofs.

### 1.1.6. Target Reader Group

The book is written for curious but critical researchers and students who want to achieve a deeper understanding of their field, and a higher throughput in their work. This includes people who want:

- to *learn* how to derive reliable, relevant and interpretable information from large tables of empirical real-world data;
- to *solve a real-world problem*, realising that it may have several causes, each being difficult to measure;
- to *explore* complex systems in a more holistic, empirical and open-minded way, but with high scientific standards;
- to achieve *reliable predictions* of quality from high-speed but non-selective measurements.

In particular, this book is written for people who, like the authors, work with one of the many aspects of *quality*: quality definition, data quality, quality assurance, quality control, quality management, etc. We want reliable, relevant and new information, in order to solve a problem in, e.g. industry, regulatory agencies or academic research. Often we have lots of data, or at least we can get lots of data, if we want to. On the one hand, mathematical statistics is not our favourite topic. On the other hand, we have found it difficult to have someone else do the data analysis for us: too many important details get lost. So we perceive the data analysis as a problem: how can we know when we have exhausted the data for relevant information? How can we know if our conclusions are reliable?

The target group for this introductory book is here termed *researchers*, as a common denominator for the explorer, the analyst, the problem-solver and the curious investigator. The book is written with a hope to reach different traditional professions, both in industry *and* academia, both among technologists *and* humanists. Common to the researchers from different areas is our desire for new, valid and useful information.

The book is also intended to be used as a textbook, for a university course in applied multivariate data analysis.

## 1.2. INFORMATION AND QUALITY

Without the human mind there would be no concept of *quality*. To gain *information* about quality is a multivariate problem. The reason is two-fold: the physical properties of an entity, e.g. a certain product or process, are seldom one-dimensional. The human perception and understanding is certainly not one-dimensional.

In research, decisions should be based on real information, not just wishful thinking or prior dogmatic opinions. The information itself must have certain qualities, which makes it useful for its intended decision-making.

Communication between disciplines about *quality* can be difficult, because the term has several meanings and is seldom defined. The *hard* technologist, physicist and chemist like to *measure* it with electronic instruments, the sensory scientist makes quantitative human *judgement* of quality, the sociologist and psychologist *evaluate* quality while the *soft* humanist has *opinions* about quality and the artist *feels* quality. Everybody *talks* about quality! But to what extent do all our measurements, judgements, evaluations, feelings and talking reflect a common understanding? Quality concerns information that people consider valuable, and therefore the definition and quantification of quality must create controversy. It is no wonder that the international effort to standardise quality assessment in, e.g. the International Standardization Organization (ISO) has required a long time, and still raises a lot of dust when about to be implemented in practice. Chapter 2 discusses this in more detail.

Likewise, it is ironic that the concept of *information* lies at the bottom of all inquiries and is used extensively, yet it has many definitions. A telecommunication engineer, working in an information technology company on the Internet, may find *Shannon's Information Theorem* helpful. But most of the information accessible from the Net, error-free as it may be, is just *noise* to him personally. An incompetent data analyst may use analysis of variance to extract statistically significant, but *meaningless* information from data. A philosopher may speculate on how the human mind allows information to self-organise, putting matter and energy *in form*. An experimental psychologist can measure the cell structures as well as the information *processing* in the brain. A physicist may think of information in terms of *negative entropy*, and then, on the way home, buy the newspaper *L'Information* to get the latest news. If she by mistake gets an old issue that she read yesterday, she will be disappointed, even though she was thrilled to be *informed* by the very same letters the day before.

An intersubjective understanding of *information* about *quality* requires rationality rooted in reality, which includes the physical reality and the psychosocial reality (the information structure within and between people). One type of rooting is to compare data from different aspects of reality. If we can find stable and interpretable relationships between them, this is a strong indication that we are about to learn something important.

## 1.3. THE ROLE OF MULTIVARIATE DATA ANALYSIS IN RESEARCH

It is important to reflect upon various aspects of information and quality in research: why measure? What to measure? How to measure and how much? Who should analyse the data, and how?

Modern advanced technical instruments (e.g. spectrophotometers and chromatographs) as well as large market analyses (e.g. questionnaires), create enormous amounts of raw data. But do we need all these data? And if so, how can we interpret them without experiencing mental overflow? How can we legitimate all the time and money spent on experiments and data collection?

Before starting a new investigation, the researchers cannot guarantee a successful experiment and a positive conclusion: there are too many unknowns for that. Even after the investigation, we cannot guarantee the correctness of the results and conclusions: that would be statistically impossible and require economically prohibitive experiments. The only thing we as researchers can guarantee is that we have tried our best to be *professional*. To be professional researchers means, among other things, that we have built on the existing knowledge in the field, planned our experiments consciously, performed the experiments diligently, and analysed the obtained data creatively and critically.

### 1.3.1. Who Should be the Data Analyst?

A central theme in this book is that the data analysis is much easier and safer if *background knowledge* about the problem is at hand. It is impossible to say which part of background knowledge becomes important for understanding a given data set, before the data analysis has been finished! Life seems to have a tension between predictability and surprise, and so does data analysis.

Therefore the person who 'owns' the problem and has the contextual background knowledge should try to do as much as possible of the data analysis, him- or herself. On the other hand, it is important to know when to call for help, like in all do-it-yourself activities.

It should be noted that a good text editing system does not make everybody a good author. Likewise, a good data analysis system does not make everybody a good researcher.

### 1.3.2. Mathematical Modelling – But Just a Little

*Models* are useful in science and technology for their ability to give compact, simplified representations of otherwise complex topics, from a certain perspective. *Mathematical models* are particularly useful: they allow quantitative interpretation and prediction. Their model *parameters* can be checked for internal and external validity. Many simple systems can be described elegantly by mathematical models.

However, too much mathematics at the wrong level of abstraction for the wrong people, is a hindrance in research. Detailed mathematical description, based on first principles, is often beyond reach in heterogeneous, non-linear dynamic systems involving life, mind, culture or economy. Complicated industrial processes are also a problem in this regard. There would be too many unknown parameters to be estimated and to be dealt with cognitively.

In such cases there are better ways for researchers to think and communicate. Mathematics is just one language among languages (albeit an efficient one, for those who speak it). Therefore it should not be given higher status than deserved in the unspoken academic pecking order. To make matters worse, there are many different mathematical cultures in science, and for non-mathematicians to cope with their different notations and terminologies is a challenge.

But a *little* mathematics can go a long way. The present book uses a few *pictogram* concepts and practical examples to explain the necessary mathematics behind *soft modelling* and the BLM method.

The book presents a pragmatic, multivariate approach to the research process, which is here referred to as *qualimetrics*, to be discussed in Chapter 2. Appendix A1 relates this to some other mathematical modelling traditions.

Professional qualimetrics includes a clear definition of purpose, a generous but cost-effective experimental planning, a critical generation of measurements, a careful quality control of the input data, and creative, yet rigorous analysis of the data and a final concluding and reporting in light of the original purpose.

The *soft modelling* approach in this book builds on data analytical developments over the last 50 years in a number of disciplines, primarily economy, experimental psychology, chemistry and sensory science. The book reflects a version of *soft modelling* that the authors personally have found to be most useful in their applied research over the last two decades. This has been research in sensory science, food science and agriculture, analytical chemistry, medical instrumentation, medical ethics, process monitoring, statistics and telecommunication signal processing.

The book advocates the use of *soft modelling* in a humble but eager spirit inspired by famous, but pragmatic statisticians like Box (1993), Deming (1987), Tukey (1977) and Coomb (1964), as well as by contemporary fellow chemometricians and sensometricians.

The BLM method is used extensively in the field of *chemometrics*: outside chemistry it is sometimes called *infometrics*; the naming depends on the field of application. The statistical methodology is related to pragmatic approaches in *chemometrics, psychometrics*, *sensometrics*, *econometrics* and contemporary resampling statistics.

The present approach is available in several well-documented commercial software packages, e.g. within the fields of chemometrics/sensometrics. Armed with such a program, the reader can start to explore complex, real-world problems without too much trouble.

## 1.4. A SMALL EXAMPLE OF MULTIVARIATE DATA ANALYSIS

A small, introductory example will now be shown: the interdisciplinary assessment of quality variation in a set of cocoa-milk drinks. The data (COCOA-I) is a subset taken from Folkenberg et al. (1999). The purpose of the study was to explore the sensory quality variations in various cocoa-milk powder mixes for automatic dispensers. To what extent can the producer predict, control and optimise the quality, as perceived by consumers?

Table 1.1 gives the input data. The three first columns show the experimental design: the percentage of the three ingredients in the dry matter (the sum is 100%). The fourth and fifth column represent physical measurements (mean of two replicates): spectrophotometric measurements of *COLOUR* (lightness) and *VISCOSITY*. The remaining five columns represent human sensory assessment of colour, odour, texture and taste of these cocoa-milk drinks.

**Table 1.1.** A table of input data (8 × 10): the quality of cocoa-milk drinks. [a]

| | | %COCOA 1' | %SUGAR 2' | %MILK 3' | COLOUR 4' | VISCOSITY 5' | Colour 6' | Cocoa-odour 7' | Smooth-txtr 8' | Milk-taste 9' | Sweet 10' |
|---|---|---|---|---|---|---|---|---|---|---|---|
| 1: Milk+ | 1 | 20.0 | 30.0 | 50.0 | 44.9 | 1.86 | 1.7 | 6.1 | 8.6 | 6.9 | 8.5 |
| 2 | 2 | 20.0 | 43.3 | 36.7 | 42.8 | m | 3.2 | 6.3 | 9.1 | 5.2 | 9.8 |
| 3: Sugar+ | 3 | 20.0 | 50.0 | 30.0 | 41.6 | 1.78 | 4.8 | 7.1 | 8.6 | 4.6 | 10.5 |
| 4 | 4 | 26.7 | 30.0 | 43.3 | 42.4 | 2.06 | 4.9 | 7.6 | 6.0 | 3.3 | 6.7 |
| 5a | 5 | 26.7 | 36.7 | 36.7 | 41.0 | 1.97 | 7.2 | 8.3 | 6.1 | 2.9 | 7.0 |
| 5b | 6 | 26.7 | 36.7 | 36.7 | 41.0 | 1.97 | 6.9 | 7.7 | 6.7 | 2.6 | 7.0 |
| 6 | 7 | 33.3 | 36.7 | 30.0 | 39.1 | 2.13 | 10.6 | 10.2 | 4.5 | 1.5 | 5.5 |
| 7: Cocoa+ | 8 | 40.0 | 30.0 | 30.0 | 38.3 | 2.26 | 11.1 | 11.3 | 3.4 | 0.9 | 3.9 |

[a] Rows: eight samples, representing seven cocoa-milk mixtures, plus a replicate of mixture 5. Columns: ten variables; three chemical composition variables (%); instrumental *COLOUR* (L = Lightness) and *VISCOSITY* (centipoise × 10, with one missing value, m), and five sensory variables: perceived *colour* intensity, *cocoa-odour, smooth-texture, milk-taste* and *sweet* taste (mean over seven assessors and three replicates).

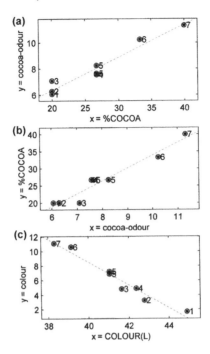

**Figure 1.2.** Three ways to model causality data from Table 1.1. The numbering represents the *names* of the cocoa mixtures *1–7*. In all three cases it is possible to make a model $y \approx f(x)$. (a) *Forward causality*: sensory *cocoa-odour* (**y** ordinate) is caused by *%COCOA* (**x** abscissa). (b) *Reverse causality*: chemical *%COCOA* (**y** ordinate) is causing *cocoa-odour* (**x** abscissa). (c) *Common causality*: sensory *colour* intensity (**y** ordinate) and instrumental *COLOUR* lightness

The original data set for this example is available at www.wiley.co.uk/chemometrics. The sensory data represent standard professional sensory analysis (the mean of seven trained sensory assessors analysing the samples in three sensory replicates, each produced and analysed in random order).

Data are given for seven different cocoa products: different mixtures of cocoa powder, sugar and milk powder in a fixed amount of water. Product *5* was assessed in two *sensory* replicates (rows *5a* and *5b*) to get a measure of the repeatability of the sensory panel.

In the following, the columns in a data table like Table 1.1 are called variables, and the rows are called samples (or, equivalently, objects). With ten input variables, there are 90 (10 × 9) different ways to plot pairs of variables. Figures 1.2 and 1.3 show a few of them, while Figure 1.4 will show a BLM overview of all of them at once.

### 1.4.1. Causality and Prediction

Considerations about *causality* is important in data analysis, because it allows better generalisations. But the researcher is free to choose how to handle causality in data analytical models.

Figure 1.2 shows three valid ways to handle causality, in plots and in data analytical modelling. Let us consider the observed changes in two sensory variables, *cocoa-odour* and *colour* darkness, and a physical instrument measurement, *COLOUR* lightness. All these are *effects*, mainly *caused* by changes in the chemical design variable *%COCOA*. But this *causality* may be expressed in different ways, as shown in Figure 1.2.

In each case, one variable from Table 1.1 is chosen as **y** and plotted (here as *ordinate*) as a function of another variable, chosen as **x** (*abscissa*), in order to model relationships of the type $\mathbf{y} \approx f(\mathbf{x})$. The numbered points represent sample names (the rows in Table 1.1). The best straight line through the points has been found by so-called linear *least squares regression* modelling (Section 4.2).

Let the direction of causality be represented by the symbols $\Leftarrow$ or $\Rightarrow$. Figure 1.2a–c may then be formalised as:

(a) Abscissa **x** (*%COCOA*) causes ordinate **y** (*cocoa-odour*): $\mathbf{y} \Leftarrow \mathbf{x}$

Data analytical model: $\mathbf{y} \approx f(\mathbf{x})$. (1.4.1a)

This is the choice of axes by default preferred by most researchers.

(b) Ordinate *y* (*%COCOA*) causes abscissa **x** (*cocoa-odour*) $\mathbf{y} \Rightarrow \mathbf{x}$

Data analytical model: $\mathbf{y} \approx f(\mathbf{x})$. (1.4.1b)

This is an equally valid modelling, and easier if *y* is to be predicted from **x** later on.

(c) Hidden phenomenon **u** (*%COCOA*) causes both abscissa **x** (instrumental *COLOUR* lightness) and ordinate **y** (sensory *colour* darkness) $\mathbf{u} \Rightarrow [\mathbf{x}, \mathbf{y}]$:

Data analytical model: $\mathbf{y} \approx f(\mathbf{x})$. (1.4.1c)

This is also a valid and useful representation. Classical deduction fundamentalists may feel uncomfortable with case (c), because it has no causal axis that the mind can use as fixed reference or *mental floor to stand on*. But, as this book will show, this plot is actually the beginning of multivariate explorative data analysis!

Note that the same forward data analytical plot formalism (**y** vs. **x**), and the same statistical regression method (regression of **y** on **x**) were used, irrespective the direction of the causality. In Martens and Næs, 1989, p.61, this topic of causality vs. prediction modelling is discussed in more detail. It should be mentioned that choosing **x** to be plotted as abscissa and **y** as ordinate is merely a convention and not a necessity.

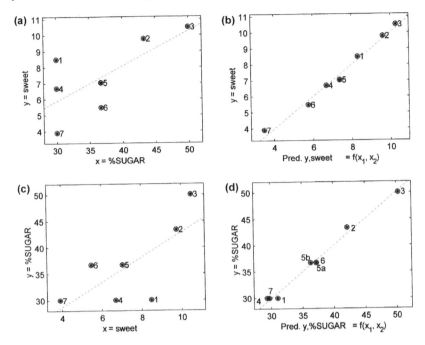

**Figure 1.3.** Solving the selectivity problem data from Table 1.1. The numbering represents the *names* of the cocoa mixtures *1–7*. Left: univariate modelling. Right: multivariate modelling. Top: forward causality. Bottom: reverse causality. (a) *sweet* taste **y** vs. *%SUGAR* **x**; (b) *sweet* taste **y** vs. predicted *sweet* taste **ŷ**, where **ŷ** $= f(\%SUGAR\ \mathbf{x}_1$ and *%MILK* $\mathbf{x}_2$); (c) *%SUGAR* **y** vs. *sweet* taste **x**; (d) *%SUGAR* **y** vs. predicted *%SUGAR* **ŷ**, where **ŷ** $= f($*sweet* taste $\mathbf{x}_1$ and *colour* $\mathbf{x}_2$).

## 1.4.2. The Selectivity Problem and its Solution

Whenever possible, univariate selectivity (one-to-one correspondence like that seen in Figure 1.2) should be striven for in research, because it is robust and simple to understand. But selectivity cannot always be attained, because the world often gives many-to-many correspondences: the observed effects are often affected by two or more causes that vary more or less independently of each other.

Figure 1.3 shows that multivariate analysis can solve the selectivity problem. The left side illustrates a problem and the right side shows its solution. The variables involved are *%SUGAR* and *%MILK* as well as sensory *sweet* taste, *colour* darkness and instrumental *COLOUR* lightness, all from Table 1.1:

(a) Incomplete causality: Abscissa **x** (*%SUGAR*) and **u** (something else) cause ordinate **y** (*sweet* taste): **y** $\Leftarrow$ [**x,u**]

Bad data analytical model in plot: **y** $\approx f(\mathbf{x})$.         (1.4.2a)

This is a dangerous situation, because we may lock onto our initial idea about causality $y \Leftarrow x$, and stop looking for improvements. The bad fit prevents us from predicting effect $y$ from measured cause $x$.

(b) Sufficiently selective causality: *Two X-variables* $x_1$ (*%SUGAR*) and $x_2$ (*%MILK*) cause ordinate $y$ (*sweet* taste): $y \Leftarrow [x_1, x_2]$

OK data analytical model:   $y \approx f(x_1, x_2)$.                                           (1.4.2b)

Plot: ordinate $= y$, abscissa $=$ predicted Y-value, $\hat{y} = f(x_1, x_2)$.

The selectivity problem has been solved by using *two different causal variables*. The response variable *sweet* taste is well predicted from *%SUGAR* and *%MILK*.

(c) Incomplete predictive correlation: Ordinate $y$ (*%SUGAR*) and $u$ (something else) cause abscissa $x$ (*sweet* taste): $[y,u] \Rightarrow x$

Bad data analytical model plot:   $y \approx f(x)$.                                           (1.4.2c)

This is a dissatisfying situation, because something is interfering with our wish to predict cause $y$ from measured effect $x$.

(d) Selective predictive relationship: Ordinate $y$ (*%SUGAR*) and $u$ (something else) cause different effects in *two X-variables*, response variable $x_1$ (*sweet* taste) and response variable $x_2$ (*colour* darkness): $[y,u] \Rightarrow [x_1, x_2]$

OK data analytical model:   $y \approx f(x_1, x_2)$.                                           (1.4.2d)

Plot: ordinate $= y$, abscissa $=$ predicted Y-value, $\hat{y} = f(x_1, x_2)$.

The selectivity problem has here been solved, by using two different *response* variables *sweet* taste and sensory *colour* darkness to predict one of the causal variables *%SUGAR*. Note that the unknown phenomenon $u$ does not have to be considered explicitly, as long as its effect is clearly different in the two measured responses $x_1$ and $x_2$!

Figure 1.3a,c may be informative for the human eye *to look at,* but they were unfit for *univariate mathematical modelling* because of the selectivity problems. However, the multivariate modelling removed the selectivity problems. Hence, from Figure 1.3b we can predict the effect-variable *sweet* taste in new products of this type if we know their causal variables *%SUGAR* and *%MILK* levels. Conversely, from Figure 1.3d we can predict the causal variable *%SUGAR* in new products of this type if we have assessed their effect-variables *sweet* taste and *colour*.

A number of examples of various types of causality will be given in later chapters in this book. But we must remember that models are only valid within a certain range.

### 1.4.3. Multivariate Data Analysis: Overview Maps

Figures 1.2 and 1.3a,c displayed the data in Table 1.1 in some of the many possible pair-wise combinations. How many such plots are necessary before we can say that we have studied the whole data table? In principle, we need to study all of them. But in practice, only a few plots are needed, if we plot the *principal components,* instead of the individual input variables!

Consider, e.g. Figure 1.3c: here two of the ten input variables were related to each other in a two-dimensional plot. If we had included a 3rd variable, e.g. *%COCOA*, we could have studied the configuration of samples in a three-dimensional plot. By straight analogy, we may well relate all ten input variables against each other. But since we cannot envision a ten-dimensional space directly, we need a little more help of the computer.

So we submit all the data in Table 1.1 to the *mother of all multivariate analyses*, the principal component analysis (PCA, Section 3.3 and Chapter 5). This is a bi-linear *soft modelling* in which the many individual input variables in a data table (ten variables in this case) are combined into a few so-called *principal components* (PCs), symbolised as *latent variables* $t_1$, $t_2$, etc. The relationships of the PCs to the *samples* (data rows) are called *scores*, and to the *variables* (data column) called *loadings*.

□ SCORE PLOT OF THE SAMPLES

The solution in Figure 1.4a shows how a whole table of variables can be summarised, with respect to information about the samples. A full 98% of the initial variance in Table 1.1 is represented by the first two *PC score vectors*, $t_1$ and $t_2$. The remaining 2% may be regarded as small enough to be ignored (probably mainly noise).

So, instead of having more than 45 different variable-pairs to look at, we now have only one pair of *super-variables*, $t_1$ vs. $t_2$. Figure 1.4a shows the resulting main pattern of samples. Again, the cocoa-milk samples are named by their mixture number, but for the three most extreme mixtures, *1, 3* and *7*, the dominant ingredient is written explicitly, *1: Milk+, 3: Sugar+* and *7: Cocoa+*. Along the first principal component, $t_1$, the cocoa-richest mixture, *7*, scores highest, and mixtures *1, 2* and *3* score the lowest. Along the second PC, $t_2$, the sugar-richest mixture, *3*, scores the highest and milk-richest mixture, *1*, scores the lowest. The two replicates for the centre mixture, *5a* and *5b,* come close to each other in the middle.

□ LOADING PLOT OF THE VARIABLES

Figure 1.4b gives interpretation meaning to the two PCs, $t_1$ and $t_2$, in terms of the corresponding pattern of the ten input variables $x_1, x_2, ..., x_{10}$, in Table 1.1, as

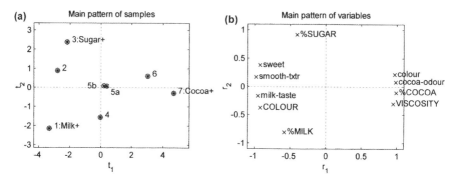

**Figure 1.4.** Main information in a data table mapped by PCA. The main variation patterns in Table 1.1, extracted by standardised principal component analysis. The first two PCs account for 98% of the total initial variance of the input data. (a) *Score plot*: pattern of relationships between the eight cocoa-milk samples. Origin = mean sample quality. Samples are identified by their *names*. (b) *Correlation loading plot*: pattern of relationships between the ten variables. Origin = no correlation. Variables are identified by their *names*.

their correlation to the first latent variable $t_1$ ($r_1$ abscissa) and to the second latent variable $t_2$ ($r_2$ ordinate).

The present example has been chosen because it has a rather obvious structure that allows the reader to follow the multivariate modelling intuitively. But it has its shares of surprises. You are hereby invited to try to interpret Figure 1.4. The data are interpreted in detail in Chapter 8.

Figure 1.4 shows that PC1 primarily corresponds to a systematic pattern where the levels of *colour, cocoa-odour, %COCOA* and *VISCOSITY* tend to increase together, e.g. from samples *1, 2* and *3* to sample *7*, and together with decreasing levels of primarily *smooth-txtr, milk-taste,*and *COLOUR*. PC2 primarily corresponds to a pattern that when *%SUGAR* increases, e.g. from sample *1* to *3*, *sweet* taste also tends to increase, while *%MILK, COLOUR, VISCOSITY* and e.g. *milk-taste* tend to decrease.

The main result is no big surprise and reflects the changes expected when the percentage of *COCOA, SUGAR* and *MILK* changes in the cocoa-milk drinks. For instance, the *cocoa-odour* is strongly correlated to *%COCOA*.

But some interesting and unexpected effects can be observed: *sweet* taste is not only correlated to *%SUGAR*; it appears also to be anti-correlated (negatively correlated) to *%COCOA*. Is this (1) because of *closure*: whenever *%SUGAR* increases, *%COCOA* or *%MILK* decrease, since their sum is 100%? Or is it (2) because there are two types of sugar involved, sucrose (represented by *%SUGAR*) and milk sugar (lactose, a certain fraction of *%MILK*), so *sweet* taste is also to some degree correlated to *%MILK*? Or is it (3) because when *%COCOA* increases, the strong *bitter* cocoa flavour masks the *sweet* taste and the *milk-taste* for the sensory assessors? These possible effects are so confounded that they cannot be fully resolved from these data alone: that would call for contextual knowledge or some more experiments.

The multivariate *soft modelling* method gives us easy visual access to the main information in a whole data table, like in Figure 1.4. This is one of the main purposes of multivariate modelling. Another main purpose is to be able to predict some variables from other variables. In that context, the selectivity enhancement illustrated in Figure 1.3 is a benefit, and so is the automatic outlier detection to be shown later. The same method can be used for a number of other purposes as well, like the statistical assessment of the effects of causal design factors, the discrimination between different classes of samples, the facilitation of multivariate statistical process control (MSPC), etc.

The present illustration concerned various aspects of the *quality* of cocoa-milk drinks: more specifically about sensory, chemical and physical quality *variations* in a *set* of cocoa-milk samples. But what is quality?

Readers eager to get into the data analytical methodology may go directly to Chapters 3–16, after having finished Chapter 1. However, it may be worth while first to read Chapter 2, which concerns more fundamental questions: what is quality, why measure it, and what to measure? The extra project-time spent on initial goal formulation and identification of informative criteria to be measured is usually earned many times over at a later stage. If the purpose of a project is clear and realistic, then the ensuing experimental work and the data analysis become much simpler.

## 1.5. GUIDELINES FOR USING THE BOOK

The main book title is *Multivariate Analysis of Quality* indicating that the overall purpose of the book is to make the reader able to start analysing large and complex data sets. It is intended to be *introductory* in the sense that it requires little or no theoretical background in mathematics and statistics. It focuses on *multivariate data,* because that is called for in modern, real-world interdisciplinary research, as discussed above.

The words *of quality* have been added, since this is a typical field of application: quality needs to be analysed and understood without loss of essential dimensionality, but also without unnecessary details and without a feeling of alienation. To this end, tools for meaningful data generation as well as data analytical information extraction and interpretation are needed, with the practical user at the controls.

The book has four main parts:

*Part One: Overview* (Chapters 1–3) motivates the book, discusses the concepts of *quality* and *qualimetrics*, and puts data analysis in a *research* perspective.

*Part Two*: *Methodology* (Chapters 4–11) explains the basic *engine* of the interactive, data-driven multivariate data analysis, the BLM method. It also describes validation methods to ensure valid conclusions in practice. Finally, it outlines ways to plan informative, yet cost-effective experiments.

*Part Three*: *Applications* (Chapters 12–16) shows the practical analysis of various real data sets. Table 3.1 in Chapter 3 gives an overview of the various examples given in this book.

*Part Four*: *Appendices* (A1–A16) gives technical details for each of the chapters in Parts One, Two and Three.

Test questions in each chapter allow the reader to check that she/he has understood the main points in the text. For those who want it, more technical details are given with small print, in appendices, and in the references.

This book uses largely the same notation, terminology and methods as the more software-specific introduction *Multivariate Data Analysis in Practice* by Esbensen (2000). A more advanced treatise, *Multivariate Calibration* by H. Martens and T. Næs (John Wiley and Sons Ltd., 1989), may serve as the next stepping stone for readers who want to become fully proficient in this kind of multivariate data analysis. The terminology concerning quality is largely taken from international standards and classical books like Juran and Gryna (1988).

Most of the illustrations in Chapters 1–11 were programmed individually in Matlab™ Version 5.2 (www.mathworks.com, 1999); the rest of the analyses and plots were made with in The Unscrambler™ Version 7.5α (www.camo.com, 1999). The plots in the book are intended to represent the quality obtained in routine data analysis. The data sets are available from John Wiley and Sons website in ASCII and UNSCRAMBLER format (www.wiley.co.uk/chemometrics).

Having read the book, the reader should be able to start carefully using experimental planning and multivariate data analysis in their own research.

## 1.6. TEST QUESTIONS

1. What is your motive for reading this book?
2. List some ways in which *quality* is assessed in different scientific fields.
3. List some ways in which the term *information* is used in different scientific fields.
4. Why is mathematics often placed on top of the unspoken academic ladder?
5. What distinguishes a mathematical model from other models?
6. What is the selectivity problem, and how can it be solved?
7. What is the advantage of multivariate overview maps?
8. What is multivariate *soft modelling*?

## 1.7. ANSWERS

1. You answer! But here are some previously obtained answers:

- I drown in data!

- I hate traditional statistics!
- I am puzzled about the many meanings of the word quality.
- I think ISO 9000 stinks, because of all the paperwork. We collect a lot of data, but so what?
- The budget for my laboratory is drastically reduced this year and I shall have to drop some of the quality control analyses. But which ones should I drop?
- I think it is important to improve the relevance of technical measurements, for instance by relating them to how people perceive quality.
- I need high-speed measurements of quality, but my process instrument has selectivity problems, i.e. I don't really know what I am measuring!

2. *Quality* may, e.g. be *measured* in the *hard* sciences, *judged* in sensory science, *evaluated* in the social sciences, *stated as opinions* in the humanities and *felt* in arts.

3. *Information* may refer to *bits* in Shannon's information theory and in some branches of statistics, to *negative entropy* in thermodynamics, to *useful facts* on the internet, to *new and interesting statements* in the press, and to a *combination of all of this* in interactive data analysis.

4. The role of mathematics is difficult and controversial. The authors have stated their opinion throughout this book. The readers may have different opinions.

5. All models represent over-simplifications for a given purpose. A mathematical model can be particularly informative, elegant and concise, when it is valid for its stated purpose. It can be particularly misleading when invalid, particularly overwhelming when too complex, and particularly confusing and alienating when forced upon someone with too much detail or at a level of too much abstraction.

6. The selectivity problem arises when a desired *target* variable cannot be determined from one single measured variable. In particular, it is evident when the measurements are affected by other, unknown phenomena in addition to the *target* variable. It can often be solved mathematically by combining the data from several measured variables.

7. The advantage of multivariate overview maps is that a lot of different variables in a lot of samples may be studied at the same time, thereby avoiding mental overflow, and with improved statistical precision.

8. Multivariate *soft modelling* is based on a few, pragmatic mathematical approximation techniques to extract the main information about relationships in data tables.

# Chapter 2

# Qualimetrics
# for Determining Quality

*This chapter focuses on the quality of information, and on information about quality. First, it discusses the uncertainty of indirectly observed information, and identifies the main quality criteria of empirical information: reliability and relevance.*

*Then multivariate data analysis is demonstrated in an empirical study of how people understand the word quality. Four different common definitions of quality are consequently discussed. Finally, qualimetrics is defined, and some cognitive aspects of data analysis are outlined.*

## 2.1. A WORLD UNDER INDIRECT OBSERVATION

In order to be able to make correct decisions, we need relevant, reliable, timely and understandable information. The best information about the real world would come from the real world itself. However, the world is under indirect observation (Jöreskog and Wold, 1982). We seldom have direct access to the *real thing*: neither with our measuring tools nor with our senses or our thoughts. In simple cases, we can count the number of screws per minute, and measure the length of a screw in millimetres. But to measure toxicity, taste or trends in a market is not that easy.

This creates methodological limitations, statistical uncertainties and cognitive interpretation problems. We may put on the reductionist's hard hat or a softer, more frivolous attire. Irrespectively, we must concede that we do not have access to error-free and complete information about the world. The best we can do as researchers is to strive to be professional in defining the needs and the opportunities, and then try to satisfy these needs as reliably, creatively and completely as possible within the given resources.

## 2.1.1. Looking for the Golden Bits of Information

Explorative research is like detective work or treasure hunting. It is not smart to take as gold everything that *subjectively* seems to glimmer; to distinguish between real gold and just glimmer calls for good eyesight and a critical attitude. Neither is it smart to look only under the lamppost, just because that is where, *objectively*, the ground is smooth and the light is strong. The real gold may lie a short way off in the shadows, waiting to be found. To search in the shadows calls for a good headlamp and an inquisitive and risk-willing attitude. But not all parts in the shadow are equally likely to hide gold. To choose where to start looking, we need some kind of map of the general landscape: an inventory. If prior knowledge is lacking, it is advisable first to look around a little, exploring more or less on pure intuition.

It is ironic, that the stronger headlamp we put on and the more focused we are, the better we can see straight ahead, but the more blinded we become to what is to the side and to valuables other than gold. The better instruments or mathematical tools we obtain, the less open-minded we may become. It is important to realise that being too precise and logical may give each of us an endorphin kick, but may also make us myopic and narrow-minded.

## 2.1.2. It is Human to Err, Even for Researchers

There is a strong subjective component to information processing in research. The researcher's tacit knowledge represents a vast resource for meaningful interpretation of information in data, if only this information can make it through the bottlenecks of mathematical modelling and sensory perception of the results.

Of course, researchers strive to distinguish valid information from noise. But it is irresponsible to be so afraid of committing Type I errors (being fooled by noise in data) that one instead commits gross Type II errors (overlooking important, real information). In this book, the classical defensive, confirmative *significance-testing* is down-played. Instead, the focus is on the more offensive discovery aspect of explorative research. Science works both by *disproving* hypotheses and by *discovering* the unexpected. Scientific discovery based on induction from empirical observations should not be underrated. Exploration is aided by good experimental planning, good information extraction models and good displays of the results. The validity of the displayed results is checked by statistical validation as well as by contextual interpretation.

Still, some errors will be made. But if the data analysis is efficient, the turn-around time to the next experiment in the project can be shortened, and many errors may then be corrected. Moreover, the society of knowledge makers in science and elsewhere constitutes one huge, cultural world-wide web, and the body of knowledge itself constitutes another web. The person who finds holes in the latter web becomes famous on the former, so the system is slowly self-correcting (Sagan, 1995). Therefore, it is also OK to make mistakes in science: you or others will eventually correct them.

Taking the human strengths and limitations into account, we shall discuss the quality of information and the information about quality in the following subchapters. In order to define quality, we shall also have a look into the multi-dimensionality of the *quality* concept in an empirical study.


## 2.2. THE QUALITY OF INFORMATION

The quality of our decisions depends at least in part on the quality of the information on which we base them. This, in turn depends on the quality of our input data as well as on the quality of the tools we use for interpreting them. This is obvious. Yet, what do we actually mean by *quality of information*?


### 2.2.1. Validity

The quality of the information which we get, and which we give, depends on several criteria, and the optimum of these criteria depends on the situation. Two important criteria are *reliability* and *relevance*, which are discussed below. In one way or another, they both concern the topic of *validity*: again a difficult concept which often is used differently across professions.

In the following we shall discuss validity in a continuum that starts from reliability in statistical terms of uncertainty, thus avoiding *ad hoc* validity. Next comes *internal* validity, implying that the information we get is in conformance with the purpose or goal for the measurements. In data analytical terms, we may talk about *predictive* validity. Methods for data analytical validation are outlined in Chapter 3.4 and described in Chapter 10. Then more ambitiously, *external* validity may be sought when we want to generalise our results to a broader population or setting, ideally based on a *causal* interpretation.

In general, reliability is a prerequisite to validity. However, we may have results that are reliable without being *relevant* for the given purpose.


### 2.2.2. Reliability

According to Juran and Gryna (1988) the term *reliability* is used in a general sense to express a degree of confidence that a part or system will successfully function in a certain environment during a specified time period. In data analytical terms, we want to minimise *uncertainty* which means doubt, and thus in its broadest sense *uncertainty of measurement* means doubt about the validity of the result of a measurement (ISO, 1995a).

Uncertainty is always an important consideration. In situations where vested interests are at stake, involving money or honour, it is particularly important. In the official assessment of a new drug, for instance, it is of utmost importance.

Then it may be useful to require formal experimental hypotheses, and to apply formal experimental design and classical confirmative hypothesis testing.

But in other types of research this traditional approach is rather inefficient, because too strict a requirement for a detailed working hypothesis can be a hindrance to discovery. It is therefore presumptuous and unrealistic to call this hypothetical-deductive approach *The scientific method.* Nevertheless, both the hypothesis confirming and the hypothesis generating explorative methods are meant to give reliable results.


❒ How to Report Reliability


Uncertainty in the input data gives uncertainty in the resulting output from the data analysis. How should reliability be reported, without distracting too much from the main message? Qualitatively, the reliability may be stated in terms of our degree of confidence in the results. For more quantitative purposes it is usually better to report the *size* of the uncertainty.

First of all, how should *uncertainty in input data* be reported? In some fields, e.g. chemistry, it is normal to use the number of printed digits as a crude indication of the reliability of measurements. When giving the result pH = 7.4, the chemist has no doubt about the *significant* digit (7), but may be somewhat uncertain about the last digit (0.4). In addition, it is common in some fields to give uncertainty bounds for the most important results, (e.g. pH = 7.4 ± 0.1). Precisely what is meant by this '±' symbol depends a little on who you are:

*The traditional chemist* would probably state that he feels 100% certain that the true, but unknown pH value lies between 7.3 and 7.5. *The statistician*, in contrast, would claim that there is no such thing as 100% certainty. If the error in this measurement 7.4 is random and normally distributed, then the best one can do is to request some replicate pH measurements and then use distribution theory. A *t*-test may, for example, provide the estimated confidence interval of 95%. *The quality control engineer*, in turn, may question the validity of too detailed statistical statements, as long as important error sources may have been overlooked (ISO 1995a). Instead, the standard error ($s$), estimated from the replicated measurements, say $s = 0.03$, as well as uncertainty ($u$) estimates from other known sources not represented in the experiment, say $u = 0.04$, are summarised in the mean-square-error (rms) sense, to a combined ($c$) estimate of the uncertainty, $u_c = 0.05$. This may be given as pragmatic error bounds, e.g. $\pm 2u_c$, in this case 7.4 ± 0.1, corresponding to approximately 95% level of confidence. 'The true, but unknown pH value probably lies between 7.3 and 7.5'.

Secondly, how should the *uncertainty of the estimated output parameters* be reported? Analytical modelling of the data may filter out most of the input noise. Still, some of the input uncertainty propagates through to give uncertainty in the model output. This book uses two approaches to quantify the output uncertainty.

1. the *average* standard uncertainty of *prediction of the input data* from the models (the root mean square error of prediction (RMSEP)) estimated by so-called cross-validation; and
2. the reliability range of the resulting model parameters estimated by so-called jack-knifing.

Reliability ranges are here defined as $\pm 2$ standard uncertainties, with reference to ISO (1995a). Moreover, the individual uncertainties, estimated by cross-validation (Chapter 10), are visualised graphically to reveal individual mistakes and outliers that are non-statistical in nature.

Information about uncertainty of the input data may be used for optimising the data analytical modelling, by giving lower *a priori* weights for the more uncertain input data (cf. Chapter 8.2). Particularly uncertain data points may even be defined as *missing values* (Appendix A5), to be filled in by multivariate modelling by imputation.

□ THE INFINITE REGRESS OF UNCERTAINTY

The levels of uncertainty of ordinary measurements are seldom known: they have to be estimated from observations. These uncertainty estimates tend to be rather uncertain themselves. This next level of uncertainty may also in turn be estimated, but again with uncertainty. This goes on in an infinite regress. In the end, we are only saved by the self-organising criss-cross of the above mentioned web of interdisciplinary scientific knowledge and of the critical web of scientific peers.

Therefore, giving too much statistical attention to uncertainty drains too much cognitive capacity, both from the provider and from the receiver of scientific information. So while not being ignored, the uncertainty assessment should play a minor role in most research; other parts of the discovery process are much more important. We should not be reckless or unethical, but we must accept the risk of making mistakes, otherwise we are just driven by fear, not by the lust for knowledge and love for truth. But how to define what is *reliable enough* information depends on the situation: the given needs and opportunities.

### 2.2.3. Relevance

Data may represent information according to Shannon's information theory, and still be irrelevant to us, because they do not concern us, or because they are incomprehensible (cf. Chapter 1.2). A bit is not a bit for us until we can see it clearly and in a meaningful context. That goes for our own input data, as well as for our conclusions, which will then serve as input data to someone else.

It is certainly not true that the more data we get, the better and more useful information we have. Too much data creates mental overflow: our ability to find meaningful information in the raw data goes down when the amount of data goes up. To describe what constitutes necessary and sufficient information is difficult. Relevance addresses both internal validity, i.e. the results should be relevant to the goal, and external validity, i.e. the degree of generalisability in relation to the scope of the project and in correspondence with the web of scientific knowledge.

The input data should be selected so as to have a good chance of bringing us something of interest. On the other hand, before we have done the experiment, and see how all the pieces of data fit together, it is difficult to know which bits of information are really important and reliable, and which are not. Therefore the present book advocates conscious planning, with problem reduction *during* or *after* the data analysis: not *before* the experiment (Munck, 1991). In the experimental planning stage, a creative and generous selection should be made for what to measure. Then, during the *soft* interactive data analysis, the important information in the data is gradually crystallised while irrelevance is removed without loss of essential quality dimensions. This process may have to pass through several research cycles before the answers are sufficiently clear.

Not all *information* is relevant. Some organisations keep on doing measurements long after they ceased to be of interest to anyone. It would have been better to use the limited resources on something else, but to discontinue a tradition can sometimes be a difficult decision.

Not all relevant information is used. In the context of some ISO standards, many companies and institutes have started to record, register and measure a number of procedures, events, properties, or qualities, *just for the record*. This may be good for traceability purposes. But perhaps these data also contain valuable information that is left unused today?

In summary, we have seen that the world is under indirect observation. Researchers must be professional. But the stronger we focus, the blinder we can become. The quality of the information that we get from our data and data analysis has several aspects, reliability and relevance being the most general validity aspects.

However, qualimetrics in its broadest sense requires that we know what we want to measure, i.e. what *quality* is. Appendix A2 describes a research field where quality is assessed with both statistical reliability and human relevance: sensory science.

Let us now consider more explicitly how people use the term *quality*. Later in the chapter, the definition of quality will be discussed more formally.


## 2.3. A STUDY OF THE QUALITY CONCEPT

What goes on in our brain when we hear the word *quality*? As we shall see, there are many definitions and uses of the word. When opening the newspaper in the morning, we find the word quality on almost each page: quality of cars, food products, paintings, etc. In society in general, we use the word in many areas: quality of health care, of school systems, and environmental quality. We even talk about quality of life! But what do life, food and cars have in common? And does quality mean the same to different people? How can industries and communities advertise for *quality products or services* if there does not exist a common or universal meaning?

Some of these questions were raised in a pilot study on the quality concept (Martens, 1996). Quality *cognition* here refers to the mental processes going on in the brain when we think about quality of an entity or item. Is there any systematic pattern that may constitute a common ground for talking about quality?

The present example will show that the word quality has different meanings depending on the context. Data (QUALITY-CONCEPT) from an actual study will be analysed, using the *soft modelling* approach which will be explained later in the book. The original data set for for this example is available at www.wiley.co.uk/chemometrics. Like all the other examples, this data set will be collected, analysed and reported as a research project cycle in six steps (cf. Chapter 3.2).

*I. Purpose.* The purpose of the study was to see what people associate with the concept of quality, and whether or not there were differences between countries.

*II. Experimental planning.* A questionnaire, based on an appropriateness scale method (Schutz, 1994), was developed in order to find out how individuals regarded various statements about quality suitable for various items. A total of 18 statements were constructed (Table 2.1) each with a given abbreviation to be used in the plots. The following seven conceptual items were chosen: *CAR, FOOD, HOSPITAL, LIFE, PAINTING, RESEARCH* and *SPORT*. The appropriateness for each statement w.r.t. each item was to be assessed on a seven-point interval scale, ranging from 1 (not appropriate) to 7 (very appropriate).

*III. Experimental work.* The questionnaire was filled out by a total of 83 respondents, with about the same number from the US, Norway and Denmark. The respondents were students and researchers from different disciplines. The questionnaire data were entered into a database and collected in a data table for further analysis.

*IV. Pre-processing and QC of data.* For simplicity, only the means over respondents for each of the three countries will be discussed here.

Figure 2.1 gives an overview of the actual table of input data. It consists of 21 rows (*samples*), representing seven items (*CAR, FOOD, HOSPITAL, LIFE, PAINTING, RESEARCH* and *SPORT*) × three countries (US, U; Norway, N and Denmark, D), and 18 variables (statements, Tables 2.1). In the figure, the printed data have been quantised to only three grey-tone levels (black:1–2.25, grey: 2.25–5.7, white: 5.7–7). Thereby, the most dominant patterns become visible, but finer nuances in the data are invisible.

**Table 2.1.** A study of the quality concept. Eighteen statements about quality. Express how *you* consider the following statements to be appropriate for perception/cognition of quality of the different items, by judging the statements along the 1–7 point appropriate scale (1, not appropriate; 7, very appropriate).

| Quality of *CAR, FOOD*, etc. | | Abbreviations used in Figures |
|---|---|---|
| 1. | is complex | *complex* |
| 2. | is difficult to measure | *difcltmeas* |
| 3. | is easy to predict | *predict* |
| 4. | varies much | *varies* |
| 5. | depends on originality | *original* |
| 6. | depends on efficiency | *efficient* |
| 7. | depends on familiarity | *familiar* |
| 8. | depends on how *new* | *new* |
| 9. | depends on price | *price* |
| 10. | depends on normality | *normal* |
| 11. | depends on appearance | *appear* |
| 12. | depends on circumstances | *circumst* |
| 13. | depends on availability | *availab* |
| 14. | depends on safety | *safety* |
| 15. | depends on liking | *liking* |
| 16. | depends on comfort | *comfort* |
| 17. | depends on pride | *pride* |
| 18. | depends on time | *time* |

At this initial stage we may note considerable agreement between the three countries, in many respects. For instance, the figure shows that the respondents from the three countries agree that the *quality* of *LIFE* is considered *complex* (*ULIFE, NLIFE, DLIFE* are white in column 1), while the *quality* of *CAR* is *not* difficult to measure (*UCAR, NCAR, DCAR* are black in column 2: *difclt-meas*).

*V. Data analysis*. The data table in Figure 2.1 was submitted to multivariate bi-linear modelling by PLS regression (PLSR), together with so-called indicator variables for the seven items. The results indeed showed patterns that could reliably be predicted between the countries. Figure 2.2 summarises the most important patterns found.

The PLSR method will be explained in Chapters 3 and 6–11. The 18 statements were used as standardised X-variables and the seven item indicators as standardised Y-variables (1/0) in a PLSR for the 21 samples (seven items × three countries). The model was cross-validated between countries (i.e. three cross-validation segments). This showed that the models with increasing number of PCs from 1 to 6 explained

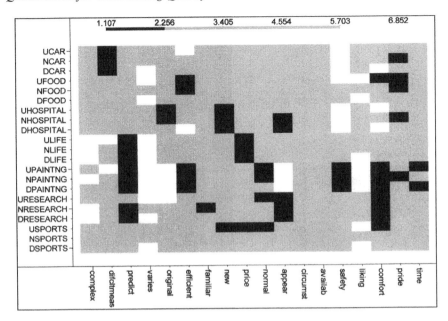

**Figure 2.1.**    A study of the quality concept: input data.

23, 35, 45, 52, 61 and 67% of the cross-validated variance in statements **X**, and 15, 30, 45, 51, 57 and 61% of the variance in item descriptors **Y**. Beyond six PCs the predictive ability in **X** flattened out and in **Y** decreased again. Hence, the first six PCs had predictive ability, but decreasingly so. For simplicity, only the first four PCs are shown explicitly here.

Figure 2.2a shows the correlation coefficient between the first two PCs and each of the X-variables (statements) and each of the Y-variables (items, connected to the origin by line segments). Figure 2.2b shows the corresponding score plot for the first two PCs. Looking towards the right in Figure 2.2a, *HOSPITAL* and *CAR* quality are associated with descriptors like *safety, comfort* and *efficiency*. But *CAR* quality is more associated with *predictability* and *price*, and is *opposite to* (i.e. *not*) *complex* and *difficult to measure*, while *HOSPITAL* quality has a *time* element (and is not associated with *appearance* and *originality*). Looking toward the left, we find *PAINTING*, whose quality apparently has to do with, e.g. *originality*. Hence, the horizontal quality dimension may be seen as a general *originality* vs. *safety* and *comfort*, as well as *complexity* vs. *predictability* axis.

The second dimension is a little more difficult to generalise, but it does seem to distinguish abstract items (*LIFE, RESEARCH*) from more concrete items

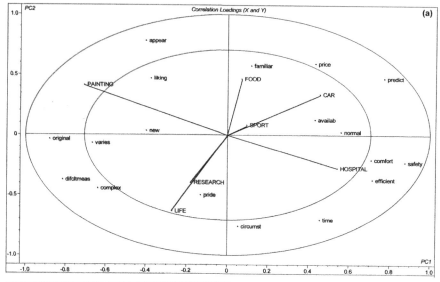

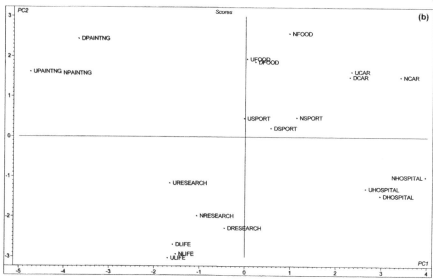

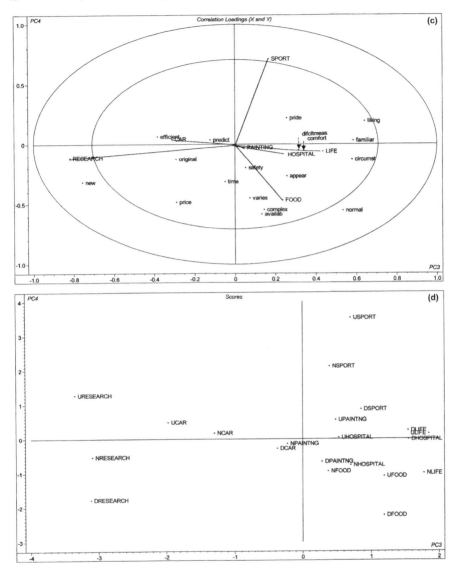

**Figure 2.2.** A study of the quality concept: multivariate overview maps. The appropriateness response variation of 18 statements for each of seven item indicators, summarised in terms of their correlations to PLSR PCs. The circles represent 50 and 100% explained variance, respectively. (a) Loading plot: PC1 (abscissa) vs. PC2 (ordinate). (b) Score plot: PC1 (abscissa) vs. PC2 (ordinate). (c) Loading plot: PC3 (abscissa) vs. PC4 (ordinate). (d) Score plot: PC3 (abscissa) vs. PC4 (ordinate).

(*PAINTING, FOOD, CAR*). *LIFE* and *PAINTING* apparently share the statements that their quality is *complex* and *difficult to measure*; both are *not predictable* (negative loadings along PC1). But where *PAINTING* quality in PC2 has more to do with *appearance*, *LIFE* quality depends more on the *circumstances* (but definitively not on *price*).

Figure 2.2b shows the corresponding score plot for PC1 and 2. It sheds further light on the meaning of these first two dimensions in the latent *quality* space. It shows that the respondents from the three countries largely agree on the assessments. Looking diagonally, the contrasts in *quality* between *PAINTING* vs. *HOSPITAL* and between *LIFE/RESEARCH* vs. *CAR/FOOD* are striking.

Figure 2.2c likewise shows the configuration of the statements and items for the third and fourth PCs, and Figure 2.2d shows how the countries score the items in these dimensions. In Figure 2.2c it is obvious that the third, horizontal dimension distinguishes the quality of *RESEARCH* (*new*, not associated with *liking*) against the others, primarily *LIFE*, *HOSPITAL* and *FOOD*. The fourth, vertical dimension mainly distinguishes *SPORT* quality, which has to do with *pride* and *liking* (and not *price*), from *FOOD* quality, which is associated with *normality, appearance, availability,* and *price* (and not *efficiency*). *Complexity* also seems to have something to do with *FOOD* quality in this figure.

Figure 2.2d shows that the countries largely agree on the third, horizontal *RESEARCH* vs. *LIFE/HOSPITAL* dimension. But in the fourth, vertical dimension the countries differ considerably. On one hand, the US respondents, on the average, have particularly strong views on the quality of *SPORT*: It is *not* associated with e.g. *new, price, normal* or *comfort*. The Danish respondents seem to take a slightly different view on *SPORT*; a check in the raw data (as summarised by Figure 2.1) shows that it is particularly important that it *varies*, in addition to their *liking* it. However, tempting as it may be, we do not proceed with a more detailed interpretation of inter-country differences here, for fear of over-interpreting this relatively small data set.

The fifth PC associates CAR quality positively with *pride* and *comfort*, while the sixth PC is small and difficult to interpret. For simplicity they are not shown here.

The above model interpretations were verified by inspection of the input data.

## VI. Conclusions of the quality concept study

- Quality cognition was found to be strongly item-dependent. Thus, it is important to make sure that we talk about the same item when we discuss quality.
- The way people associated statements about quality with different items gave good agreement between the three independent countries, so the data seem informative and the results reliable.
- Quality appeared to be an ambiguous concept. A quality paradox seems to exist when we can talk about quality as simultaneously having to do with both safety and originality.

More data is required before final conclusions with more nuances can be drawn. But the results point to the difficulty in giving a simple description of what quality is.

On the other hand, the results were not exactly shocking and unexpected: the so-called *WOW! factor* was not very high. This points to the fact that people use the word quality constantly in the daily conversations with the feeling of understanding each other quite well. In other words, as long as speaker and listener share a common cultural context, the word quality is an efficient carrier of information between them.

We see that quality is a complex concept. How can we then define quality to make the concept operative?

A classical definition: Quality consists of freedom from deficiencies (Juran and Gryna, 1988), does not seem to cover the active use of the word *quality* today.

## 2.4. FOUR DEFINITIONS OF THE WORD QUALITY

This subchapter gives a more formal specification of the word quality in terms of *quality of an entity*. An *entity* may be

a product, an activity or a process, a service, an organization or a person or any combination thereof (ISO, 1995b).

Another word for entity in this book will be an object, alternatively called an item or a sample (i.e. in the chemical, not statistical or commercial usage of the word sample).

### 2.4.1. Overview of Various Quality Definitions

There are at least four common definitions of quality (Table 2.2). In the following, these Quality Definitions (QD) are named QD1–QD4.

**Table 2.2.** Four definitions of quality taken from daily life, dictionaries and international standards, to be specified in the text.

QD1: *QUALITAS:* quality as *qualitas*, i.e. essential nature, inherent characteristic, property
QD2: *EXCELLENCE/GOODNESS:* quality as an expression for the intuitively evaluated *excellence/goodness*
QD3: *STANDARDS:* quality as a practical interaction between inherent characteristics of an entity, and stated or implied human needs; '*needs*' should be *satisfied*, requirements should be fulfilled
QD4: *AN EVENT:* quality as a *subjectively* experienced *event*!

The first and second definitions have traditionally constituted the two extremes between the *objective*, rationalistic and product-oriented approach

(QD1), vs. the *subjective*, phenomenological every-day approach to quality (QD2). When they are used at the same time, and undefined, there is room for misunderstanding.

The third definition of quality (QD3) pragmatically aims at linking the two extremes, in an *explicit* relation that can be used for practical purposes. This appears to be a basis for international quality standards like ISO, to be discussed below. The fourth definition of quality (QD4) is the subjective, existential or artistic definition, *implicitly* combining knowledge about what *is* with feelings about what is *best* into an event (an action). As such, we shall see that the two latter approaches, being so different, have something in common.

In Tables 3.1 and 3.2, the reader will be guided to examples in this book where the different quality definitions are understood and used in practice.

Next, let us discuss the four quality definitions in more detail, to prepare for various multivariate data analytical approaches.

❏   QUALITY AS QUALITAS (QD1)

---

Webster's New 20th century dictionary: QUALITY, n.; pl. QUALITIES [Lat. QUALITAS]; that which belongs to something and makes or helps to make it what it is; basic nature, characteristic attribute

---

QD1 reflects that human beings are *describing* the world. Quality as *qualitas* has main focus on the inherent properties of the entity. These qualities constitute the *potential* information to be discovered and explored by human beings. In general, the goal at this level is to describe the properties of an entity (e.g. a product) as objectively as possible.

There is an almost unlimited number of different qualities in physical objects that can be measured, be it in fundamental research, applied development or quality control. Stored and made easily available in a database, many of these measured qualities can contribute to the scientific body of knowledge, at least *in the long-term*. The human genome project is one example of this.

However, *in the short-term* and for a given purpose, only some properties are relevant, while many measurable properties are *irrelevant*. For instance, with sufficient work it is possible to detect, identify and quantify thousands of volatile chemical compounds in a set of whisky samples. But how many of these compounds are necessary in order to distinguish between the flavour of Scotch and Irish whisky, to predict the changes in consumer preference of whisky, or to be used in product quality control? Certainly not all of them!

In summary, there are potentially *many* characteristics to be measured in the world around us, but which ones are relevant?

❏  QUALITY AS EXCELLENCE/GOODNESS (QD2)

---

Webster's New 20th century dictionary: quality is any character or characteristic which may make an object good or bad, commendable or reprehensible

---

QD2 reflects that human beings are *evaluating* the world, in a cultural context. Quality as *degree of excellence* is the most common approach to using the term quality in daily life, yet the most confusing one. It assumes a common understanding of what is good and bad. But is good for me the same as good for you?

Examples of meaningless, but common, conversations

*Conversation 1: circular communication*
– How do you like this car?
– It is good
– Why is it good?
– Because I like it

*Conversation 2: quality ill defined*
– In our company we are producing quality products
– Oh, that's interesting. Please, tell me what do you mean by that?
– Don't you know what quality is?
– Not really...
– But I cannot tell you that. Quality is a personal, subjective thing. Taste cannot be discussed...
– Well, good luck with producing quality products

Within the humanities and social sciences the focus is on the culture of human subjects, not on the objects or products. This approach has consequences for the use of *quality* which goes back to the antique, to the *man of virtue* who aristocratically and with authority *knows* what is good and bad. Today this attitude is reflected, e.g. in the role of the master of wine and the critique of theatre and music. In his book 'The Varieties of Goodness' von Wright (1964) makes us aware of the mixture of uses of what is *good for...*, what is *good at* and what is just *good* in itself. Some uses of *good* have an instrumental or extrinsic value, i.e. good in conformance to a certain goal, and others have intrinsic value, i.e. we do not have to be conscious of why we assess it as good.

Therefore, it may be risky to use *degree of excellence* in the classical meaning, as a quality goal in practice, based on only one person and used as one single term (cf. ISO warning to QD3 below). Thus, today, e.g. large market studies based on *many* representative people (e.g. consumers) are carried out when quality as *degree of excellence* is at stake. We search for *intersubjective* patterns of responses. The quality concept study in Chapter 2.3 illustrated this, and another example is given in Chapter 13.

In summary, there seems to be *many* ways to understand *excellence/goodness*. Our perception of *excellence/goodness* reflects how *well* our 'needs' are satisfied.

> The authors are well aware of different approaches to define **goodness** and **needs**, especially from a philosophical point of view. That goes beyond the scope of the present book.

## ❏ QUALITY AS STANDARDS (QD3)

The International Standardization Organization (ISO) defines quality the following way:

Quality is the totality of characteristics of an entity that bear on its ability to satisfy stated or implied needs (ISO, 1995b), or the recent version:

Quality is the ability of a set of inherent characteristics of a product, system or process to fulfil requirements of customers and other interested parties (ISO, 1999)

QD3 is a quality definition intended to be useful in operational practice. In a technological, commercial or organisational context quality definitions are often goal-oriented. The goal is to satisfy certain needs, expectations and wants, in other words, to fulfil certain requirements. QD3 reflects that human beings are relating the *description* of an entity to an *evaluation* of good or bad. Thus, quality becomes a *relational* concept. Let us have a look at the *content* of the two parts of the ISO-standard separately.

*...a set of inherent characteristics of an entity...*

Here the translation from the latin *qualitas* = characteristic attribute is included, and constitutes an objective, descriptive part of the definition, QD1.

*...that bear on its ability to satisfy stated or implied needs/fulfil requirements*

Here the subjective *excellence/goodness* constitutes a normative, culture- and value-dependent part of the quality definition, QD2.

The ISO standard definition of quality aims at bridging together the two parts for practical purposes. Sometimes this is expressed as *conformance to requirements* or *fitness for use*. The product characteristics shall satisfy customers' needs or requirements, where *customer* means an organisation or a person that receives a product (e.g. consumer, client, end-user, retailer).

Thus, the quality standards may represent a vehicle for bringing forward both the understanding of inherent characteristics, and the ability to *satisfy/needs*, as illustrated in Figure 2.3.

The work of implementing and maintaining the quality standards is carried out by people, who will find the burden meaningful, only if it brings them somewhere. Unfortunately, the quality standard vehicle has no motor in itself: It requires some kind of locomotive to move it. Standing still, it does not take long for a quality certification to get rusty. But the organisations that push the quality concept actively, can gain considerable momentum from their increased ability to solve problems and to improve communication. This is a characteristic of good quality management.

To make the quality standards operative, two phases can be identified, to be illustrated later (Table 2.3):

(3a) *Finding* quality criteria, i.e. inherent characteristics of an entity that are relevant for the stated or implied needs. This calls for explorative data analysis.

(3b) *Using* these criteria as *specifications with control limits* in quality assurance and control. This calls for predictive or confirmative data analysis.

In summary, the quality standards call upon active uses of the *many* characteristics measured in, e.g. a product, and relating these to the *many* perceived or

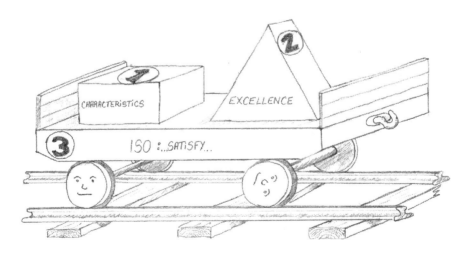

**Figure 2.3.** Quality standard. The ISO standard definition of quality (QD3) as a vehicle for satisfying needs by combining and carrying forward both an *objective* description of characteristics (QD1), and a *subjective* evaluation of excellence/goodness (QD2). Ilustration by Jens Strøm.

required needs that are to be fulfilled. Being *operative* means to act, not only talk. That leads us to the last approach to *quality*.

◻  QUALITY AS AN EVENT (QD4)

---

Quality is not a thing, it is the event at which awareness of subject and object is made possible

Quality...you know what it is, yet you don't know what it is. But that's self-contra-dictory. But some things are better than others, that is, they have more quality. But when you try to say what the quality is, apart from the things that have it, it all goes poof!! There's nothing to talk about. But if you can't say what quality is, how do you know what it is, or how do you know that it even exists? If no one knows what it is, then for all practical purposes it doesn't exist at all. But for all practical purposes it does exist. What else are the grades based on? Why else would people pay fortunes for some things and throw others in the trash pile? Obviously some things are better than others...But what's the 'betterness'?...So round and round you go, spinning mental wheels and nowhere finding any place to get traction. What the hell is quality? What is it? (Pirsig, 1974)

---

The fourth definition of quality is the subjective, existential or artistic definition: Implicitly combining knowledge about what *is* with feelings about what is *best*. This is a subjective balancing *act* required in every step we make, as illustrated in Figure 2.4. Educated, thinking persons on the move in a confusing and over-whelming world constantly experience moments of this nature. A good researcher can handle this balance.

This last approach to quality, is often considered as the ultimate subjective and phenomenological view, opposite to the objective, technological, goal-oriented and realistic approach. Being in the world is understood as something intrinsically good, having value in itself without any external or instrumental goal. Focus is changed from a third-person perspective (an observer) to a first-person existential perspective. Quality as such, is experienced implicitly, as an event that can only be lived. When grasped, it may be communicated to others in terms of the phenomenological concept *qualia* expressing conscious experiences.

Again, interdisciplinary communication can be difficult; in the ears of a *hard* scientist the word *qualia* may sound like a mispronounciation of the word *qualitas*. The authors have sufficient respect for the verbal capabilities in post-modern humanities to avoid further discussion on the subject.

### 2.4.2. Relationships Between the Four Definitions

'What the hell is quality? What *is* it?' Determining quality is a confusing activity

**Figure 2.4.**   Lived quality. The individually experienced event of balancing descriptive and evaluated perception of quality. Illustration by Jens Strøm.

as long as the various definitions are mixed up. Nevertheless, the various definitions may be linked together in a continuum from goals, with focus on the object (product), to focus on the subject (person), or to focus on the interaction between the two parts, resulting in action.

Figure 2.5 shows how the definitions may be linked together. An interaction between an object's properties (QD1) and the perceived *goodness* (QD2) results in definition of needs and explicit quality standards (QD3) with its two parts: (a) finding quality criteria, and (b) using these criteria as specification in, e.g. quality control. For each individual involved, the process is implicitly experienced as an event or an action (QD4).

In summary, we may talk about a third-person response as a descriptive, verbal response, as well as a *report* of experiences, emotions and behaviour. QD1, 2 and 3 are linked in a triad from a third-person perspective. QD4 is quality as if it occurs from a first-person perspective, doing away with the product–person, object–subject dichotomies.

How the four definitions may be used in an actual project context is shown in Table 2.3: in this case the investigation of the quality of cocoa-milk drinks from Chapter 1.4.

**Figure 2.5.**    Four definitions of quality. The definitions related to each other.

**Table 2.3.** Four definitions of quality: examples as applied in the book, e.g. the cocoa-milk study in Chapter 1.4.

---

QD1: *QUALITAS:* find the characteristic properties that describe the essential nature and variations in cocoa-milk drinks

QD2: *EXCELLENCE/GOODNESS:* find which cocoa-milk product the consumers like the most

QD3: *STANDARDS*
3a: *Quality criteria:* find which sensory and/or chemical quality criteria that are relevant for predicting consumers' preference within a set of cocoa-milk samples
3b: *Using criteria:* during production and/or quality control of the given cocoa-milk products, use the quality criteria and check that the product is within its control limits for the quality specification (e.g. accepted/ not accepted)

QD4: *AN EVENT:* drink hot cocoa-milk and enjoy it!

---

As will appear throughout the book, the various definitions are more or less in focus depending on the goal addressing different data analysis (cf. Table 3.2).

## 2.5. INFORMATION ABOUT QUALITY

### 2.5.1 Qualimetrics

Chapter 2.1 stressed the importance of scientific exploration in a world under indirect observation. Chapter 2.2 discussed the quality of information. Chapter 2.3 showed some of the dimensions of the quality concept. Chapter 2.4 attempted to structure the different meanings of quality into four definitions. Now we shall discuss how data analysis can give more information about quality.

Typically the bi-linear model of PCA may be used for summarising a number of property aspects of *qualitas*, QD1. More importantly, bi-linear regression by PLSR is a powerful tool for extracting *relevant* quality criteria, i.e. relating a set of inherent characteristics to a certain set of needs, QD3. Together these two versions of the bi-linear method provide cognitively accessible overviews of quality of an entity, without loss of essential dimensionality. Quality and the art of quantitative measurements are inseparable, and may be expressed in the discipline of *qualimetrics*.

---

*Qualimetrics* as used in this book, is the discipline dealing with quantitative methods of quality evaluation (EOQC, 1989). This includes the development and use of mathematical and statistical methods for measuring and understanding quality in a broad context. The *broad context* indicates awareness of both the physical product quality and the perceived quality.

---

Juran and Gryna (1988) describe how the word *qualimetry* was adopted as a name for the science of evaluating quality. According to a title '15 Years of Qualimetry: Problems and Prospects' by Glichev (1985):

> ...progress has been made in design of algorithms and models and also in identification of quality parameters. However, practical application has lagged because of various limitations, such as evaluation of sensory qualities, evaluation of long-life parameters, variations in product use and environments, and consumer perception. Future directions appear to involve reliance on market research to secure consumer's evaluations of consumer goods, use of panels of qualified experts to evaluate quality of commercial goods, and computerised data collection and analysis

Molnár (1983) introduced qualimetric methods that included both sensory aspects and data analysis in a quality index for food products.

The word qualimetrics is also sometimes used in a more narrow, technical way, referring to a set of modelling and validation methods for extracting information about quality from data, as in the extensive 'Handbook of Chemometrics and Qualimetrics (Massart et al., 1997). The present book is written to conform with the former, broader context.

### 2.5.2. Communicating Quality Grade and Confidence

Quantitative determination of quality in its simplest form may be expressed by grades. *Quality grade* is a category or rank given to entities having the same functional use but different requirement for quality, e.g. classes of airline tickets and categories of hotels in a hotel guide (ISO, 1999). Grade reflects a planned and recognised difference in requirement for quality. The emphasis is on the functional use and cost relationship.

As such, quality standards are intended to facilitate and stabilise communication between countries, between companies or between disciplines within a company. Therefore it is important to simplify the quality message as much as possible. When describing quality grade, two types of *scales* are recommended by ISO (1995b).

Grade 1 (highest), 2, 3, 4

Grade **** (highest), ***, **, *

Even this may have to be simplified further. As long as an entity is *within specifications*, its quality is often defined as high ( = OK) and the customer is happy.

The quality measure is often *quantised* very strongly: The decision *OK/Not OK* is fundamental. We cannot sit and stand at the same time. *OK/Not OK* represents just one bit of information (1/0). Multilevel quality grading is a little more ambitious, but still simplifies communication: *Grade 1, 2, 3 or 4* represents a quantisation to slightly higher resolution; it represents two bits of information.

It is interesting to note the analogy between classifying an entity into quality grades 1–4, or ****, ***, **, * and the traditional graded statistical reliability into significance levels ***, **, * or n.s. (*not significant*). Both methods intend to convey important conclusions with very few bits of information (see Appendix A16 for more details).

By the way, when using traditional statistical hypothesis testing, it should be noted that a statistically *significant* effect is not necessarily a *meaningful* effect. Moreover, for effects that are *not* found to be statistically significant, the approximate uncertainty bounds of one's results ought to be stated. It is not enough to report *n.s.*, because then it would be possible, e.g. to draw a *desired* negative conclusion just by doing *bad* enough research: 'Our product has shown no significant detrimental health effect' could mean: '*We used very few people in the test, and did sloppy measurements, in order to ensure that all statistical tests came out as non-significant*'.

In this book we generally recommend giving slightly more information about the quality of the results (cf. Chapter 2.2). Approximate *reliability ranges* or *confidence regions* and visualisation of the model stability are more informative. But they do require more bits of information, and fundamental OK/Not OK decisions sooner or later have to be taken by someone, based on the results.

In the future, cognitive research may provide a better insight into how different individuals and cultures perceive risk, and understand uncertainty. Meanwhile, communicating risk and uncertainty is a challenge in research. The international effort for quality standardisation is a major step in the right direction.

### 2.5.3. The Word *Quality* as Carrier of Information Between People

What happens when a speaker **X** utters a statement to a listener **Y** about quality in a daily connotation, *goodness*? Here is a possible metaphor, inspired by the communicative language model of Nørretranders (1991). A complex, high-dimensional set of intentions, perceptions, memories and opinions concerning an entity has in the speaker's mind been filtered, combined and compressed, in light of the context, into a low-dimensional latent variable of mind **X**: a quality scale of goodness. This quality value has further been quantised into language with a low number of quality grade-words (e.g. very bad, bad, good, very good, i.e. two bits of information), and person **X** may briefly say: 'The quality of this entity is *good*'. Upon hearing this, the listener **Y** will mentally unpack and expand this compressed verbal information, in terms of his or her contextual understanding, and will, if **X** and **Y** share the same cultural context, get the intended message.

Interestingly, this conceptual **X–Y** communication via compressed words has a clear analogy to the quantitative modelling of **X–Y** relationships in data via latent variables, the fundamental data analytical method in this book. In Figure 2.2 this bi-linear *soft modelling* was used for seeing what people in different countries seem to mean by the word quality. In Chapters 3–16 the same method will be used for visualising the main **X–Y** relationships in a number of different data sets, applying PLSR to reveal which variations in **X** are *relevant* to **Y**.

### 2.5.4. Cognitive Aspects of Multivariate Data Analysis

What goes on in our brain when we get confronted with a lot of data? Cognitive science concerns investigation of the structure and function of mental processes which account for human knowledge and behaviour (Ellis and Hunt, 1993). As was shown in Figure 1.1, data analysis is a cognitive discipline. This includes sensation, perception and cognition (cf. Appendix A2). During data analysis, we are bombarded with a multitude of visual, auditory and other sense inputs which are more or less filtered in the sensory registers of the brain, and matched with previous patterns from the memory. Our visual cortex has specialised neurons for detecting straight lines and other prototype sub-patterns. When this pattern recognition processing is completed, complex patterns are recognised, and ascribed meaning. In other words, perception as *information processing* integrates outer, physical stimuli which are sensed (bottom-up data) with internal, mental stimuli from memory (top-down data) to result in a human response.

This information processing is influenced by many factors, like attention, emotion, allocated mental capacity and context. The contemporary problem of defining *consciousness* is highly relevant for understanding of these responses, although beyond the scope of this book (see e.g. Dennett, 1991).

Let us go a little deeper into what is meant by *limited capacity* and *context*. The human brain has *limited capacity* for what can be processed, and our mind apparently has to choose between the many input data competing for our attention. In general, it is often said that we can consciously handle 7 ± 2 chunks of information simultaneously as a limit. This is not a magical number but depends on how we organise and elaborate upon the input data. Working with visual representation we can handle many more. That is why graphics are so important in data analysis.

Some antropologists use a word *capta* instead of *data* to indicate that there is only a selected amount of all the possible data that can be captured by one person (Caplan, 1988). From a multitude of data, the capta are brought into mental processing, resulting in meaningful information.

What we perceive as meaningful information is dependent on *context*. Take a look at Figure 2.6 (Bugelski and Alampay, 1961) and first decide what the last drawing in the upper row represents. Then change attention to the last drawing in the lower row and make up your mind what that illustrates.

**Figure 2.6.**    Same plot, two different meanings. What is the last figure in each row? A mouse or an old man? Copyright 1961. Canadian Psychological Association. Reprinted with permission.

**Figure 2.7.** Two different plots, same meaning. The two symbols mean the same (more or less).

How can it be that we have physically the same last drawing in the upper and lower row, respectively, but different meanings? This illustrates that we may have *one* pattern, but *two* different meanings depending on the context.

Now let us turn to Figure 2.7 which illustrates that we may have *two* different patterns, but *one* meaning. This exemplifies some of the mental mechanisms at play also when we look at data!

### 2.5.5. Quality and Qualimetrics Bridging Technology and the Humanities

For a long time it has been difficult to find a common ground for cross-scientific discourse between natural scientists and technologists on one hand, and the humanists on the other hand. The former may be realists who believe they measure an object independent of the person who measures. The latter may be phenomenologists who focus only on a person's subjective experience of an object (*qualia*).

Since the words *quality* and *information* are used in so many contexts ranging from philosophy and social sciences to physics and technology, they may represent a platform for rational discourse between the professional cultures and between academia and industry. The ongoing debate for a non-reductive theory of consciousness based on psychophysical principles and cognition (e.g. Chalmers, 1996; Núñez and Freeman, 1999) is interesting in this connection. According to Chalmers (1996) '...information links the physical world to the experiential world...an answer to controversies in science and the humanities.'

Thus, one approach to bridging that gap between the two cultures, may be to use *soft modelling* for finding which combinations of properties or physical stimuli (**X**) are relevant for interpreting human responses (**Y**) in a given set of objects. Or oppositely, finding which human responses (**X**) are relevant for perceiving and understanding the physical quality properties (**Y**). In both cases a latent structure in an information space is revealed, between predictability and surprise, and between objectivity and subjectivity.

The concluding argument in this chapter is that qualimetrics offers a useful set of concepts and tools for extracting and interpreting an underlying, latent information space common to the physical and experiential worlds; thus revealing a common ground for communication about quality.

## 2.6. TEST QUESTIONS

1. What is meant by the statement 'The world is under indirect observation'?
2. What determines the quality of information?
3. What were the main steps in the study on the quality concept (Chapter 2.3)?
4. Name four different ways to define quality
5. Is the quality of a cadillac better than quality of a bicycle?
6. Is a success product (i.e. large sales volume) automatically a quality product?
7. Is a *good wine* for you dependent on your *knowledge* about the name, price or the chemical content of the wine?

## 2.7. ANSWERS

1. 'The world is under indirect observation' is here meant to indicate that neither our senses nor our instruments give us direct access to the phenomena we want to observe; our measurements convey information, via transfer functions, of psychological, chemical, electronic or computational nature.
2. The quality of information mainly depends on its reliability and relevance.
3. The main six steps in the quality concept study consist of (i) define a *purpose*; (ii) set up questionnaire and an *experimental plan* for which people to address; (iii) do *experimental work*, i.e. find the individuals and get them to fill out the questionnaire; (iv) enter the data in the computer and do *preprocessing and quality control of data*; (v) do the *data analysis*, i.e. by computing the average responses for each country and analysing the data by cross-validated bi-linear modelling, and (vi) draw *conclusions* and report this.
4. The four ways to define quality in this book are: (QD1) inherent descriptors (qualities). (QD2) Excellence/goodness. (QD3) Quality standards: (3a) choose quality criteria by finding the descriptors that are important for a given need and setting specifications for these descriptors: (3b) quality control of new products according to the chosen quality criteria. (QD4) Lived quality.
5. Not necessarily: that depends on which quality definition you use and what your needs are. Without petrol, you are better off on a bicycle.
6. A successful product (i.e. large sales volume) may automatically be considered a quality product for the producer if he makes money on it, but not necessarily for the consumer.
7. Up to you to decide!

# Chapter 3

# A Layman's Guide to Multivariate Data Analysis

*This chapter guides the reader through six steps in the research project cycle in which multivariate data analysis is considered to be important. Qualimetrics is shown to link questions and goals to answers and conclusions, by using cross-validated, soft bi-linear modelling (the BLM method) as a tool. Choice of data analytical models, validation tools and plots are demonstrated and used in analysing a small data set (a consumer study). The appendix lists the many alternative data analytical methods that can be replaced by the present method.*

## 3.1. RELIABLE AND RELEVANT INFORMATION AS A GOAL

A good goal formulation is the first step to successful results in research. The *overall goal* is expressed by the purpose and motivation of the activities. The operational *project goals* and purposes express a problem to be solved or an investigation to be performed. The overall goal is often to get *information about quality*. The operational goals usually entail a search for new information, and this information must be of *sufficient quality*: it must be reliable and relevant (cf. Chapter 2.2). The present Layman's Guide to Multivariate Data Analysis is built on an earlier version (Martens et al., 1983).

### 3.1.1. Examples in this book

Table 3.1 lists the ten data analytical examples in this book. Table 3.2 outlines the quality goals which these examples illustrate, addressing the various quality definitions in Chapter 2.4, and the appropriate BLM method used. All the examples (except the first one) follow the present layman's guide.

No two research projects are alike. No two researchers are alike, either. To set strict rules of how research should be conducted would be futile at best: researchers are like cats and make their own choices. But the following section

**Table 3.1.** Overview of examples with data sets analysed in this book. The original data sets are available at www.wiley.co.uk/chemometrics

| No. | Example | Data set | Chapter |
|-----|---------|----------|---------|
| 1 | Causal vs. predictive modelling of quality | COCOA-I | 1.4 |
| 2 | The confusing word *quality* | QUALITY-CONCEPT | 2.3 |
| 3 | Predicting consumer quality perception | COCOA-III (cont. 1.4) | 3.5 |
| 4 | Quantitative information from light | LITMUS-I, -II, -III | 4–7 |
| 5 | Interdisciplinary study of product quality | COCOA-II, -III (cont.1.4, 3.5) | 8–11 |
| 6 | High-speed determination of quality | WHEAT-NIR | 12 |
| 7 | Quality of the working environment | JOB-QUALITY-I, -II | 13 |
| 8 | Health quality with respect to toxicity of pollutants | TOX-QUANTCHEM | 14 |
| 9 | Multivariate statistical process control | MSPC-SUGAR | 15 |
| 10 | Food safety quality | MOULD-DESIGN-I, -II | 16 |

gives some advise on how data may be generated and analysed, based on the experiences of the authors.

## 3.2. QUALIMETRICS AS MEANS

To reach the quality goals, certain means are necessary. Qualimetrics in a broad sense (cf. Chapter 2.5), including multivariate planning and data analysis of experiments, represents a powerful set of means. The approach taken in this book is to focus on *data-driven modelling*, instead of, or before more traditional *theory-driven modelling*. Therefore, the input data must be sufficiently *informative*. What does that mean?

For efficient and safe research in complex, incompletely understood systems, the best way forward is to generate and analyse empirical data. Existing knowledge should be used, both during experimental planning and in the data analysis. But the data should not be forced into too narrow theoretical mathematical models: the data should be allowed to speak for themselves, at least initially. On the other hand, some modelling is required (cf. Chapter 1.3): the mind cannot otherwise grasp the contents of large tables of data.

### 3.2.1. From Measurements to Data

The researcher starts with a universe of *potential* observations from which he/she has to choose what *actually* to record or measure (Coomb, 1964). Before the recorded observations can be analysed, they must be put into data tables to become compatible and comparable with each other. A data table has rows

**Table 3.2.** Quality goal for the examples in this book giving typical applications and with links to some quality definitions (QD1–3) in Chapter 2.4.

| Quality goal | Typical application | BLM method | Example no.[a] |
|---|---|---|---|
| **Find inherent quality characteristics QD1** | | | |
| Sensory descriptors | R&D | PCA, PLSR | 1, 5 |
| Chemical and physical properties | Process monitoring | PLSR | 4, 5 |
| Microbiological growth | Lab. Experiment | APLSR | 10 |
| **Explore subjective quality perception QD2** | | | |
| Quality cognition | Communication | PCA, DPLSR | 2 |
| Consumer liking | Marketing | PLSR | 3 |
| Job satisfaction | TQM[b] | APLSR | 7 |
| **Quality monitoring and control QD3** | | | |
| Economical and nutritional quality | Agricultural QC[c] | PLSR | 6 |
| Environmental and health quality | Pollution control | PLSR | 8 |
| Technological processing quality | Multivariate SPC[d] | PLSR | 9 |

[a] The example numbers refer to Table 3.1.
[b] Total quality management.
[c] Quality control.
[d] Multivariate statistical process control.

and columns, which we call samples (or objects) and variables. Figure 3.1 outlines some principles involved in choosing which samples to study and which variables to measure. The lower part of Figure 3.1 shows the *basis* for planning an experiment so that the data, when they finally are obtained, will have a good chance of giving the *necessary information:*

❐ SAMPLES (OBJECTS)

First, the samples to be studied must be relevant to the purpose. To ensure this, it is usually an advantage to try to describe explicitly what type of samples would be of interest, i.e. the *target* population.

This does not have to be a population in the strict statistical meaning, because real-world situations make it difficult to draw sharp limits between what is inside and what is outside the population. But it can be useful to make some restrictions, (e.g. *here we are only concerned with hard red winter wheat; only samples measured in this instrument;* or *only from this sugar beet factory*). More details

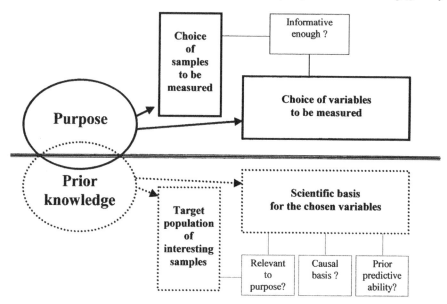

**Figure 3.1.**    Using prior knowledge as a basis for choosing samples and variables.

on conscious choice of samples to get a sufficiently informative sample set (experimental design) will be given in Chapter 11.

❐  VARIABLES

The choice of variables to be measured should also be relevant to the purpose and have a plausible scientific basis. This basis could either be a more or less strong theoretical evidence of a *causal* relationship between the measured variable and the effects of interest, be it forward, reverse or common-cause relationships (cf. Chapter 1.4). Alternatively, it may be a well-documented *empirical* correlation between the variables and the effects of interest. If there is neither a good causal basis nor a good empirical basis for the variables chosen, the experiment is a risky one.

❐  NUMBER OF SAMPLES AND VARIABLES

The input data must have the necessary variation to allow the data-driven soft model to become informative and trustworthy. This means, first of all, that the chosen set of variables must have the ability to describe the interesting phenomena in the system to be studied. For example, a set of shape- and size-measurements of the heads of adult humans is insufficient for describing their intelligence or personal character, even though that was believed a century ago.

Moreover, the chosen set of samples must be diverse enough to span each of

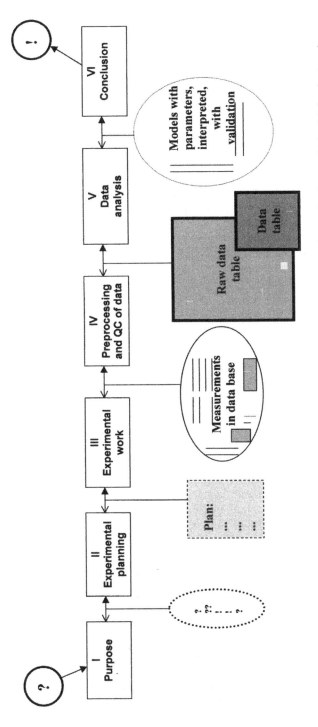

**Figure 3.2.** (a) Six steps in one research project cycle from question to answer, and how they relate to the data analysis; (b) the six steps in more details.

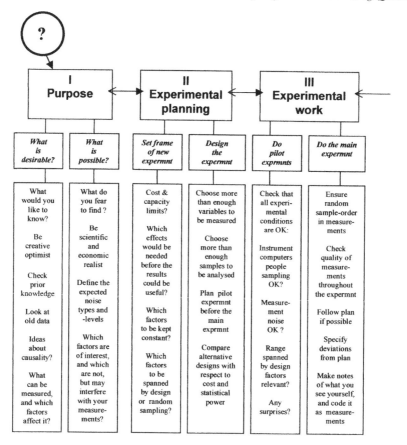

| I<br>**Purpose** | | II<br>**Experimental planning** | | III<br>**Experimental work** | |
|---|---|---|---|---|---|
| *What is desirable?* | *What is possible?* | *Set frame of new expermnt* | *Design the expermnt* | *Do pilot exprmnts* | *Do the main expermnt* |
| What would you like to know?<br><br>Be creative optimist<br><br>Check prior knowledge<br><br>Look at old data<br><br>Ideas about causality?<br><br>What can be measured, and which factors affect it? | What do you fear to find ?<br><br>Be scientific and economic realist<br><br>Define the expected noise types and -levels<br><br>Which factors are of interest, and which are not, but may interfere with your measurements? | Cost & capacity limits?<br><br>Which effects would be needed before the results could be useful?<br><br>Which factors to be kept constant?<br><br>Which factors to be spanned by design or random sampling? | Choose more than enough variables to be measured<br><br>Choose more than enough samples to be analysed<br><br>Plan pilot expermnt before the main exprmnt<br><br>Compare alternative designs with respect to cost and statistical power | Check that all experimental conditions are OK:<br><br>Instrument computers people sampling OK?<br><br>Measurement noise OK ?<br><br>Range spanned by design factors relevant?<br><br>Any surprises? | Ensure random sample-order in measurements<br><br>Check quality of measurements throughout the expermnt<br><br>Follow plan if possible<br><br>Specify deviations from plan<br><br>Make notes of what you see yourself, and code it as measurements |

**Figure 3.2(b).** (*continued*)

the interesting variation phenomena in the system. For example, a set of samples consisting of young men and old women only is unable to distinguish the *confounded* effects of age and gender.

Finally, if the noise level in the input data is high, compared to the variation range of the interesting phenomena in the system, then the number of samples required increases before precise enough soft modelling can be obtained. However, with the use of BLM as a modelling method, there is no longer a problem with having more variables than samples, like it was in many traditional multivariate statistical methods.

More details on informative, yet cost-effective choice of samples (experimental design) will be given in Chapter 11. The rest of Chapter 3 provides an overview of how the soft modelling approach may be applied in research projects, and then outlines the BLM method and its versatility. The practical details of the BLM method will then be explained in subsequent chapters.

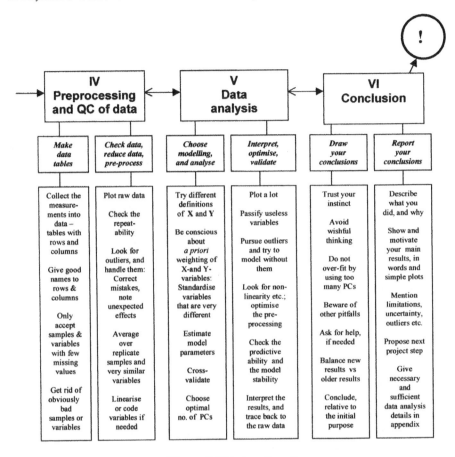

| IV Preprocessing and QC of data | | V Data analysis | | VI Conclusion | |
|---|---|---|---|---|---|
| *Make data tables* | *Check data, reduce data, pre-process* | *Choose modelling, and analyse* | *Interpret, optimise, validate* | *Draw your conclusions* | *Report your conclusions* |
| Collect the measurements into data – tables with rows and columns | Plot raw data | Try different definitions of X and Y | Plot a lot | Trust your instinct | Describe what you did, and why |
| Give good names to rows & columns | Check the repeatability | Be conscious about *a priori* weighting of X-and Y-variables: | Passify useless variables | Avoid wishful thinking | Show and motivate your main results, in words and simple plots |
| Only accept samples & variables with few missing values | Look for outliers, and handle them: Correct mistakes, note unexpected effects | Standardise variables that are very different | Pursue outliers and try to model without them | Do not over-fit by using too many PCs | Mention limitations, uncertainty, outliers etc. |
| Get rid of obviously bad samples or variables | Average over replicate samples and very similar variables | Estimate model parameters | Look for non-linearity etc.; optimise the pre-processing | Beware of other pitfalls | Propose next project step |
| | Linearise or code variables if needed | Cross-validate | Check the predictive ability and the model stability | Ask for help, if needed | Give necessary and sufficient data analysis details in appendix |
| | | Choose optimal no. of PCs | Interpret the results, and trace back to the raw data | Balance new results vs older results Conclude, relative to the initial purpose | |

**Figure 3.2(b).** (*continued*)

### 3.2.2. One Research Project Cycle: *Steps I–VI*

A research project spans from questions via data to answers. Figure 3.2a outlines six steps in a research project cycle, that are associated with multivariate data analysis. More details on each of these steps are given in Figure 3.2b.

The six steps are the following:

*I. Purpose.* The cycle begins with thorough thinking about the purpose and overall goal of the project, in increasing levels of detail. This results in the assessment of what would be relevant and desirable, what is technically possible, what is known and what is unknown, what are the most likely outcomes and what type of phenomena or factors are expected to affect the intended measurements and their interpretation. This project purpose guides the research cycle.

*II. **Experimental planning***. A detailed plan for a new experiment or investigation is made. This involves setting an operational framework for the new experiment, identifying the different factors expected to have effects and deciding how to treat each of them. It also involves choosing a generous set of *variables* to be measured, and setting up a generous, but cost-effective *experimental design*: a plan for obtaining *samples*, as discussed above in relation to Figure 3.1 (cf. Chapter 11 for more explanation).

*III. **Experimental work***. The actual experiment or data collection is conducted. The details of this important work are project specific, and will not be discussed here. It is often a good idea to perform preliminary pilot or screening experiments, in order to simplify and optimise the main experiment. The many measurements or registrations obtained, are stored in a database.

*IV. **Pre-processing and quality control (QC) of data***. The measurements are organised as raw data in a large table that allows systematic comparisons between samples (rows) and between variables (columns). Quality control of data means ensuring that the raw data are correct and available. Thus, the raw data table is briefly checked with respect to repeatability (Chapter 10). If gross errors are detected, their cause is pursued; their data are either corrected or just noted and replaced by the symbol for *missing values*. Remember that *missing values* mean *missing information* and this should be less than 10% of the data. By averaging over repeatability-replicate samples and over adjacent or otherwise near-identical variables, a more compact, although still overwhelming, table of input data is obtained.

*V. **Data analysis***.The input data table is submitted to multivariate data analysis for quantitative interpretation, as illustrated in Chapters 12–16 and explained in Chapters 4–11. This has several steps:

1. *Choosing a suitable model*: the researcher selects a way to relate different parts of the data set to each other. When two or more *types of variables* are available for the same samples (e.g. design variables and measured response variables), it is useful, both cognitively, statistically and for practical purposes, to analyse them as *two blocks of variables*, here called **X** and **Y**. With BLM the model choice is primarily determined by the definition of **X** and **Y** variables and of training set (*calibration set*) of samples. The main alternatives for choosing a model will be given later (Figure 3.4).
2. *Estimating the model parameters*: the variables are weighted according to approximate *a priori* assumptions about their precision (Chapter 8.2.3). The bi-linear model is fitted to the data by the principle of weighted least squares (Chapter 4.3). The main information content from the data may thus be extracted as compact, interpretable models with a few parameters that summarise all the samples and all the variables.

3. *Model validation*: the stability and the predictive ability of the models are checked by cross-validation (Chapter 10), at the chosen level of data analytical validity (repeatability, reproducibility etc.). Chapter 3.4 gives an overview of some useful validation tools, and choice of optimal model.

4. *Graphical interpretation*: each model is plotted in various ways and inspected in light of the available background knowledge, in search for anomalies and for patterns with plausible causal explanations. An overview of the main types of plots will be given in Figures 3.5 and 3.6.

5. *Handling outliers and non-linearities*: if the model validation reveals the presence of abnormal outlier-samples (Chapter 9.3) or apparently useless variables (Chapter 10.4), these may be pursued separately, and the modelling repeated without them. If the modelling indicates a need for non-linear model expansion or handling heterogeneous data set, this is tried (Chapters 8.2 or 9).

6. *Quality control of the results*: the main findings from the modelling are traced back to the raw data, to guard against mistakes or misinterpretation in the data analytical process.

*VI. Conclusions*. Finally, conclusions are drawn from the experiment, checked with respect to the original purpose, and reported as detailed in Figure 3.2b.

Generally, the six steps follow each other sequentially. But it may sometimes be useful to go back and improve previous steps, as the two-way arrows in Figure 3.2a,b indicate. At every stage it is important to be *creative*, pursuing one's intuition and applying one's contextual knowledge. But it is also important to be *critical*; beware of wishful thinking. Finally, one should avoid becoming a *mental prisoner* of the computer programme, the input data or the data analytical results at hand: use common sense and do not lose track of your purpose.

## 3.3. SOFT BI-LINEAR MODELLING (BLM) AS A TOOL

### 3.3.1 Choosing the BLM Method: a Versatile Engine

In the research project cycle, choosing a suitable model (*Step V.1*) is of utmost importance. The soft modelling approach chosen in this book is based on one single "work horse": the so-called bi-linear modelling method. This is a general multivariate method for finding the main information from data tables. It is called *bi-linear* because it is linear in two ways, with respect to both the variables (columns) and the samples (rows) in data tables. It belongs to *soft modelling* because it requires little or no causal assumptions to be built into the model: the data are allowed to *speak for themselves*, outside the cage of prior theory.

In brief, the BLM method consists of trying to focus the relevant and reliable information in the many input-variables called $\mathbf{X}=[\mathbf{x}_1,\mathbf{x}_2,\mathbf{x}_3,\mathbf{x}_4,\mathbf{x}_5,\ldots]$ into a smaller sequence of highly informative 'latent variables' or *Principal Compo-*

*nents (PCs)*, $\mathbf{T}=[\mathbf{t}_1, \mathbf{t}_2, \mathbf{t}_3, \ldots]$ via a mathematical function, here called $w(\ )$: $\mathbf{T}=w(\mathbf{X})$. These PCs are, in turn, used for modelling the X-variables themselves. This additive bi-linear model is a sum of an informative, systematic part $p(\mathbf{T})$ and a less informative, residual error $\mathbf{E}$: $\mathbf{X}=p(\mathbf{T})+\mathbf{E}$, cf. Chapters 5–7.

If (optionally) some of the input variables have been defined as Y-variables $\mathbf{Y}=[\mathbf{y}_1, \mathbf{y}_2, \mathbf{y}_3, \mathbf{y}_4, \ldots]$ instead of in $\mathbf{X}$, then they are now modelled just like the X-variables, as a sum of a systematic part $q(\mathbf{T})$ and a residual part $\mathbf{F}$: $\mathbf{Y}=q(\mathbf{T})+\mathbf{F}$. This is explained in Chapters 8 and 9.

Cross-validation is used for determining $A_{Opt}$, the optimal number of PCs $\mathbf{T}=[\mathbf{t}_1, \mathbf{t}_2, \ldots, \mathbf{t}_{AOpt}]$ to be used in the modelling of $\mathbf{X}$ and $\mathbf{Y}$. Moreover, the cross-validation allows us to assess the quality of the obtained model. Cross-validation consists in re-computing the models $w(\ )$, $p(\ )$ and $q(\ )$ many times, each time temporarily treating a different part of the available sample set as 'secret' test-samples, as explained in Chapter 10.

The models $w(\ )$, $p(\ )$ and $q(\ )$ can then be combined in order to relate the Y-variables directly to the X-variables: $\mathbf{Y}=b(\mathbf{X})+\mathbf{F}$. The obtained model parameters and residuals may be inspected graphically for interpretation, and used for quantitative prediction of the Y-variables in new samples from the X-variables.

Figure 3.3 illustrates the most typical application of BLM: a practical *engine* for extracting and displaying the systematic essence from two input data sources $\mathbf{X}$ and $\mathbf{Y}$, while throwing aside their unsystematic errors $\mathbf{E}$ and $\mathbf{F}$.

Figure 3.4 gives an overview of different models to choose from, within the BLM method (cf. ***Step V.1*** in Chapter 3.2.2).

***1. BLM of one data table X:*** extracting the main variation patterns from one single data table $\mathbf{X}$ for a set of samples and a set of variables. This is here done by *principal component analysis (PCA)*.

> Simple example: let table $\mathbf{X}$ represent various *body measurements* (variables) from a group of children (samples) of different ages. When children grow, all their body parts grow: that constitutes a dominating pattern of variation between the children, i.e. a principal component (PC).

PCA is explained in Chapter 5. Although PCA is the 'mother of all multivariate methods', it is here primarily regarded as a special version of the more general two-table BLM method, without the second table ($\mathbf{Y}$).

> The basic method behind PCA has been re-invented again and again in different fields of science and therefore goes by many names: PCA (statistics), singular value decomposition (SVD, numerical analysis), Karhunen Loeve transform (signal processing). It can be computed in a number of different ways. Its *squared form* is called eigen analysis (the extraction of eigenvalues and eigenvectors), which is used extensively, e.g. also for determination of electron orbitals in atoms and for analysis of harmonics in music: it is a central mathematical tool in many sciences.

**Figure 3.3.** The way to extract essence from data. A researcher who has learnt to use the information search engine BLM, with direct dual data injection and double garbage expulsion. Illustration by Jens Strøm.

***2. BLM of two data tables X and Y***: extracting the main variation patterns from one data table **X** that have relevance also for another data table **Y** from the same samples. This allows us to interpret the structures within and between **X** and **Y**, and the obtained model may be used for predicting **Y** from **X** in future samples (cf. title page). This is here done by partial least squares regression (PLS regression, or just PLSR).

Simple example: From the size and shape of the children, i.e. the body measurements **X** we can predict their age and language abilities **Y**.

The partial least squares regression method is sometimes just called *PLS* in the chemometrical literature. The term *PLSR* is here used in order to avoid confusion with other types of PLS modelling, e.g. the original multiblock path modelling still used in econometrics (Jöreskog and Wold, 1982; Tenenhaus and Morineau, 1999). The original PLSR method (Wold et al., 1983a) has here been modified slightly to make it even more robust and versatile, cf. Appendix A6.8.

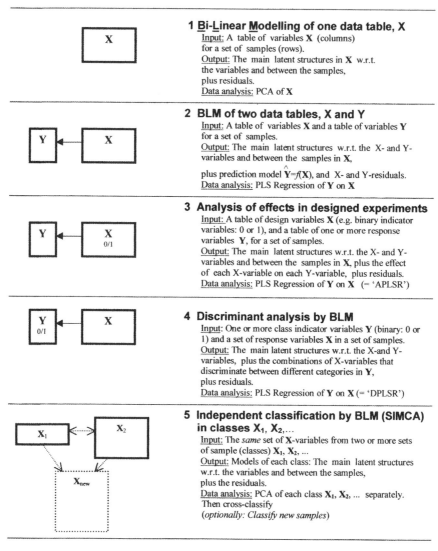

## 1 Bi-Linear Modelling of one data table, X

Input: A table of variables **X** (columns) for a set of samples (rows).

Output: The main latent structures in **X** w.r.t. the variables and between the samples, plus residuals.

Data analysis: PCA of **X**

## 2 BLM of two data tables, X and Y

Input: A table of variables **X** and a table of variables **Y** for a set of samples.

Output: The main latent structures w.r.t. the X- and Y-variables and between the samples in **X**,

plus prediction model $\hat{\mathbf{Y}} = f(\mathbf{X})$, and X- and Y-residuals.

Data analysis: PLS Regression of **Y** on **X**

## 3 Analysis of effects in designed experiments

Input: A table of design variables **X** (e.g. binary indicator variables: 0 or 1), and a table of one or more response variables **Y**, for a set of samples.

Output: The main latent structures w.r.t. the X- and Y-variables and between the samples in **X**, plus the effect of each X-variable on each Y-variable, plus residuals.

Data analysis: PLS Regression of **Y** on **X** (= 'APLSR')

## 4 Discriminant analysis by BLM

Input: One or more class indicator variables **Y** (binary: 0 or 1) and a set of response variables **X** in a set of samples.

Output: The main latent structures w.r.t. the X-and Y-variables, plus the combinations of X-variables that discriminate between different categories in **Y**, plus residuals.

Data analysis: PLS Regression of **Y** on **X** (= 'DPLSR')

## 5 Independent classification by BLM (SIMCA) in classes X₁, X₂,...

Input: The *same* set of **X**-variables from two or more sets of sample (classes) $\mathbf{X}_1$, $\mathbf{X}_2$, ...

Output: Models of each class: The main latent structures w.r.t. the variables and between the samples, plus the residuals.

Data analysis: PCA of each class $\mathbf{X}_1$, $\mathbf{X}_2$, ... separately. Then cross-classify

(*optionally: Classify new samples*)

## 6 Extension of BLM: Find relation between M-way Y and N-way X

Input: An *M*-way table **Y** (here: 2-way) and *N*-way table **X** (here: 3-way) for a set of samples (rows).

Output: Same as for 2) above, or as 3) if **X** is binary, or as 4) if **Y** is binary, but the Y- and X-patterns are given for each of the *M* and *N* ways, respectively.

Data analysis: N-way PLS (N-PLS) Regression of **Y** on **X**

**Figure 3.4.** Some common data analytical goals and choices for a suitable BLM method.

PLSR is explained in Chapter 6. These two bi-linear methods, PCA and PLSR, may be used in many different manners, and for many different kinds of samples and variables.

Some special uses of the BLM method need to be mentioned specifically. If certain *facts* are available about the samples, these facts may be coded as *indicator* variables. A typical case is the interpretation of designed experiments: If the design *factors* are of categorical nature (e.g. *Boy* or *Girl*), they may be brought into the modelling by coding each factor as a binary *indicator* variable (e.g. 0 = *Boy*, 1 = *Girl*). A binary variable is a special type of variable often used in classification and in experimental design, quantised to only two levels (eq. 8.3).

In general, this type of representation may be used for qualitative variables with level 0 = *false* or *no*, and 1 = *true* or *yes* (cf. Chapter 13). Other levels of such *quantisation* (stepwise representation) may also be used, e.g. $-1$ = *Low*, 0 = *Medium* and 1 = *High*. This is illustrated in Chapters 11 and 16.

Depending on how these indicator variables are used, we may name the PLSR modelling differently.

***3. Analysis of effects in designed experiments by BLM (APLSR):*** the effect of each level of each design factor may be assessed when $\mathbf{X}$ = design variables e.g. binary (0/1) indicator variables and $\mathbf{Y}$ = one or more response variables, with the model parameters estimated by the PLSR. The reliability of the observed effects is assessed by cross-validation, as an alternative to traditional statistical analysis of variance (ANOVA). To mark that $\mathbf{X}$ consists of indicator variables, this is here called APLSR or ABLM. APLSR is explained in Chapter 11.3 and applied in Chapter 16 and in the literature (Aastveit and Martens, 1986; Martens and Martens, 1986a; Martens et al., 1987).

> Simple example: Let table $\mathbf{X}$ represent *known* information about the above-mentioned children, e.g. $x_1$ = gender (0 = boy, $+1$ = girl), $x_2$ = age (no. of years), and let $\mathbf{Y}$ represent both the body measurements and the language abilities measured. The APLSR will then show how each of the Y-variables, on the average, changes with X-variables gender and age.

***4. Discriminant analysis by BLM (DPLSR):*** if the samples belong to different classes or categories, we may learn how to distinguish between them by defining $\mathbf{Y}$ = binary (0/1) indicator variables for the categories, and $\mathbf{X}$ = a set of other variables, e.g. measured responses. The PLSR model reveals what distinguishes the categories, and allows us in the future to categorise new, unknown samples of the same general type, from their X-data. This method of discriminant analysis is here called DPLSR or DBLM. But like for APLSR, the PLSR itself is the same as in 2. DPLSR is explained and applied in Chapter 9, and in the literature (Wold et al., 1983b, 1984).

Simple example: Let table **Y** represent *known* information about the children, e.g. $y_1 =$ gender ($0 = $ boy, $+1 = $ girl), $y_2 = $ age (no. of years), and **X** represent both the body measurements and the language abilities measured. The DPLSR will then reveal the systematic gender- and age-related variations in the measured X-variables, and show how, and to what extent, the gender and age may be predicted from the body- and language-measurements.

**5. Independent classification by BLM (SIMCA):** the BLM-based soft independent modelling of class analogy (SIMCA) is a multiblock BLM-based classification method applicable if the samples belong to classes that are too different from each other to fit into one simple bi-linear model. Instead, each class is modelled separately by PCA (as in Wold et al., 1983b). Later, each sample is fitted to each of the independent PCA models to check the class separation; this may also be done for classifying future unknown samples. SIMCA analysis can also be based on PLSR instead of PCA, as explained in Chapter 9.

Simple example: let table $X_1$ represent the body- and language-measurements for the *boys*, and $X_2$ the *same* measurements for the *girls*. Similarly, let $X_{new}$ represent the same measured variables for new children of unknown gender. Tables $X_1$ and $X_2$ are first analysed separately by PCA, and the obtained models compared to each other in order to see what really distinguishes boys and girls from each other. Then each new child may be fitted to each of the two models, for gender classification.

**6. Extensions and modifications of BLM:** related, but more advanced models exist, which allow data analytical *experts* to enhance the modelling even further. For instance, the BLM method may be extended and modified in a number of ways, but this goes beyond the scope of the present introductory book.

If there is a need for it, the two-block bi-linear PLSR may be extended to *multiway* data tables: this is illustrated for a two-way Y-table and three-way X-table. Different extensions reflect different professions (cf. Appendix A1). The advantage of this (instead of the alternative of unfolding **X** into a conventional two-way matrix) is simpler model interpretation and better noise rejection (Bro, 1996, 1998). The two-block PLSR may also be extended to multiblock path modelling in latent variables: the original version of the PLS modelling principles developed in econometrics (Wold, 1985). The original econometrics version of PLS (Wold, 1975) allowed three or more blocks of variables. For more advanced path modelling, this is still a very useful tool. But for most practical purposes the advantage of this is more than offset by the increased complexity in interpreting and validating the models. Multiblock situations may also be modelled by the present two-block BLM approach.

Likewise, the one-block bi-linear PCA, which is suitable for analysis of a single two-way data table, may be extended to analyse one multiway data table (tri-linear Tucker or PARAFAC analysis) as well as several two-way data tables (multiblock generalised procrustes analysis, Gower, 1975).

Additional knowledge may be included inside the bi-linear parameter estimation. For instance, if the data are known to reflect some smooth process (smooth time series or smooth spectra), mathematical smoothing may be built into the PCA/PLSR algorithm, to make the model more robust against non-smooth noise. More details are given in Appendix A6.14 and by Martens and Næs (1989).

From an application point of view we may regard the multivariate bi-linear modelling with *no* Y-variable (PCA), *one* Y-variable or *many* Y-variables (PLSR) as being one single method, here called the BLM method. Appendix A3.1 lists a number of different statistical and computational methods that may be replaced by this simple, unified approach. This shows how versatile the BLM method is.

### 3.3.2 What is X and What is Y?

The choice of what to use in **X** and in **Y** as part of *Step V.1* in the research project cycle depends on the researcher and on his/her goal. This allows some control over what the *soft modelling* should particularly look for. That freedom can be a little confusing in the beginning, but the choice is usually non-critical, as illustrated in Chapter 8. The choice of what is **X** and what is **Y** does not have to follow unspoken tradition that **X** = cause and **Y** = effect.

> In classical regression theory **X** are called *independent variables* or *regressors*, and **Y** are called *dependent variables* or *regressands*. These names may cause a confusion between the underlying causality and how they are used in data analytical modelling. We learnt to crawl on the *x*-axis as babies; only later we started to lean on the *y*-axis. Does that have anything to do with how we spontaneously choose **X** and **Y**?

In the case of multichannel analytical instruments, **X** is usually defined as the output of the instrument while **Y** is the quality or property to be calibrated for (e.g. the concentration of a chemical compound, the values of a set of interesting quality criteria, or a set of class indicators).

In assessment of designed experiments, **X** may be the known *causes* and **Y** the observed *responses*: this is the simple and safe APLSR approach. But the opposite choice, DPLSR, is often more interesting, if many response-variables have been observed.

In all other cases it is best to let **X** be the block of variables that spans the most information (e.g. with the highest number of informative variables or with the highest general precision), and let **Y** be all the other potentially interesting variables. The choice of **X** and **Y** will be illustrated and discussed several times in Chapters 4–16, and particularly in Chapter 8.

Generally speaking, if two blocks of variables reflect similar information, it does not matter which is **X** and which is **Y** in BLM. But if, upon swapping **X** and **Y**, very different solutions are obtained, that disagreement can be very informative.

The set of objects may be split into two or more classes. Conceptually, this book uses two ways to split the samples: (1) Calibration vs. prediction samples: the samples available NOW are used for calibration; samples arriving in the

FUTURE are treated as prediction samples; (2) Modelling of heterogeneous systems by SIMCA (cf. Figure 3.4) or multiclass PLSR (cf. Chapters 9 and 14).

## 3.4. SOME USEFUL DATA ANALYTICAL VALIDATION TOOLS

Since the combination *chance correlations* + *wishful thinking* forms a dangerous trap for the researcher's mind, data analytical models and conclusions must be *validated*. Proper reproducibility tests require independent experiments, performed at another time with other instrumentation and even other people. While being strongly recommended, this is beyond the scope of the present book.

But even during the analysis of one given data set, validation is important. Statistical significance and predictive ability within one data set will ensure *internal* validity, while causal interpretability based on expert scientific knowledge gives a simple type of *external* validity (cf. Chapter 2.2). In this book the techniques of cross-validation and jack-knifing are central. They will be explained in Chapter 10 and applied in all the examples in Chapters 12–16.

During the actual analysis of a given data set, the following steps, within *Step V.3* in the research project cycle, are useful:

*Model validity check*: The first step in interpreting the modelling results is to check if the model makes sense at all, statistically. It is only safe to start interpreting its detailed parameters if the model shows at least some predictive ability. An exception is when we have too few observations to be able to check predictive ability, e.g. when using strongly reduced experimental designs such as fractional factorials (Chapter 11.3). If the results correspond to what we, as experts, already know or can easily understand, they can still make sense to us.

*Model rank assessment*: The second step is to determine the optimal number of latent variables (PCs) in the bi-linear model, here called $A_{Opt}$, e.g. by internal cross-validation at some chosen level of validity (repeatability, reproducibility, etc). It is important to use as few PCs as possible, in order to simplify the interpretation and to stabilise the results against errors in the data.

*Predictive ability*: Once the optimal number of PCs has been chosen, the cross-validation also shows the estimated prediction error in **Y**, usually expressed as the root mean square error of prediction, RMSEP(Y), at the chosen validity level. This is important in order to assess how successful the modelling has been and if it can be relied upon in practice. Part of this apparent prediction error will be due to errors in the reference data **Y**: the actual predictive ability may be better, as long as the model is not used outside its range of applicability.

*Stability of the model parameters*: The stability of the estimated model parameters (regression coefficients, sample scores and variable loadings for the $A_{Opt}$ PCs) may be assessed during the cross-validation. This is sometimes called *jack-knifing*.

*Interpretation of model parameters*: One should always at least *try* to interpret the detailed modelling results contextually. This is done by graphical inspection of the model parameters and the residual statistics (prediction error etc).

## 3.5. SOME USEFUL PLOTS

### 3.5.1. The General Types of Plots

The *soft modelling* approach in this book combines one modelling method – the BLM method – with some simple statistical concepts of validation as described above, and finally, with computer graphics. Figure 3.5 illustrates some the most important types of plots to choose between in *Step V.4*, i.e. plots used in this book when monitoring input data, QC of data, interpreting models, checking results from data modelling against raw data, etc.

*Data table plot*: Figure 3.5a shows a bird's eye view of an actual data table. This map is useful for envisioning the data matrices involved in the data analysis, as well as for identifying gross errors. The samples (objects) are shown as rows, the variables as columns, and the values of the different data elements are quantised to a few intensity levels (here printed in three grey tones, instead of the more informative interactive multicolour coding).

This particular table concerns input data for the application in Chapter 14 giving values on toxicity and quantum chemical descriptors of various pollutants.

*Curve plot*: Figure 3.5b shows a way to gain more detailed view of a whole table of data: a one-way plot of a set of curves. This is very useful when the variables follow each other smoothly; the eye can then make local comparisons as well as general pattern recognition. The individual curves in the plot may be individual samples (*a spectrum*) or individual variables (e.g. *a time series*).

The present plot concerns light spectra in the near-infrared (NIR) wavelength range for ten wheat samples: these data will be analysed in Chapter 12.

*Bar plot*: Figure 3.5c shows another version of one-way plot, called *bar plot*. It is similar to the one-way curve plot in Figure 3.5b, but often allows more quantitative read-out from the plot. This advantage comes at the expense of increased visual complexity. The bar plots may be used for individual variables, as well as for samples, as in c.

The plot in Figure 3.5c compares three samples, in this case the most extreme cocoa-milk samples, for six of the 15 sensory variables in Table 1.1 (Chapter 1.4).

*Two-way plot*: Figure 3.5d shows one of the most important graphical techniques in soft modelling: a two-way *map* with text information. This type of plots is very useful for discovering patterns in data, either in the input data or in obtained bi-linear model parameters (see below).

In Figure 3.5d the two-way plot is used for studying two empirical variables against each other (abscissa = *colour*, ordinate = *cocoa odour*), with factual

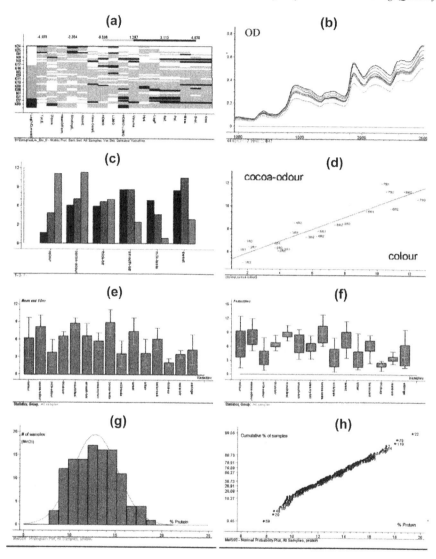

**Figure 3.5.**    Some useful plots for inspection of input data. (a) Data table map: dark/light = high/low. Toxicity and Quantum descriptors for 51 toxins. (b) One-way curve plot. NIR spectra of 10 wheat samples in the 1000–2500 nm range. (c) One-way BAR plot of three samples: *1* (most milk), *3* (most sugar), *7* (most cocoa). (d) Two-way plot of two variables: *Cocoa odour* vs. *colour* for 21 cocoa-milk samples. (e) Mean ± SD for each of 15 sensory variables, taken over a set of 21 cocoa-milk samples. (f) Percentiles: summary of 15 sensory variables over a set of 21 cocoa-milk samples. (g) Histogram of a variable: distribution of protein percentage in a set of 121 wheat samples. Curve = normal probability distribution. (h) Normal probability plot of a variable: distribution of protein percentage in a set of 121 wheat samples, plotted as if they were normally distributed.

names for each individual sample point, the seven cocoa-milk drinks from Table 1.1, measured in Replicates 1, 2 and 3 (Chapter 10).

***Statistical summary, many variables***: Figure 3.5e,f shows two ways of summarising data tables statistically. In (e) a set of variables are shown in terms of their means (bars) ± their total initial standard deviation. This gives a quick overview of their general levels and variation ranges. In (f) a set of variables are shown with respect to the 0 (minimum), 25, 50 (median), 75 and 100% (maximum) percentiles. This gives slightly more statistical detail, and reveals, e.g. variables that are strongly non-normally distributed and therefore might have an outlier or need a linearising pre-treatment prior to the multivariate analysis.

Figure 3.5e,f shows the distribution of 15 sensory variables over the 21 cocoa-milk samples from above. In (e) we see that the variables differ in mean value; hence it makes sense to mean-centre the variables prior to analysis. Furthermore, some variables (e.g. the first variable, *colour*) have much higher variation range than the others (e.g. third last, *thick-texture*); hence we may wish to standardise the variables, forcing all of them to have the same total initial standard deviation, prior to the multivariate analysis. In (f) we see, e.g. that the last variable (*astringent*) is somewhat asymmetrical: There are relatively fewer samples at the higher end than at the lower end of its distribution.

***Detailed statistics, one variable***: Finally, Figure 3.5g,h shows two detailed statistical ways to study an individual variable. These are usually not required, but may sometimes be helpful for identifying the nature of an error revealed by other means.

Figure 3.5g shows a bar plot histogram: the abscissa shows the range of values of the variables segmented into a certain number of bins, and the ordinate shows the number of samples that fall inside each bin. The curve superimposed on this distribution histogram is the normal probability distribution obtained from the mean and standard deviation of the variable. In Figure 3.5h the sorted data have been integrated into the cumulative distribution and plotted on non-linear normal-probability scale. If the histogram had perfectly fitted the normal probability curve in (g), then all the sample points would have fallen along a straight line in (h). These two plots are only useful if there is a sufficient number of samples to give meaningful distribution histograms.

Figure 3.5g shows the distribution of protein content in the set of 121 wheat samples from which the curves in (b) were taken. Figure 3.5h shows these samples to be near normally distributed; no abnormal outliers are evident. Samples 23 and 56 represent the most extreme values of protein, but they do not fall far away from the normal trend in the figure.

### 3.5.2. Example of BLM Plots: a Small Consumer Study

Figure 3.6 summarises a practical example of the ways this book uses plots during the multivariate data analysis. The reader does not have to understand the plots in detail at this stage: the figure just serves to give an impression of the thinking process and graphics in the multivariate analysis. The example is a continuation of the cocoa-milk study in Chapter 1.4 (from Folkenberg et al., 1999), and it follows the six steps in the research project cycle.

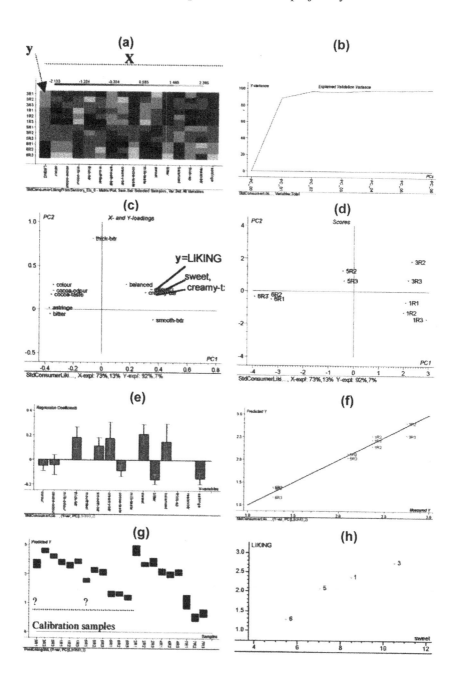

*I. Purpose*. The example concerns the modelling of consumer preference of ready-made cocoa-milk drinks from descriptive sensory analysis. To what extent can we use the available data to develop a BLM, which then allows us to predict consumer *LIKING* even in other samples, from their sensory quality descriptors?

The data analytical purpose in this compact representation is just to demonstrate some relevant BLM plots.

*II + III. Experimental planning + work*. In this small pilot experiment, only four of the cocoa mixtures from Figures 1.2–4 were submitted to a consumer study (mixtures *1*, *3*, *5* and *6*). Sixty consumers gave their score of *LIKING* on a four-point hedonic scale. The same samples had previously been evaluated in three replicates, using 15 sensory variables in a sensory descriptive analysis (cf. Chapter 10).

*IV. Pre-processing and QC of data*. One single y-variable, consumer *LIKING* (averaged over 60 consumers); is to be regressed on 15 sensory variables from Figure 3.6a. The average *LIKING* results for mixtures *3*, *1*, *5* and *6* were 2.8, 2.4, 2.1 and 1.3, respectively, i.e. sample *3* was liked best (file COCOA-III).

Figure 3.6a characterises the input data tables for **y** and **X** of the 12 cocoa-milk samples: there are three copies of the y-value for each of the four mixtures, one for each of the three sensory replicates. Each variable has been standardised to a total initial standard deviation of 1.0, to ensure that each variable has equal chance of contributing to the modelling.

*V. Data analysis*. PLSR analysis was applied for exploring the relationships between the remaining ten sensory variables, **X,** and the consumer *LIKING*, **y**.

Figure 3.6b. shows how much of the original y-variance was explained by the PLSR model as the number of PCs increased. After one PC 90% of the total variation in consumer LIKING **y** was explained, while after two PCs 97% was explained: beyond this there was little or no change. Therefore, the two-PC solution is chosen.

---

**Figure 3.6.** Some useful BLM plots, illustrated for a small consumer study of cocoa-milk drinks. (a) Input data tables (standardised): consumer *LIKING* (**y**) and 15 descriptors (**X**) for 12 cocoa-milk samples. (b) Explained variance in **y** vs. number of PCs, **y** = consumer *LIKING*. (c) Main pattern of variables: X- and Y- loadings $p_1$, $q_1$ (abscissa) vs. $p_2$, $q_2$ (ordinate) for ten active sensory variables in **X** and for **y** = *LIKING*. (d) Main pattern of samples: scores for first two PCs, $t_1$ (abscissa) vs. $t_2$ (ordinate). Cocoa-milk samples *1*, *3*, *5* and *6* in three replicates. (e) Final model: Regr. coeff. $\hat{b} \pm 2\,\hat{s}(\hat{b})$ for prediction of **y** = consumer *LIKING*.(f) Predicted $\hat{y}$ vs. measured **y** *LIKING* in cal. Samples. (g) Predicted $\hat{y}$ vs. sample no. Pred. *LIKING* in both cal. samples and new samples. (h) Final check in input data **y** = *LIKING* vs. sensory *sweet* taste.

A preliminary analysis showed two of the samples (3R1 and 5R1) in Figure 3.6a to be erroneous outliers which were eliminated from the main data analysis. A new preliminary modelling showed that some of the sensory variables had no predictive value, and these were eliminated.

Figure 3.6c,d shows the pattern of co-variation between the variables (loading plot) and samples (score plot), respectively. Consumer *LIKING* is positively correlated to *sweet* taste and *creamy texture*, and negatively correlated to *astringent* and *bitter* taste. In (d) the sensory replicate samples are reasonably close to each other. Further, the best liked sample *3* on the average, as found in *Step IV*, is situated to the right in the score plot, corresponding to *LIKING* in the loading plot. On the opposite, sample *6* is confirmed to be the least liked among cocoa-milk drinks tested.

Figure 3.6e shows the regression coefficients $\hat{b}$ obtained using two PCs, with approximate reliability range bars ( $\pm$ 2 estimated standard uncertainties). It shows that high consumer *LIKING* in this small pilot experiment appears to be associated with high levels of *milk-odour, smooth texture, creamy texture, sweet* taste and *balanced*, and low levels of *cocoa-taste, bitter* taste and *astringency*.

The consumer *LIKING* **y** was predicted from the sensory variables **X** within the ten good calibration samples via the coefficients in Figure 3.6e, and the *predicted LIKING* plotted in Figure 3.6f as ordinate against the *measured LIKING*. A reasonably linear relationship is seen, indicating the model to be satisfactory.

The obtained calibration model (Figure 3.6e) was then applied to the sensory descriptions of *new* cocoa milk mixtures (the remaining ones that had not been included in this small consumer study, namely mixtures *2, 4* and *7*). The resulting predictions of consumer *LIKING* are shown in Figure 3.6g for all the samples, both the calibration samples, as well as the new cocoa-milk mixtures *2, 4* and *7* not yet submitted to consumer study. Also the two samples, initially found to be outliers, 3R1 and 5R1 now could be predicted, marked by (?). The bars represent the predicted values with an approximate uncertainty range that is outlier-sensitive, estimated as described in Chapter 10.

Figure 3.6g shows that the first two of the untested samples (*2* and *4*) receive high *predicted LIKING*, while the last untested one (*7*) receives very low values of the same. In addition, sample *7* is tagged as a new outlier: particularly its first replicate falls outside the applicable range of the calibration model.

Finally, the importance of tracing the results from the multivariate data analysis back to raw data should be demonstrated. Figure 3.6h gives a two-way plot of *LIKING* vs. *sweet* taste, indeed, showing that *sweet* taste is strongly positively correlated with consumer *LIKING*, as expected from Figure 3.6c.

**VI. Conclusions from the small consumer study**. Consumer *LIKING* of cocoa-milk drinks could be predicted well from a sensory descriptive laboratory test within the available samples. The model indicated which sensory quality characteristics might contribute the most for liking of the product. But the study was too small to allow us to generalise the results beyond the actual study. On the other hand, more information from the study could have been extracted if more detailed data analysis had been applied to various segments of the consumer population (e.g. age or gender segmentation).

## 3.6. GETTING STARTED WITH MULTIVARIATE DATA ANALYSIS

When working with extremely expensive, dangerous or ethically difficult issues, we recommend that you let a professional statistician take over both the planning and the data analysis.

But in most other cases, it is best if you plan and analyse your own experiment. It is fun, and it is effective. In the beginning it will be difficult, with all the new and unfamiliar concepts. Later, it will still be difficult, but for a different reason: research in "dirty" real-world systems *is* difficult. Nevertheless it is valuable, because it can give results that are *relevant*.

To get started, the reader is advised to try to understand the methods in Chapters 4–11, at least conceptually. Look also at the illustrations in Chapters 12–16. Then, most importantly, start to analyse your own data. Get hold of a professional software package for this type of analysis; several good ones are commercially available. Choose one that is well documented and has concrete user guidance (e.g. Esbensen (2000)). Start with the analysis of a small data set that you already know, because that helps you understand the modelling process. Make sure that you have someone to ask for help when needed, be it over the internet or next door: pay for it, if necessary.

In the beginning, trust only modelling results that are both interpretable *and* show good predictive ability and stable model parameters in the cross-validation. Later on, this can be relaxed a little. But you should always remain sceptical to interpretations from a model that does not predict well in the cross-validation. Likewise, you should always be sceptical to predictions from a model that is difficult to interpret.

To get started in this new type of data analysis is much easier with management support. An example of quality management systems in the food industry where huge amount of data may be collected is given by Early (1995). But a conservative professor, research director or quality manager may fear this new and pragmatic *soft modelling* approach to be "insincere black magic without scientific merit". Therefore it is important to find a good topic to start with: one that has a high chance of success. Here are some suggestions:

***Better planning***: In some institutes, companies and scientific disciplines there is not a strong tradition for conscious, systematic design of experiments. A good idea is then to start by selecting an important, but ill understood problem. Study it, step by step, by a sequence of factorial designs (Chapter 11) in order to investigate several design factors by the use of many measured variables, analysed as described in this book, e.g. Chapter 16.

*Better use of existing data*: Some organisations, especially in industry, have long traditions of measuring many things, but make little use of the data in spite of its high cost. Then it is a good idea to systematise compatible measurements, from lying disconnected in a database into data tables that can be meaningfully analysed. The BLM analysis first of all probably reveals that some of the analytical methods are unreliable and others are unnecessary. Not everybody will like that. Secondly, the modelling will probably reveal some familiar pattern, but it may also show unexpected patterns with high *WOW! factor*. Thirdly, discussions of the results may reveal that some very important phenomena are not yet being measured at all.

*Bridge-building*: A continuous quarrel between different professions may be observed in some organisations, e.g. between different branches of science in a university, or between the production and marketing section in a company. A good idea is then to propose a joint project where each profession is allowed to describe or measure the qualities they find important and in their own way, but where everybody uses the same well-designed set of samples or products. BLM will then map the professions together.

## 3.7. TEST QUESTIONS

1. List some quality goals for which qualimetrics may be relevant to use.
2. What were the six steps relevant to data analysis in a research project cycle?
3. What determines the choice of a suitable data analytical model?
4. Why is the choice of number of samples vs. number of variables not so critical in BLM?
5. How to explore relationships between two data tables, **X** and **Y**?
6. Why is it important to validate?
7. Why is it important to plot?
8. Why is the BLM method considered to be versatile and powerful?

## 3.8. ANSWERS

1. Examples of goals may be to study how products satisfy the customers' needs, wants and requirements, to find quality criteria of an item, or to set up quality specifications for monitoring and control of an industrial process. Table 3.2 lists some examples, referring back to the quality definitions (QD1–4) in Chapter 2.4.
2. The six steps are *I Purpose; II Experimental planning; III Experimental work; IV Pre-processing and quality control (QC) of data; V Data analysis; and VI Conclusion*, as outlined in Figure 3.2a,b.
3. The choice of a suitable model should reflect the goal and data at stake. For studying latent structure in one data table, PCA is suitable, while for two general data tables, PLSR should be chosen. If effects of various design factors are to be investigated, then APLSR and DPLSR are relevant. If the goal is to find which variables discriminate between sets of samples, DPLSR

and SIMCA are useful tools. Figure 3.4 shows the main choices within the BLM method.

4. The BLM method is based on collinearity between the variables, thus reducing the rank problems which some classical statistics fight with. In the latter it is a requirement that e.g. the input X-variables in a regression should be independent of each other, and that is seldom the case in real-world data!

5. Relationships between two data tables **X** and **Y** may be explored by PLSR. By modelling **X** and **Y** in a set of calibration samples, **Y** in future samples can be predicted from **X** by using the calibration model (cf. title page).

6. First, data analytical validation is important to guard against *ad hoc* solutions and wishful thinking. Some tools for this is mentioned in Chapter 3.4, with cross-validation as a central technique. Checking the model results vs. the raw data and interpreting the results with respect to the original purpose, ensures internal validity, i.e. that you have measured what you were supposed to measure. Verifying the results with respect to other types of measurements or giving causal interpretation, strengthen external validity, i.e. your results may be generalised to a larger population (cf. Chapter 2.2).

7. ''Plot a lot'' gives you an overview of the main tendencies of variation in the data, and points out gross outliers. Figure 3.5 lists some useful plots, while the main BLM plots are exemplified in a small study in Figure 3.6.

8. The BLM method can be fitted to many different quality goals, with data from various fields, and of different nature. The soft modelling was invented to *let the data speak* without locking the data into a deterministic model. Table 3.3 in Appendix A3.1 gives a short overview of which classical statistical methods can be replaced by the BLM method. For a layman in data analysis, with expert knowledge in another field, it is considered better to spend time on the main problems, and just learn to use a few data analytical methods.

# PART TWO:
# METHODOLOGY
# (Chapters 4–11)

This is the book's theoretical method part. It uses as few formulae as possible. Instead, it relies on examples from two data sets, each with rather obvious structures, to give a *visual* understanding of the mathematics behind the bi-linear modelling (BLM) method.

Chapter 4 presents the basic notation, terminology and univariate least squares regression methodology. Chapter 5 presents the BLM method for one single data table **X**, the principal compontent analysis (PCA). Chapter 6 expands this to BLM of two data tables, **X** and **Y**, by partial least squares regression (PLSR). Chapter 7 illustrates how the combination of PCA and PLSR modelling can be used in practice for cost-effective calibration and prediction purposes. Chapter 8 shows BLM used for interdisciplinary, explorative purposes. Chapter 9 presents ways of analysing heterogeneous sample sets. Chapter 10 gives some important validation tools to ensure reliability and relevance in conclusions. Finally, chapter 11 outlines how to plan experiments.

Chapters 4–7 use a smooth spectral data set LITMUS for algebra visualisation. Chapters 8–11 (and chapter 1.4.1) use the data set COCOA for interdisciplinary illustration.

The corresponding technical appendices in Part Four cover the following:
A4 defines some elementary algebra and common statistical expressions (means, standard deviations, etc.), for the interested reader. It also provides more details on linear univariate regression.

A5 explains how the bi-linear modelling uses linear univariate regression in two different ways – over samples and over variables. Then it provides more details about PCA, concerning scaling, rotation and plotting of the bi-linear PCs, and about how to deal with occasional *missing values* in the input data.

A6 explains in detail the main BLM method – the stabilised PLSR method.

A7 summarises an example of how to *model the unknown*, based on lots of empirical data – as a contrast to classical causal modelling in science.

A8 describes how to model *non-linear relationships* by the BLM method. It also gives a detailed account of more advanced ways to use *a priori* knowledge about uncertainty in the input data to improve the data analysis.

A9 summarises the two BLM-based *classification* methods SIMCA and DPLSR, and gives some formulae for detecting *outliers*.

A10 explains how to use the cross-validation to study *different sources of uncertainty* in the data, and gives various formulae for model reliability assessment.

A11 explains the Monte-Carlo based assessment of statistical *power* of experimental designs.

# Chapter 4

# Some Estimation Concepts

*The basic notation and terminology is first defined. Then the problem of selectivity, and how it may be solved by bi-linear modelling (BLM), is demonstrated on the data set, LITMUS. Finally, the basic building block of the multivariate BLM method is explained: the univariate linear least squares regression.*

## 4.1. BASIC NOTATION AND TERMINOLOGY

With a lot of measurements stored in a *database*, we need mathematical modelling to extract the systematic information from these measurements. However, the measurements themselves are not directly suitable for multivariate analysis. We must first collect them in a data table (a data *matrix*), with tidy *rows* and *columns* that allow systematic analysis, as discussed in Chapter 3.2.

Data tables may then be analysed by the *soft modelling* approach, in particular by BLM. This chapter defines the notation and terminology needed. The formal notation and terminology follows Martens and Næs (1989).

### 4.1.1. Some Reserved Letters

Letters **x** and **y** (or **X** and **Y**, if they are multivariate) represent input variables. Their corresponding residuals (errors) are named **e** and **f** (or **E** and **F**).

Principal components (PCs) from **X** are called **T**. Other important model parameters are named **B**, **P**, **Q** and **V**, and will be explained later. At a more detailed technical level, letters **W**, **s**, **h** and **I** (the identity matrix) will also be reserved.

Index $i = 1, 2, ..., N$ represents the samples. Indices $k = 1, 2, ..., K$ and $j = 1, 2, ..., J$ represent X- and Y-variables, respectively. Index $a = 0, 1, 2, ..., A$ represents PCs.

### 4.1.2. Matrix Notation

The *rows* in a data *matrix* usually represent the individual physical entities or objects for which data are available: the *samples*. The *columns* represent the different types of data available: the *variables*, which describe the various qualities or properties by which the samples have been characterised.

A two-way data table (a *matrix*) is represented by *upper-case boldface letters*, e.g. **X**. The dimensions of a matrix **X** with $N$ rows and $K$ columns are sometimes symbolised by **X** $(N \times K)$.

One-way tables of data are called *vectors*, denoted by *lower-case boldface letters*. There are two types of vectors, *column vectors* and *row vectors*. For instance, the input data from one single sample is a row vector; from a single variable it is a column vector. If not specified otherwise, vectors are usually defined as column vectors.

An individual data element is called a *scalar* and denoted by *lower-case italics*. (But in tables and plots, scalars are written as normal letters, not in *italics*, due to software limitations.) When the *scalar* is an element in a *vector*, it has an index. When a scalar is an element in a *matrix*, it has two indices.

For simplicity, indices in subscripts are not written in italics; element number $i$ in vector **y** is written $y_i$. When vector or matrix symbols are used as adjectives, they are written in normal boldface letters, for example, vector **y** is a Y-variable, and matrix **X** represents a set of X-variables.

Matrix **X** $(N \times K)$ hence contains row vectors (samples) $\mathbf{x}_i$, $i = 1, 2, ..., N$ and column vectors (variables) $\mathbf{x}_k$, $k = 1, 2, ..., K$, and elements $x_{ik}$, $i = 1, 2, ..., N; k = 1, 2, ... K$, (counting downwards and rightwards). Row vector $\mathbf{x}_i$ has elements $x_{ik}$, $k = 1, 2, ..., K$. Vector $\mathbf{x}_k$ has elements $x_{ik}$, $i = 1, 2, ..., N$.

*The transpose of a matrix or a vector:* The use of the *transpose* operator$'$ is sometimes necessary in order to make matrix expressions clean and tidy. The transpose of a matrix changes its columns into rows and its rows into columns. The transpose of a column vector changes it a row vector, and vice versa.

Some texts use the transpose notation $^{T}$, as in $\mathbf{y}^{T}$, in other texts the transpose notation$'$ is used, as in $\mathbf{y}'$. The latter is used here, in order to avoid confusion with **T** as matrix name.

### 4.1.3. Other Symbols

The mathematical symbol $\wedge$, (pronounced *Hat*), as in $\hat{\mathbf{y}}$ (*y Hat*) is used, when it is necessary to distinguish the values in a matrix, vector or scalar that have been *estimated* statistically, from values that have been measured directly, or from values not yet estimated as model parameters. (In some plots it is replaced by its verbal name, e.g. *yHat*.)

The mathematical symbol $^{-}$ represents the average (the arithmetic *mean*). Example: $\bar{\mathbf{x}}$ (*row vector x mean*), which contains the mean of each column in matrix **X**.

Mathematical *functions* are usually denoted by lower-case *italics*. For instance, $\hat{\mathbf{y}} = f(\mathbf{X})$ is pronounced 'vector **y** is estimated *as a function of* matrix **X**'.

Sometimes, the usual symbol for *multiplication*, $\times$ as in $\mathbf{t} \times \mathbf{p}'$, is replaced by the computer-equivalent symbol, *, as in $\mathbf{t}*\mathbf{p}'$, or simply taken as being implicit, as in $\mathbf{tp}'$.

The symbol, $\equiv$, means 'identically equal to'.

The symbol $\Sigma$ means *the sum of...*:

$$\bar{y} = \Sigma y_i/N \equiv \sum_{i=1}^{N} y_i/N$$

The symbol I means *given that...*, as in $\beta|\alpha = 0.05$: 'the value of beta, given that alpha equals 0.05'.

### 4.1.4. The Only Matrix Algebra Really Needed in this Book

The matrix notation enables us to employ *linear matrix algebra*. In this book, only a simple subset of algebra operators is going to be used.

1. *The sum or difference*: Vectors or matrices may be added or subtracted, element by element, just like ordinary numbers, as long as their dimensions are the same. Examples: $\hat{\mathbf{f}} = \mathbf{y} - \hat{\mathbf{y}}$, and $\mathbf{y} = \hat{\mathbf{y}} + \hat{\mathbf{f}}$
2. *The product of vectors or matrices:* This is an important notation because it allows a very compact description of the various structure models used in this book. Examples:

$$\hat{\mathbf{y}} = \mathbf{X} \times \hat{\mathbf{b}} \text{ or for simplicity, } \hat{\mathbf{y}} = \mathbf{X}\hat{\mathbf{b}}$$

$$\hat{\mathbf{X}} = \mathbf{T} \times \mathbf{P}' \text{ or for simplicity, } \hat{\mathbf{X}} = \mathbf{T}\,\mathbf{P}'$$

Vector- and matrix-multiplication is an extension of scalar multiplication. Its meaning will soon become evident. More technical details may be found in Appendix A4.

### 4.1.5. Names of samples and variables

*Names* of samples and variables are written in *italics* (except in figures). The choice of informative, yet compact names is very important for the graphical data interpretation. Short names of data sets and methods are written in normal uppercase letters. When, e.g. samples are named by *numbers*, e.g. *5a, 5b, 6, 7*, and these numbers are not identical to index *i*, then even these *numbers* are written in *italics*.

### 4.1.6. A Matrix (Table) of Data

In a *data table*, the *samples* follow below each other (rows), and the *variables* follow beside each other (columns).

**Table 4.1.** Litmus solutions, part of input data with matrix concepts illustrated.[a]

| Names # | | 416 nm 1' | 424 nm 2' | 432 nm 3' | 440 nm 4' | 448 nm 5' | 456 nm 6'... | 792 nm 48' | % Litmus 49' |
|---|---|---|---|---|---|---|---|---|---|
| 100 | 1 | 0.390 | 0.412 | 0447 | 0.460 | 0.480 | 0.502 | −0.000 | 100 |
| 100 | 2 | 0.416 | 0.436 | 0.454 | 0.480 | 0.503 | 0.529 | −0.020 | 100 |
| 93 | 3 | 0.361 | 0.375 | 0.388 | 0.396 | 0.431 | 0.449 | −0.004 | 93 |
| 84 | 4 | 0342 | 0.357 | 0.379 | 0.395 | 0.408 | 0.430 | −0.015 | 84 |
| 67 | 5 | 0.294 | 0.301 | 0.311 | 0.330 | 0.335 | 0.357 | −0.030 | 67 |
| 50 | 6 | 0.250 | 0.254 | 0.249 | 0.260 | 0.267 | 0.280 | −0.019 | 50 |
| 33 | 7 | 0.206 | 0.208 | 0.187 | 0.190 | 0.199 | 0.202 | −0.007 | 33 |
| 17 | 8 | 0.106 | 0.107 | 0.098 | 0.102 | 0.110 | 0.109 | −0.004 | 17 |
| 0 | 9 | 0.039 | 0.046 | 0.043 | 0.047 | 0.048 | 0.043 | −0.021 | 0 |
| 0 | 10 | 0.007 | 0.007 | 0.009 | 0.015 | 0.022 | 0.016 | −0.001 | 0 |
| 17 | 11 | 0.172 | 0.186 | 0.183 | 0.188 | 0.196 | 0.201 | −0.014 | 17 |
| 33 | 12 | 0.304 | 0.325 | 0.323 | 0328 | 0.345 | 0.359 | −0.008 | 33 |
| 33 | 13 | 0.235 | 0.271 | 0.282 | 0.304 | 0.324 | 0.342 | 0.003 | 33 |
| 50 | 14 | 0.370 | 0.411 | 0.436 | 0.465 | 0.499 | 0.523 | −0.007 | 50 |
| 67 | 15 | 0.475 | 0.523 | 0.563 | 0.617 | 0.660 | 0.696 | 0.002 | 67 |
| 67 | 16 | 0.506 | 0.551 | 0.590 | 0.627 | 0.674 | 0.704 | −0.016 | 67 |
| 95 | 17 | 0.583 | 0.654 | 0.723 | 0.784 | 0.804 | 0.865 | 0.005 | 95 |
| 97 | 18 | 0.608 | 0.660 | 0.741 | 0.793 | 0.888 | 0.970 | −0.004 | 97 |
| 100 | 19 | 0.692 | 0.771 | 0.846 | 0.898 | 0.963 | 1.034 | 0.004 | 100 |
| 67 | 20 | 1.572 | 1.543 | 1.569 | 1.579 | 1.581 | 1.595 | 1.216 | 67 |
| 33 | 21 | 0.764 | 0.788 | 0.797 | 0.822 | 0.832 | 0.861 | 0.534 | 33 |
| 33 | 22 | 1.605 | 1.622 | 1.627 | 1.607 | 1.617 | 1.628 | 1.408 | 33 |
| 33 | 23 | 1.005 | 1.012 | 1.011 | 1.012 | 1.017 | 1.020 | 0.766 | 33 |
| 0 | 24 | 1.361 | 1.366 | 1.368 | 1.358 | 1.331 | 1.348 | 1.248 | 0 |
| 0 | 25 | 0.092 | 0.098 | 0.073 | 0.071 | 0.071 | 0.061 | 0.083 | 0 |
| 0 | 26 | 0.612 | 0.609 | 0.602 | 0.599 | 0.584 | 0.591 | 0.563 | 0 |

[a] Samples (rows): Rows 1–10: pH >7, (blue colour), 11–20: pH ≤ 7, (red or purple colour), 21–26: with ZnO powder (whitish colour). Variables (columns): 1–48, OD at 48 different wavelength channels; every 40 nm from 416 to 792 nm; 49, relative analyte concentration, named *%Litmus*.

Table 4.1 shows an example of a data table, namely a part of the LITMUS data set to be analysed below. The samples represent various known amounts of litmus dissolved in water with various additives. The variables represent different light colours.

The present data set concerns the colour of a chemical colorant or dye called litmus. This LITMUS data set will be used throughout Chapters 4–7 for illustrating the mathematics behind BLM.

The reported wavelength range of the colour variables is from blue light ($\approx 400$ nm) via green, yellow and red light to near infrared light >700 nm).

The light intensity response was measured at 48 wavelength channels. Each variable's *name* is given as its wavelength (in nanometers, e.g. '*440*'). For simplicity the table shows only the first six and the last one.

The rightmost column shows the vector of concentrations of litmus in the solutions. The variable's *name* is *%Litmus*. The sample *names (100,...,93,...,0)* also reflect this colorant concentration. The litmus concentration is given in percent of a certain reference concentration (15 g/l; Geladi and Martens, 1996a,b).

## 4.2. SOLVING THE SELECTIVITY PROBLEM: THE LITMUS DATA SET

Good data analysis implies that we need a good understanding of the relationship between the problem to be solved, and the data analytical tools. Although one and the same data analytical *method* will be applied to a variety of data types in this book, each *data set* deserves to be analysed in its own right. This means that the reader is presented with a small amount of background knowledge for each example used in this book.

The present data set (LITMUS-I, -II, -III) is *per se* uninteresting for most readers, and has only been chosen because its smooth sequences of samples and of variables simplifies the visualisation of the BLM algebra. But the reader is urged to take the data set as an example of how to increase the reliability and relevance of quality measurements, be it light spectra, sound spectra, images or a set of more traditional measurements, for quality control, medical research, process monitoring or whatever.

The selectivity problem was first presented in Chapter 1.4 encouraging the necessity of multivariate data analysis. Here we discuss the problem more extensively by showing how to extract relevant information from colour measurements of various mixtures, focusing on relevant concepts and principles in this area.

### 4.2.1. Background, LITMUS Data Set

Before proceeding with the analysis of the LITMUS data set, readers unfamiliar with chemistry will first get a brief introduction to how light measurements are used.

Our eyes are sensitive to light in the visible spectral range, with wavelengths ranging from about *400* nm (blue colour) to about *700* nm (red colour). Changes in the light spectrum define how we see the colours of the world. The same light-changes provide quantitative information in the field of analytical chemistry.

Litmus was traditionally used for testing acidity, because it changes colour with pH, from blue in alkaline water solutions (pH > 7) to red in acidic solutions (pH < 7). In addition, of course, the colour intensity of the solutions depends on the amount of dye in the solutions, the *dye concentration*.

In the present example, litmus will be treated just like any other compound in analytical chemistry: our main quality criterion is the amount (*concentration*) of the dye, which we would like to measure from light measurements.

The present data concern the measurement of how light at different wavelengths is modified when it passes through differently coloured water solutions. Colourless glass and pure water let the light pass through more or less unchanged in the visible wavelength range. But when a sample has colour, it is usually because it contains pigments, etc., which steal (*absorb*) some of the light. This is true both for solid objects (coloured glass, human skin or a red car) as well as for our present samples of dye-containing water solutions.

A bright, saturated colour means that more light has been stolen at *some* wavelengths than at *others* in the visible range. Therefore the relative amount of light that survives through the sample and reaches our eyes (or an electronic light detector) will be different at different wavelengths; we have *spectral* information. The relative variations in surviving light are transformed in our brain (or in a computer) into colour spectra, which in turn lead to inference about sample quality.

◻ OPTICAL DENSITY

Instrumentally, light spectra may be measured and expressed in various ways. In analytical chemistry it is common to measure the relative amount of light transmitted through a solution, and then to convert the measured transmittance $T$ into *optical density* (OD) by non-linear transformation

$$OD = \log(1/T) \tag{4.1}$$

OD expresses how much light has been *lost* in a sample. An OD of 0 means that no light has been lost, while an OD of 1, 2 or 3 means that 10, 1 and 0.1% of the light has survived, the rest has been lost. The loss is caused by the dye, etc.:

Dye concentration $\Rightarrow$ OD

As discussed in Chapter 1.4, such a causal relationship may be inverted, to determine dye concentration of OD measurements:

Dye concentration $\approx f$(measured OD)

◻ DIFFERENT EXAMPLES OF LIGHT MEASUREMENTS

In the present example, the OD is calculated by measuring the visible light that is *transmitted* through more or less clear water solutions. The same OD unit will be used in Chapter 12, although in Chapter 12 it is based on near-infrared light that is *reflected* from dry powders. In Chapter 15 a different type of measurement is involved, based on ultraviolet light *emitted* from samples by auto-fluorescence.

### 4.2.2. The Selectivity Problem: One, Two and Three Causal Phenomena

Figure 4.1 illustrates how the OD may be related to the dye pigment concentration, if only the selectivity problem can be overcome by multivariate data analysis. The upper row of subplots shows the colour spectra of three sets of litmus solutions. In each plot, the input OD is given at 48 *named* wavelength channels in the wavelength range of *400–800* nm (more specifically *416–792* nm, with 8 nm intervals).

Figure 4.1a represents ten first examples in Table 4.1: clear *alkaline* water solutions (pH $> 7$), with litmus at relative concentrations ranging from 0 (clear water) to 100% (deep blue).

Figure 4.1b shows the same ten samples as in (a), together with the data from nine clear *acidic* (pH $< 7$) litmus solutions in the same concentration range (red in colour, except one which was purple, pH $\approx 7$).

Figure 4.1c finally adds the last seven samples, all of them more or less whitish (turbid) litmus solutions, made by mixing various amounts of the white powder zinc oxide into various concentrations of litmus in water at various pH levels.

Hence, the three subplots display data sets with one, two and three types of systematic variation. The middle row of subplots shows the performance of traditional data analysis (univariate calibration) for the three data sets. The bottom row shows how the BLM method can improve this performance (multivariate calibration).

☐ SELECTIVE INPUT DATA: ONLY ONE PHENOMENON

The leftmost subplots represent the near-ideal condition in analytical chemistry, where concentration variations in one single constituent uniquely define the variation in a measured response variable.

In Figure 4.1d there is therefore a well-defined relationship between dye concentration (abscissa, %*Litmus*) and OD measurements at the *best* colour wavelength, (ordinate, taken at the OD peak at *568* nm, $OD_{568}$). Hence, the litmus concentration can be uniquely determined from measuring OD at one single wavelength, even though a slight non-linearity may be observed.

If we, instead of using just one wavelength channel, use a linear function of *all* 48 channels (a weighted average of them), we can make even better determinations. This is demonstrated in Figure 4.1g, where the *known* quality %*Litmus* (abscissa) is plotted against the quality predicted from the OD measurements, %*Litmus*$_{FromOD} = f(OD_{416}, OD_{424}, ..., OD_{792})$ (ordinate). Even the non-linearity has now been corrected!

☐ Non-selective Input Data: Two Phenomena

The middle column of subplots represents a problem for the univariate calibration, because two different phenomena affect the OD data. This is commonly seen, e.g. when the desired quality *changes its appearance* (in this case the analyte litmus changes its colour with acidity).

Figure 4.1e shows that the *best* channel from Figure 4.1a, $OD_{568}$, no longer provides a good measure of the concentration of litmus, due to the selectivity problem caused by the colour change.

However, the multivariate calibration, using all 48 wavelength-channels (Figure 4.1h), has little or no problem in yielding good quality determinations, thanks to the multivariate data analysis.

A traditional analytical chemist might protest and claim that the univariate

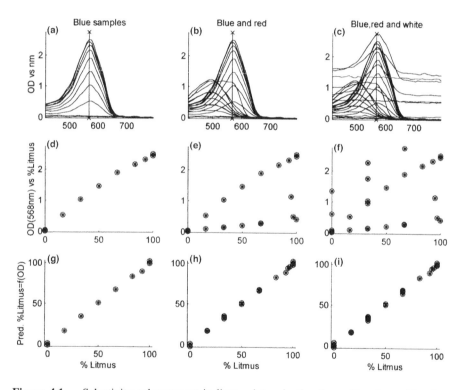

**Figure 4.1.**    Selectivity enhancement in litmus determination by multivariate calibration. Rows: top, input data **X,** OD spectra at 48 wavelength channels; middle, univariate calibration, best OD channel (*568* nm), vs. measured *%Litmus*; bottom, multivariate calibration, predicted *%Litmus* (from 48 OD channels), vs. measured *%Litmus*. Columns: left, ten samples of litmus at high pH (blue, cf. Table 4.1); middle,+nine extra samples of litmus at low pH (red) and one at neutral pH (purple); right,+ seven extra samples containing white light scattering powder (ZnO).

calibration could also work, if we had just chosen a better colour wavelength, e.g. the *isospestic point* around *512 nm*, where the red and blue litmus at the same concentration gives the same OD. This may be correct, but consider the following:

❐ NON-SELECTIVE INPUT DATA: THREE PHENOMENA

In Figure 4.1c the turbidity variations also affect the measurements. With two uncontrolled variations, colour and turbidity, in addition to the quality variation of interest (*%Litmus*), there is no way a single colour wavelength could uniquely describe the litmus concentration, and certainly not $OD_{568}$, as shown in Figure 4.1f.

However, when all of the 48 instrument channels are used, the multivariate calibration makes it possible automatically to correct for these interferences, as the bottom right Figure 4.1i shows.

How was this done? This will be explained at the end of Chapter 7.

### 4.2.3. Three Challenges in the LITMUS Data Set

This data set represents some general selectivity problems that any quantitative measurement may encounter, and that must be taken seriously when, e.g. planning what to vary and what to keep constant, during the design of an experiment (Chapter 11).

*A multi-component category*: First of all, the analyte *litmus* is not a *single, pure thing*: not a single chemical molecule species. It is a natural product, extracted from certain lichens, and contains several similar brightly-coloured chemical constituents. If these different molecule types behave in the same way, it would be OK from some perspectives to regard them as one single *chemical* with one single name. But what if they do not? Multivariate models can reveal such phenomena, and can usually be modified to compensate for them automatically.

*A graded category:* Secondly, litmus is not a *stable thing*. The traditional use of litmus as a pH indicator was possible because its molecules change colour with pH. This means that if we want to determine the concentration of litmus from OD measurements, we must sometimes expect to find blue colour, and sometimes more red colour (cf. Figure 4.1b). That creates a problem for the (unspoken) ideal of direct selectivity in science. There is no simple one-to-one correspondance between the desired property or quality (litmus concentration) and a single measurement (OD at one colour channel). With sufficient data available, multivariate analysis can usually detect and correct for this automatically.

*Foreign interferences:* A third kind of problem often arises because the measurements themselves create additional selectivity problems. In on-line industrial process monitoring and control, there is no time to purify the samples prior to analysis, in the way required by traditional analytical chemistry. The measurements in Figure 4.1c were made directly in *dirty samples* with uncontrolled turbidity variations, with an industrial process analyser: a real-world quality monitoring situation. Again, multivariate analysis can handle many such problems, as long as there are enough empirical data available.

Other causality/selectivity problems were discussed in Chapter 1.4. Like the COCOA example (see also Chapters 8–11), the present LITMUS data set has been chosen because it has a clear and simple structure. Still, the data set has its share of surprises. Moreover, the LITMUS set illustrates the type of data analytical problems that had to be overcome before diffuse near-infrared spectroscopy (Chapter 12) could achieve its present, widespread use for high-speed quality measurements.

### 4.2.4. A Law is Not THE Law

Scientific *laws* summarise empirical knowledge about causal relationships. They are often formulated both in common language and in compact mathematics, which is economical for the mind. Laws allow science to be remembered and communicated more easily (cf. Chapter 2.5). Keeping various *laws* in mind, researchers deduce critical, testable hypotheses and creatively induce possible explanations for unexpected observations.

Under ideal conditions the OD of a solution is proportional to the concentration of light-absorbing molecules. Therefore OD is used in physical sciences for quantitative analysis, e.g. in analytical chemistry for concentration determinations. The quantitative description of this has traditionally been called *Beer's law* in analytical chemistry.

Scientific *laws* usually apply only under ideal conditions. The problem is that these ideal conditions may be hard to attain, except in over-simplified model systems or after a lot of extra laboratory work. Therefore, the Swedish chemometrician Svante Wold in 1986 said

> Don't call it 'Beer's law', call it 'Beer's approximation', or preferably, 'Beer's experience'

Measurements outside these ideal conditions may still contain valuable quantitative information. But the resulting data may have some non-linear or non-selective characteristics that may require multivariate data analysis and a conscious validation of the results.

The LITMUS data set comes from one such experiment (Martens and Næs, 1989; Geladi and Martens, 1996a,b). The measurements were made with an industrial process-instrument via fibre optics, and some of the samples are turbid, so Beer's law certainly does not apply. Still, Beer's law, based on OD transformation (eq. (4.1)), is a useful approximate pre-processing tool.

The data analytical validation in this example will be rather simple, because the structure of the data is visually obvious, and because the proper statistical validation tools will only be introduced in Chapter 10.

## 4.3. LINEAR LEAST SQUARES REGRESSION

Bearing the mathematical and pragmatic concepts in mind, we shall now use this in explaining the basics of the BLM method. The BLM method consists of mathematically drawing lines and planes through *swarms* of empirical data points in a conceptual data space. The modelling relies on *two* elements:

(1) The mathematical *model specification* is a formula that allows lines and planes to be constructed through the data.

The only model type used in this book is the BLM, which, in turn, may be summarised by a linear model (LM). In the simplest case, the bi-linear model *is* the linear model. Therefore, we start the method discussion by considering the linear regression model for the simplest case. This mathematical model speci-fication has some model *parameters*.

(2) The *values* of the model parameters are estimated statistically from the data. If needed for clarity, the parameters with estimated values are distin-guished from the more abstract model parameter symbols. For instance, $\hat{b}$ is the estimated value of parameter $b$ in the mathematical model described below.

There are many ways to fit lines and planes to data. In this book we rely on the most commonly used principle: *least squares (LS)*. This means that an algorithm (computation method), implemented in a computer program, is used for estimat-ing a set of model parameters that *minimise the sum of squares (SS) of residuals* between the mathematical model and the data.

Figure 4.2 illustrates ordinary LS estimation of the linear regression model in a univariate example. The more powerful multivariate bi-linear modelling, to be used later in the book, may be implemented in an algorithm (NIPALS) that only consists of a series of such simple univariate regressions.

*Input data:* Figure 4.2a repeats the OD spectra of some of the red, blue and whitish litmus samples in Figure 4.1c. To make the illustration understandable, only $N = 5$ samples were included. The *isospestic* wavelength, *512* nm, which traditionally would be considered the *best possible* for determining the amount of litmus irrespective of the acidity, is marked by the vertical line.

The samples $i = 1, 2, ..., 5$ were originally numbered $[8,14,17,20,21]'$, and this is kept as their sample *names*. Their measured OD responses (o) at *512* nm, $OD_{512}$, are now called $\mathbf{y} = [0.24, 0.59, 1.15, 1.98, 0.96]'$. Their known concentrations of *%Litmus* are called $\mathbf{x} = [16.5, 50, 95, 67, 33]'$.

□ REGRESSION PROBLEM: HOW TO FIT A LINE THROUGH THE INPUT DATA?

Figure 4.2b shows $\mathbf{y}$ ($OD_{512}$) vs. $\mathbf{x}$ (*%Litmus*) for the five samples. If we want to

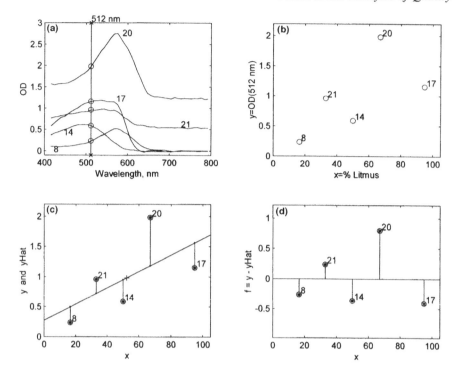

**Figure 4.2.** Linear univariate least squares regression. (a) Regressor data $x$:OD spectra for five litmus samples from Figure 4.1c, with the *isospestic* wavelength *512* nm marked. (b) Input data to univariate regression: $N = 5$ litmus samples described by $\mathbf{x} = \%Litmus$ and $\mathbf{y} = OD_{512}$ (c) LS regression line: regression of $\mathbf{y}$ on $\mathbf{x}$. The vertical line segments represent the residuals in $\mathbf{y}$, $f_i$, i = 1, 2, ..., $N$, whose sum-of-squares has been minimised by LS. (d) The residuals in $\mathbf{y}$, $f_i$, i = 1, 2, ..., $N$.

draw a straight line through this mini-swarm of five points, where should we draw it?

The linear univariate model for Figure 4.2b may then help us. It is written as

$$y_i = b_0 + x_i \times b + f_i \text{ for } i = 1, 2..., N \tag{4.2a}$$

where $i$ is the sample number, $N$ is the number of samples (5), $x_i$, $y_i$ are the data for sample number $i = 1, 2, ..., 5$ (named *8,14,17,20,21*), $b$ is the regression coefficient, which defines the slope of the line, $b_0$ is the offset parameter which defines the value of $y$ for $x = 0$, $f_i$ is the residual, due to errors in $y_i$, errors in $x_i$, and/or errors in the model specification.

Using vector notation, the model may be written more compactly

$$\mathbf{y} = b_0 + \mathbf{x}b + \mathbf{f} \tag{4.2b}$$

The values of the slope and offset parameters $b$ and $b_0$ define the line. How should we estimate them? That depends a little on what we want to use the line for. Let us say that we intend to use the line for predicting the OD ($\hat{y}_i$) from known *%Litmus* ($x_i$) in future samples $i = 6, 7, \ldots$. Then we simply estimate the parameters by regressing **y** on **x** by LS regression.

*LS linear regression solution:* The solid line in Figure 4.2c shows the solution obtained by estimating model parameters $b_0$ and $b$ by fitting the model in eq. (4.2a,b) to the data **x**, **y** ( $= x_i$, $y_i$ for samples $i = 1, 2, 3, 4, 5$) by LS.

The line in Figure 4.2c is defined by the estimated slope $\hat{b} = 0.014$ and offset $\hat{b}_0 = 0.27$. The vertical lines show the residuals obtained, $f_i = y_i - \hat{y}_i$, where the fitted y-values are defined by the model line

$$\hat{y}_i = \hat{b}_0 + x_i \times \hat{b}$$

$$= 0.27 + x_i \times 0.014 \qquad (4.2c)$$

This LS solution is the line that minimises the sum-of-squares of the residuals $f_i$ (the vertical lines in Figure 4.2c). Hence, in regression, the regressand **y** residuals after vertical projection on the regressor **x**, are minimised. That is important to remember. But the reader does not have to worry about how the estimation algebra and computer algorithm for the LS regression work. That is taken care of by the software.

With the obtained result (eq. (4.2c)), we may later predict the y-values $\hat{y}_i$ in new samples $i = 6, 7, \ldots$ from their x-values $x_i$. So we would now have reached our present goal, if it had not been for the fact that the resulting univariate model is seen to be pretty bad for these data. The correlation coefficient between **x** and **y** is only $r_{xy} = 0.62$. This means that only 38% ( $= 100\% \times r_{xy}^2$) of the total initial variation (variance) in **y** can be described by **x** (correlation coefficients and variance will be defined in Appendix A4.2).

Closer inspection of Figure 4.2b shows that samples *8, 14* and *17* (containing blue, red and purple litmus at different concentrations) lie nicely on a straight line. This extends to the origin (0,0), as predicted by the isospestic theory. But samples *20* and *21*, which contain white zinc-oxide powder, destroy the nice picture! In other words, the selectivity problem in the OD data stops us from being able to find a one-to-one relationship between **x** and **y**, even at this isospestic wavelength channel **y** (*512* nm).

Later on we shall learn how to use multivariate calibration to solve this problem, like it was done in Figure 4.1i.

◻ RESIDUALS FROM THE LINEAR REGRESSION

The LS regression of **y** on **x** may be seen as an orthogonal projection of vector **y** on vector **x**. Figure 4.2d shows the obtained Y-residuals **f** = **y** − $\hat{\mathbf{y}}$, plotted against **x**.

The residuals **f** have several useful properties. (i) The sum of the residuals equals zero: $\Sigma f_i = 0$. (ii) The residuals **f** are orthogonal (i.e. 90°) to the variable **x** in the training set of samples. Hence **x** and the residuals after projection on **x** are uncorrelated: $r_{xf} = 0$, cf. Appendix 4.5.

These useful properties allow the ensuing multivariate BLM method to be based on a sequence of such simple, univariate LS regressions.

## 4.4. TEST QUESTIONS

1. Does it matter what we use as abscissa and as ordinate in a two-way plot?
2. What is the difference between regression and correlation?
3. What is the point in using the criterion of least squares in regression?

## 4.5. ANSWERS

1. Apart from the fact already mentioned, that most of us crawled on the abscissa as babies, and still feel shaky when the floor moves, it does not matter.
2. Regression shows the relation. Correlation shows how good that relation is.
3. Least squares in regression? If you had asked the authors when they were still students, it was a deck of about 50 punch-cards that you also could borrow. If you ask us today, we would say: the main point of least squares is that it lets you replace $=$ with $\approx$ in most formulae and then forget the residuals, because (a) it treats negative and positive residuals equally and (b) it lets large residuals count more (a lot more!) than small residuals. Moreover, least squares regression has very nice theoretical properties, as sometimes alluded to in small print in the appendices.

# Chapter 5

# Analysis of One Data Table X: Principal Component Analysis

*This chapter explains how the bi-linear soft modelling works for the analysis of one single matrix of input data, which we may call* **X**. *First, it shows that the mathematical bi-linear model has the same structure as the every-day model: 'The more we have of a phenomenon, the clearer we can see it'. Secondly, it demonstrates how the bi-linear modelling process peals off layer after layer of systematic co-variations from the data, in terms of latent variables (principal components (PCs)). Finally, it shows how the first few PCs give informative maps of the main phenomena in the input data.*

## 5.1. WHAT IS A PRINCIPAL COMPONENT?

### 5.1.1. A PC Shows a Systematic Pattern of Variation in a Data Set

A PC is a special type of variable: namely a *latent* variable. To be *latent* in statistics means that the variable is not *manifest*: it cannot be measured directly. Instead, the latent variables are computed as linear combinations of a set of manifest input variables. They are called *principal* because they are particularly dominant or relevant.

As used in this book, *the PCs summarise the systematic patterns of variation between samples*. They are computed from a table of input data, based on a certain mathematical structure model specification: *the bi-linear model*.

This bi-linear model is fitted to the data with a least squares criterion (analogous to Chapter 4.3), and it describes as much as possible the variations in the data table in as few PCs as possible. The PCs reflect the largest *eigenvalues* of the co-variances in the data. Thereby, the *first few PCs* usually give an adequate description of the whole data table. All *redundancy* (repeated information in the variables and the samples) is thereby summarised. This simplifies the graphical interpretation of the data as well as their quantitative use. The resulting PC model is more compact and statistically stable than the individual input data.

### 5.1.2. PCA: Bi-linear Modelling of One Single Data Matrix X

Figure 3.4 (p. 62) outlined the various ways in which the bi-linear modelling can be applied to one, two or several data tables. This chapter is concerned with the analysis of *one single set of input variables*, **X**. In this case the bi-linear method is called *principal component analysis* (PCA). Technical details on the PCA are given in Appendix A5. In Chapter 6 the bi-linear model will be extended to *two* sets of input variables, **X** and **Y**.

Figure 1.4 (p. 20) showed an example of how bi-linear modelling by PCA can give a visual overview over a whole data matrix **X**.

### 5.1.3. The Physical Meaning of a PC

PCs reveal the systematic, dominant types of variations between samples. Usually, each individual PC does not correspond to an individual physical phenomenon. But seen together, the first few PCs usually *span the important information space* of how the physical phenomena vary. Analogy: the east/west longitude (PC1) and the north/south latitude (PC2) may not be too meaningful *per se* for describing the towns and villages in a local region. But plotted together, they give a nice 2D road map of the region. This book will give a variety of examples of how the first few PCs give insight into otherwise over-whelming tables of input data.

In special cases, there will be a one-to-one correspondance between an individual PC and an individual physical phenomenon. The simplest case is when the data in fact contain only *one* physical variation phenomenon. This will be the case in the present example, intended to reveal what a bi-linear PC is, more or less sculpturally.

Another case is when the relative size of the different physical variations in the data are so different that they are automatically separated into subsequent PCs, one after another.

### 5.1.4. A Simple Model For Simple Data: Different Concentrations of Blue Litmus

The leftmost set of curves in Figure 5.1a show data table **X** (LITMUS-I): the 10 OD spectra at 48 wavelength channels from Figure 4.1a in a 3D perspective. These samples contain litmus at high pH (hence, those that have colour are all blue), but at different concentrations. A smooth change can be observed from the highest to the lowest litmus concentrations (100–0%).

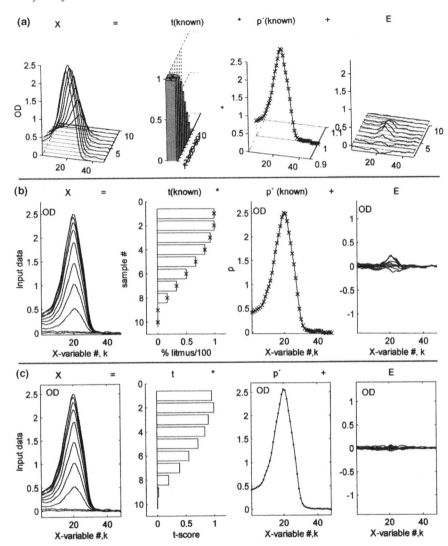

**Figure 5.1.** The bilinear model (BLM) for a one-component system, for data $\mathbf{X} = $ OD spectra of ten blue litmus samples (Figure 4.2 top left). (a) Known constituent concentration $\mathbf{t}$ and known signal $\mathbf{p}$, as 3D overview of the matrix algebra. (b) Same as (a), but as 2D view. (c) Unknown $\mathbf{t}$ and unknown $\mathbf{p}$, estimated by PCA of $\mathbf{X}$.

❏ THE ONE-CONSTITUENT BI-LINEAR MODEL WITH KNOWN PARAMETERS

First, let us express the relative concentration of litmus as $1 = 100\%$, $0 = 0\%$, and let the column vector of relative amounts of litmus in the ten samples be called $\mathbf{t} = [t_i, i = 1, 2, ..., 10]'$. Let us likewise give the typical OD spectrum of a 100% solution of blue litmus a name, column vector $\mathbf{p} = [p_k, k = 1, 2, ..., 48]'$.

*Beer's law* (cf. Chapter 4.2) for a one-constituent system says that we can express the OD in $\mathbf{X}$ as the product of the amounts $\mathbf{t}$ and the characteristic spectrum $\mathbf{p}$ of the constituent, in this case blue litmus.

If we had known both the relative amounts $\mathbf{t}$ and the typical spectrum $\mathbf{p}$, and Beer's law were followed by the OD data, then the whole data table $\mathbf{X}$ could have been summarised as shown in Figure 5.1a: $\mathbf{X}$ = the outer vector product of column vector $\mathbf{t}$ and row vector $\mathbf{p}'$, plus a residual matrix

$$\mathbf{X} = \mathbf{t}(known) \times \mathbf{p}'(known) + \mathbf{E} \tag{5.1}$$

where the unknown residual matrix $\mathbf{E}$ has the same dimensions $(N \times K)$ as $\mathbf{X}$.

This is a bi-linear model of $\mathbf{X}$, with parameters $\mathbf{t}$(known) and $\mathbf{p}$(known). The model splits the input data $\mathbf{X}$ into a parameter contribution $\mathbf{t}$(known) $\times$ $\mathbf{p}'$(known) and a residual $\mathbf{E}$. The model is said to be of a *bi-linear* type, since $\mathbf{X}$ is linearly dependent on both of its parameters $\mathbf{t}$(known) and $\mathbf{p}$(known).

The unknown residual may of course be computed from

$$\mathbf{E} = \mathbf{X} - \mathbf{t}(known) \times \mathbf{p}'(known)$$

Figure 5.1b shows the same as Figure 5.1a, but in 2D instead of 3D. For these particular data, the residual matrix $\mathbf{E}$ has values so small that we may choose more or less to ignore them. Thus, all the important information in the $10 \times 48$ table $\mathbf{X}$ has been simplified into a main structure, consisting of $10 \times 1$ column vector $\mathbf{t}$ and $1 \times 48$ row vector $\mathbf{p}'$.

This case, with both $\mathbf{t}$ and $\mathbf{p}$ in eq. (5.1) known, is not terribly exciting, since this is how we expect well-behaved data from a series of increasing concentrations of one single analyte to look. At least we have been able to confirm that the OD data follow the single-constituent *Beer's law* relatively well. But we do note that the residuals $\mathbf{E}$ display some small, but systematic features, proving that the theoretical model with known parameters is not perfect.

We note that the OD measurements go much higher than the recommended applicability range of *Beer's law* (OD < 1 ). And we already noted a slight curvature in the OD's relationship to the known litmus concentration in Figure 4.1d). Still, this model, with *both* types of bi-linear parameters known, is pretty good, although not particularly interesting.

When only *one* of the parameters $\mathbf{t}$ or $\mathbf{p}$ is known, this linear one-component model becomes more interesting: Then we can estimate the other parameter statistically from $\mathbf{X}$ and the known parameter $\hat{\mathbf{t}} = f(\mathbf{X}, \mathbf{p}(known))$ or $\hat{\mathbf{p}} = f(\mathbf{X}, \mathbf{t}(known))$ as explained in the technical Appendix A5.1.

### 5.1.5. The Same Bi-linear Model With Unknown Parameters: PCA

When *both* of the parameters $\mathbf{t}$ and $\mathbf{p}$ are unknown, then the model really becomes exciting, because that is where multivariate modelling starts. The model can then be written as

$$X = tp' + E \tag{5.2}$$

where both **t**, **p** and **E** are unknown. Figure 5.1c shows that a PCA of **X** in this one-component case yields estimated parameter values for **t** and **p** that look almost identical to the known values.

Judging from the size and shape of the residual **E**, this *reverse engineering* of **X** by PCA seems to give a better description of **X** than the *known* model did! Apparently, the multivariate consensus structure in the 10 × 48 OD measurements **X** tells us that our presumed *knowledge* was not quite correct!

> Upon closer scrutiny, even the PCA residuals in **E** in Figure 5.1c display a small, but systematic variation, which in turn may be modelled, if needed.
>
> A renewed PCA on **E** (instead of **X** itself, not shown here) yields a small second PC that primarily reflects OD response-curvature differences between the different X-variables. This is discussed in more detail at the end of Chapter 7.

We have now seen that a PC is a mathematical representation of a systematic type of variation in the data. In the present example the first PC corresponded more or less directly to a physical phenomenon. That was due to the fact that this phenomenon was so dominant. In Chapter 3.3.1 another example of a PC was mentioned, concerning childrens' growth. The pattern of how different body parts (variables) grow at the same time is common to most children (samples) and may be regarded as a PC.

Let us now make the litmus data more complicated, in order to demonstrate the power of PCA, and to explain how PCA is actually carried out in practice.

## 5.2. HOW ARE PRINCIPAL COMPONENTS OBTAINED?

### 5.2.1. The Mean-centred Multi-component Bi-linear Model of PCA

Figure 5.2 shows what happens when we extend the data set (LITMUS-II) of the ten blue litmus samples (high pH) in Figure 5.1 (and Figure. 4.1a) to include also data from nine red or purple litmus samples (lower pH, Figure 4.1b), and analyse them by PCA.

The 19 × 48 input data matrix **X** is shown in Figure 5.2a. In visual appearance, the bottles containing these samples ranged from clear water to deep red, purple and deep blue solutions.

With the present smooth spectral data, at least two types of curve shapes (blue litmus and red litmus) can be identified by just looking at the input data in Figure 5.2a). But if adjacent X-variables had not followed each other smoothly like they do here, the naked eye would not have been able to detect this systematic structure. But the PCA would.

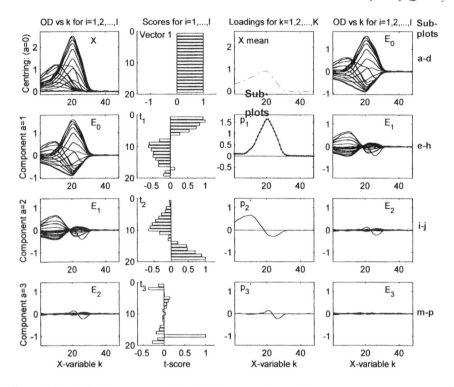

**Figure 5.2.** PCA illustrated: peeling off information-layers from input data table **X**. Rows: $a = 0$: mean-centring: subtraction of the mean $\bar{x}_k$ from each of the $K = 48$ X-variables. $a = 1, 2, 3$: estimation and subtraction of PC number $a = 1, 2, 3$. Columns: (1) data **X** to be modelled by PC number $a$; (2) samples' score vector $\mathbf{t}_a$ for PC number $a$; (3) X-variables' loading parameters $\mathbf{p}_a'$ for PC number $a$; (4) residual $\mathbf{E}_a$ after PC number $a$.

*Beer's law* for a multi-constituent system says that we can express the **X** data as the *sum* of the bi-linear product of amounts × the characteristic spectrum from the different constituents. In this case we have a suspicion that there might at least be two constituents (blue and red litmus). So, we extract at least 3 PCs, just to be sure.

The four rows of subplots illustrate how the PCA *peels off* four successive layers of information from the input data **X**. The leftmost column of subplots shows the X-data to be modelled at each stage, the rightmost column the X-residuals after having peeled off one more layer of information. The two middle columns show the model parameters. The PCA may be summarised as:

The first row (called subplots a–d) represents the centring operation. The centred data matrix, $\mathbf{E}_0$, represents the interesting between-samples variation in **X**. Row 2 (subplots e–h) shows how $\mathbf{E}_0$ is modelled in terms of the first principal component number $a = 1$ (PC1), in analogy to the one-component PCA modelling of $\mathbf{X}$ in Figure 5.1c). Row 3 (subplots i–l) shows the same for $a = 2$ (PC2) and row 4 (subplots m–p) for $a = 3$ (PC3).

◻ MEAN-CENTRING

In multivariate data analysis based on *soft modelling*, we are usually interested in the *patterns of variations between the samples*, not in absolute numbers. Therefore, it is common practice to mean-centre the variables prior to bi-linear modelling.

This means that the input data $\mathbf{X}$ are first modelled in terms of the mean (row vector $\bar{\mathbf{x}}$) for the variables, plus an initial residual $\mathbf{E}_0$. This is computed by

$$\mathbf{E}_0 = \mathbf{X} - \mathbf{1}\bar{\mathbf{x}} \tag{5.3}$$

(For algebraic consistency, the mean spectrum $\bar{\mathbf{x}}$ is pre-multiplied by the 19 × 1 vector $\mathbf{1} = [1\ 1\ 1,...,1]'$ in order to be equally applied to each of the 19 samples.)

◻ BI-LINEAR MODELLING

The residual $\mathbf{E}_0$ is now decomposed into *a few PCs* plus *noise*. The structure model for PC number $a$ is

$$\mathbf{E}_{a-1} = \mathbf{t}_a\mathbf{p}_a' + \mathbf{E}_a \tag{5.4}$$

Column vector $\mathbf{t}_a = [t_{i,a}, i = 1, 2, ..., N]'$ contains the so-called *scores* of the samples. It represents their *amounts* of PC number $a$. Row vector $\mathbf{p}_a' = [p_{k,a}, k = 1, 2, ..., K]'$ contains the corresponding so-called *loadings* for the X-variables, representing the *characteristic signature* of this PC.

In PCA, the values of $\mathbf{t}_a$ and $\mathbf{p}_a$ are defined so that as much as possible of the variation in $\mathbf{E}_{a-1}$ is explained by the product $\mathbf{t}_a\mathbf{p}_a'$, and hence as little variation as possible remains unmodelled in $\mathbf{E}_{a-1}$ in the least squares sense.

The two vectors in the expression $\mathbf{t}_a \times \mathbf{p}_a'$ may be scaled differently without affecting their product, much like $2 \times 6 = 3 \times 4 = 1 \times 12$. For the present visual presentation, each PC has here been scaled so that $\mathbf{t}_a$ has a maximal absolute value of 1.

Let us see how the PCA model develops for PCs $a = 1$–3 in Figure 5.2.

*First PC: $a = 1$*   The mean-centred data $\mathbf{E}_0$ is moved from subplot (d) to (e) and decomposed into the bi-linear product of the first principal component, $\mathbf{t}_1 \times \mathbf{p}_1'$, plus a residual $\mathbf{E}_1$.

The first score vector $\mathbf{t}_1$ in (f) displays a relatively smooth development from sample to sample. The strongest positive scores are seen for samples 1 and 2 (which actually had the deepest blue colour), and the strongest negative scores are for samples 9 and 10 (the two water samples). Thereafter, the score value returns again to 0 for the last samples (the red coloured ones).

The first loading $p_1$ in (g) in this case resembles the nice spectrum of pure blue litmus. Thus, $p_1$ in this data set appears to resemble an *average* difference spectrum between high and low levels of blue litmus! But because the spectrum of red litmus to some degree overlaps that of blue litmus, the red coloured samples (11–19) also obtain non-zero scores.

The residual after one PC, $E_1$ shows a clear unmodelled structure in Figure 5.2h.

*Second PC: $a = 2$*   Let us now move $E_1$ to (i) and seek to approximate this residual by a second PC. The score vector $t_2$ in (j) shows the strongest negative value for the water samples (10 and 11) and the strongest positive value for sample 19 (the sample with the deepest red colour). The second loading $p_2$ resembles a spectrum of red litmus, but it has some negative values in the region where blue litmus also gives a signal. A chemist cannot interpret it directly as the signal of one given chemical compound; it appears to be an *average difference spectrum* between strongly red and weakly blue litmus samples.

Why did the PCA choose this shape? Simply because that was the dominant remaining pattern unmodelled in $E_1$ (just look at it!).

In contrast, the residual $E_2$ shows very small values, except for one sample ($i = 17$).

*Third PC: $a = 3$*   The third principal component shows that $E_2$ may be largely decomposed into the contribution of an outlier ($i = 17$) plus some very small residuals $E_3$. The outlier is apparent in that the score vector $t_3$ has one single strong element ($t_{17,3} \approx 1$). The corresponding loading $p_3$ shows a smooth little spectrum, similar to the major feature that remained in $E_2$.

Outliers are not necessarily erroneous. In fact, an outlier may be the most informative sample! But an outlier should be given extra attention. It should be corrected or removed if we find that it probably reflects errors. In the present case, sample *17* is considered to be particularly interesting, and it does not seem to destroy the modelling. Therefore it is accepted among the other samples, instead of being left out.

> The reason why sample *17* is an outlier is not yet clear. It has a pH of about 7 (purple). One hypothesis is that since litmus is a group of several similar chemical compounds, this intermediate pH allows some of these dyes to stand out from the rest of the group, effectively causing an extra variation phenomenon.

In summary, Figure 5.2 shows that two principal components are clearly valid, and necessary in order to model the present two-phenomena spectra (*blue and red* litmus samples). The third factor was spanned by one unexpected, but interesting outlier. After $A = 3$ PCs, the residuals are so small that we ignore them as measurement errors. Chapter 10 will give additional tools for deciding the optimal number of components, $A_{Opt}$.

## 5.2.2. The Full Bi-linear Model of X

Summarising, the bi-linear structure model of the input data table **X**, with $A$ PCs, may be written in vector algebra

$$\mathbf{X} = \mathbf{1}\bar{\mathbf{x}} + \mathbf{t}_1\mathbf{p}_1' + ... + \mathbf{t}_A\mathbf{p}_A' + \mathbf{E}_A \tag{5.5a}$$

Ignoring the trivial vector **1**, and representing the bi-linear contributions in more compact matrix form, this may be written

$$\mathbf{X} = \bar{\mathbf{x}} + \mathbf{TP}' + \mathbf{E}_A \tag{5.5b}$$

where $\mathbf{T} = [\mathbf{t}_1, ..., \mathbf{t}_A]$ and $\mathbf{P} = [\mathbf{p}_1, ..., \mathbf{p}_A]$.

Thus, the input data **X** can be written as a sum of a modelled reconstruction plus residual

$$\mathbf{X} = \hat{\mathbf{X}}_A + \mathbf{E}_A \tag{5.5c}$$

where $\hat{\mathbf{X}}_A = \bar{\mathbf{x}} + \mathbf{TP}'$ and where the elements in $\mathbf{E}_A$ are small. So what good has that brought us?

## 5.3. HOW PCs GIVE MEANINGFUL MAPS OF THE DATA

In Figure 5.2 the ability of PCA to *peel off layers of information* from an input matrix **X** was demonstrated. But why should we want to do that?

It is often important to study how different variables are correlated with each other. But in modern science it is common to get many input variables that are more or less intercorrelated, and this creates a problem for conventional univariate data analysis (what should we look at?) as well as many traditional methods for multivariate data analysis (*collinearity problems*).

The main advantage of the BLM method, which includes the least squares tool PCA, is that it provides informative maps of the main information of interest in data table **X**. In Chapter 1.4 this was demonstrated for a small data set. Figure 5.3 illustrates this for a larger data set:

### 5.3.1. Too Many Raw Data Plots

Figure 5.3a shows the input data again for the 19 blue and red litmus samples. The spectrum outlined by a series of squares is the mean of the 19 individual spectra, $\bar{\mathbf{x}}$.

◻ ONE PAIR OF X-VARIABLES

With 48 input variables, there are $48 \times 47/2 = 1128$ different pairs of variables

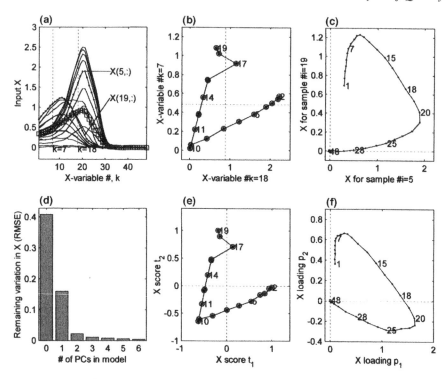

**Figure 5.3.** Multivariate modelling: inspecting *some* of the data vs. *all* the data. Top: raw data. Bottom: PCA summary based on the bilinear model: $\mathbf{X} = \mathbf{1}\bar{\mathbf{x}} + \mathbf{t}_1\mathbf{p}_1' + \mathbf{t}_2\mathbf{p}_2' + \mathbf{E}$. (a) Input OD spectra $\mathbf{X}$ ($K = 48$) of $N = 19$ red and blue litmus samples (Figure 4.1b). (b) The samples seen for OD at *two* of the 48 X-variables. Abscissa: $k = 18$ vs. ordinate: $k = 7$. Some samples are numbered. (c) The X-variables seen for *two* of the 19 samples. Abscissa: $i = 5$ vs. ordinate: $i = 19$. Some X-variables are numbered. (d) PCA modelling: average (rms) varia- tion remaining in X as a function of the number of PCs subtracted. (e) PCA *score plot* of the first two PCs. The samples mapped for *all* 48 X-variables in the first two latent variables, $\mathbf{t}_1$ (abscissa) vs. $\mathbf{t}_2$ (ordinate). (f) PCA *loading plot* of the first two PCs. The variables mapped for *all* 19 samples in the first two latent variables' loading vectors, $\mathbf{p}_1$(abscissa) vs. $\mathbf{p}_2$ (ordinate).

that we could look at. Two of the variables, $k = 7$ and $k = 18$, are marked by vertical dotted lines in Figure 5.3a. Figure 5.3b shows the 19 samples plotted for this pair of variables (abscissa $k = 18$ vs. ordinate $k = 7$). The sample number is shown for some of the samples, and consecutive samples are connected by line segments. This type of *bivariate plot* is useful, because it provides a *map* of the samples. But which of the 1127 other pairs of variables should we look at?

◻ ONE PAIR OF SAMPLES

In Figure 5.3a the spectra of two of the 19 samples are marked, a blue litmus sample, $i = 5$, and a red litmus sample, $i = 19$. These two spectra are plotted against each other in Figure 5.3c, with the number given for some of the X-variables. There are $19 \times 18/2 = 171$ different such sample pairs to plot.

Together, plots (b) and (c) give an interesting insight. But there are so many aspects of the input data which we have not yet studied! How can we know that our analysis is complete?

### 5.3.2. Compact Overview Plots From PCA

Figure 5.3d shows how the PCA, by peeling off layers of information in **X**, reduces the level of unmodelled variation between the samples in **X**. The bars represent the average OD variability in the **X** data after 0,1,2,...,6 PCs. The total initial variability in the centred data $\mathbf{E}_0$ is about 0.4 OD units. This is reduced to about 0.02 OD units after 2 PCs. Beyond PC2 the reduction in the remaining variation is very small. Thus, our chosen model of **X** is

$$\mathbf{X} = \mathbf{1\bar{x}} + \mathbf{t}_1\mathbf{p}_1' + \mathbf{t}_2\mathbf{p}_2' + \mathbf{E}_2$$

Variabilities are averaged in the root-mean-square (rms) sense. Details on how to compute the unmodelled residual are given in Chapters 9 and 10.

This 2 PC model can now be inspected:

◻ SCORE PLOT: A MAP OF THE MAIN RELATIONSHIPS BETWEEN SAMPLES

Figure 5.3e shows the so-called *score plot* of $\mathbf{t}_1$ (abscissa) vs. $\mathbf{t}_2$ (ordinate). As before, the scores have been scaled to have a maximum value of 1.0 in each PC. We know that sample *10* is pure water, sample *2* is 100% blue litmus (alkaline) solution and sample *19* 100% red (acid) litmus solution. Sample *17* is purple in colour (neutral acidity) and contains a 95% litmus solution.

Armed with this background knowledge, it is easy to recognise the set of blue litmus samples, ranging from sample *2* via *5* to *10*, and the red litmus samples ranging from sample *10* via *11* and *14–19*.

Sample *17* lies more or less on the line from 100% red litmus towards 100% blue litmus, indicating that this pH-neutral sample contains mostly red, but some blue litmus: which is what a chemist would expect.

$\mathbf{E}_2$ in Figure 5.2 shows this sample to have a small, but surprising residual, indicating that it is not only a mixture of red and blue litmus. But this effect is small and therefore ignored here.

Score vectors $t_1$ and $t_2$ represent the axes in the score plot. These PC score vectors have the same shape and function as each of the 48 individual variables in $X$, but they are *super-variables*, in the sense that they *summarise* the systematic variations in $X$.

Together, the score plot of $t_1$ vs. $t_2$ represents a very compact *geographical map* of the samples in this set of input data. Since we centred the data before the PCA, the origin in Figure 5.3e represents the average sample quality, $\bar{x}$. But what do the different directions in the mapped *landscape* of $X$ mean?

☐ LOADING PLOT: A MAP OF THE MAIN RELATIONSHIPS BETWEEN THE VARIABLES

Figure 5.3f shows the loadings $p_1$ (abscissa) vs. $p_2$ (ordinate). They have been scaled to correspond to scores with a maximum value of 1. The loading plot may be used for interpreting the score plot.

Since all the elements in $p_1$ are positive or zero ( $= non\text{-}negative$), this means, e.g. that sample 2 (100% concentration, blue colour, far right in Figure 5.3e) has generally higher X-values than average (squares in Figure 5.3a), but particularly so in the range of variables $k = 15\text{--}25$, with a maximum near $k = 20$.

Sample *19* (100% concentration, red colour) seems to have higher values of $X$, compared to the average $\bar{x}$, in the range $k = 1\text{--}15$, where $p_2$ has positive values, and lower values in the range $k = 20\text{--}48$, where $p_2$ has negative values.

There are several different ways to display the scores and loading information from bi-linear models. As discussed in conjunction with Figure 3.6, they are useful for different purposes. This is illustrated in Figure 5.4 for the loadings.

A one-way loading plot, Figure 5.4a, shows loadings $p_1$ and $p_2$ as functions of X-variable number $k = 1, 2, ..., 48$, in analogy to how the input data for each individual sample were displayed in Figure 5.3a and Figure 5.4c.

This way of displaying the loadings is particularly suitable for smooth input data, where peaks and valleys in the loadings can give the user ideas for causal interpretation. Since the first few PCs usually summarise information from many samples, their loadings are often statistically more stable than the input data.

A two-way *map*, Figure 5.4b, repeats Figure 5.3f. The reader should note that by varying the aspect ratio of the plot (length of the abscissa relative to that of the ordinate), the appearance of the plot may differ a little. More detail on perception of graphical displays may be found in Cleveland (1985).

☐ CORRELATION LOADINGS: MAKING THE TWO-WAY LOADING PLOT SCALE-INVARIANT

The loadings in the upper two subplots in Figure 5.4 are given in the same unit as the X-variables submitted to the PCA. In the present case all the variables are given in the same unit, OD, so that is not a problem.

But in many other cases we have to amplify variables with small variation

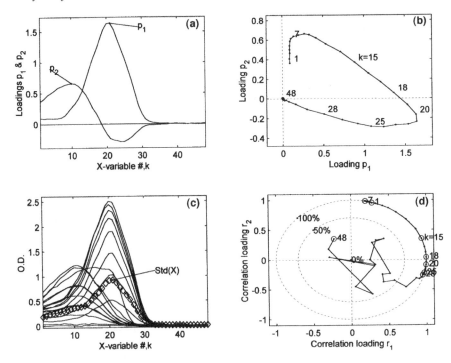

**Figure 5.4.** Three ways to plot the variables' loadings. (a) Line plots of loadings $\mathbf{p}_2$ and $\mathbf{p}_1$ as *difference spectra*. (b) Bi-variate plot of $\mathbf{p}_2$ and $\mathbf{p}_1$. (c) Input data $\mathbf{X}$, and the standard deviations of the X-variables, $\mathbf{s} = \text{Std}(\mathbf{X})$ (diamonds, eq. (4.5)). (d) Correlation loadings $r_{ak}$ (eq. (5.13)) between each of the input variables $\mathbf{x}_k$, $k = 1, 2, ..., 48$ and latent variables $\mathbf{t}_a$, for $a = 1, 2$.

relative to variables with big variations (*making mice and elephants compatible*), for instance when modelling many different types of variables given in many different units, in the same matrix. This *a priori* weighting will be explained below, and later demonstrated with a different data set (Chapter 8).

The lower right subplot in Figure 5.4d shows how the loadings can be made scale-invariant and independent of the choice of *a priori* weighting, by transforming them to correlation coefficients between the input variables and the PCs.

Figure 5.4c shows the total initial standard deviation $s_k$, of each variable (diamonds), superimposed on the data $\mathbf{X}$ as submitted to the PCA. It demonstrates that the variability beyond X-variable $k = 30$ is very small. We *expect* this range to reflect mainly noise. But we do not *know*, because there may be systematic variability too small to be visible in Figure 5.4a,b. However, by a simple rescaling of each loading element $p_{ak}$, we can express it as the ordinary correlation coefficient $r_{ak}$ between the latent *super-variable* $\mathbf{t}_a$ and each input variable $\mathbf{x}_k$.

The resulting correlation loading plot of $\mathbf{r}_1$ (PC1) vs. $\mathbf{r}_2$ (PC2) (eq. (5.14) p. 381) is shown in Figure 5.4d. The squared sum of $\mathbf{r}_1$ and $\mathbf{r}_2$ represents the relative amount of variance explained, $r^2 = r_1^2 + r_2^2$. The dotted ellipsoids represent 100 and 50% correctly modelled variance ($r^2 = 1$ and 0.5, respectively). Some of the variables are marked and numbered, for illustration. Variables in the range $k = 1$–7–15–18–28 are seen to lie close to the $r^2 = 1$. From $k = 28$–48 the $r^2$ falls drastically and erratically towards 0, confirming our assumption that in this baseline-range there is no useful variation, only the type of chance correlations expected from random measurement noise.

### 5.3.3. Reconstructions From the Score Plot

The score plot is a map of the landscape of samples. The score plot $t_1$ vs. $t_2$ in Figure 5.3e summarised the configuration of the $N = 19$ samples. Figure 5.5 explains this score plot as a 2D geographical map of the landscape in which these samples are situated.

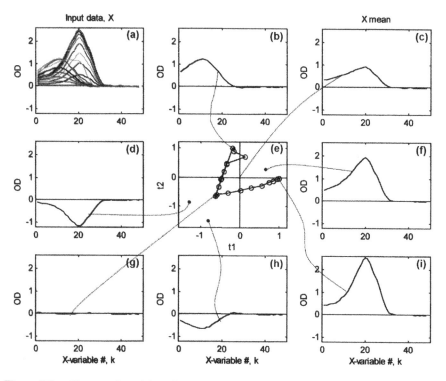

**Figure 5.5.**    The meaning of the PCA score space. (a) The input OD of the 19 litmus samples from Figure 4.1b. (e) Their score plot from Figure 5.3e: $t_1$ (abscissa) vs. $t_2$ (ordinate). The surrounding subplots show the OD spectral shape of different positions in the score plot. (c) The mean spectrum $\bar{x}$. (b,g,i) Spectra corresponding to three of the observed samples. (f) Spectrum of a possible, but unobserved sample quality. (d, h) Spectra of impossible sample qualities (negative OD).

◻ THE DATA AND THE MODEL

Figure 5.5a again shows the blue and red litmus input spectra **X**. The central subplot shows the score plot **t**$_1$ vs. **t**$_2$. Each point $(t_{i1}, t_{i2})$ in the score plot represents one real or potential X-spectrum reconstructed from the model:

$$\hat{\mathbf{x}}_i = \bar{\mathbf{x}} + t_{i1}\mathbf{p}_1' + t_{i2}\mathbf{p}_2' \tag{5.6}$$

where in this case the mean spectrum $\bar{\mathbf{x}}$ is taken from Figure 5.3a and loadings $\mathbf{p}_1$ and $\mathbf{p}_2$ from Figure 5.3f (= Figure 5.2c,g,k).

The curved lines connect different locations in the score plot, $(t_{i1}, t_{i2})$, to different possible realisations of reconstructed spectra $\hat{\mathbf{x}}_i$:

◻ SOME RECONSTRUCTED SAMPLES

Figure 5.5c represents the average sample quality, $\bar{\mathbf{x}}$. Subplots (b), (g) and (i) represent red litmus (sample *19*), clear water (sample *10*) and blue litmus (sample *2*). This illustrates how a bi-linear mathematical model may be used for reconstructing the input data of individual samples.

◻ A HYPOTHETICAL SAMPLE

More importantly, Figure 5.5f represents a possible, but not measured sample quality. It comes from a location near the outlier, sample *17*, but with more blue and less red litmus (conceivably obtained at slightly higher pH). This shows that the bi-linear model may also be used for reconstructing unknown sample qualities.

◻ PHYSICALLY IMPOSSIBLE SAMPLES

Finally, subplots (d) and (h) represent two physically impossible locations: Since they lie beyond the clear water, they must have negative concentrations of litmus, resulting in negative OD values in their hypothetical spectra $\hat{\mathbf{x}}_i$. This shows that not all sample reconstructions are physically meaningful.

## 5.4. WHEN AND HOW

### 5.4.1. When to Use PCA

PCA, the version of bi-linear modelling for one single data matrix **X**, has several uses in multivariate data analytical processing.

◻ PRELIMINARY INSPECTION OF INPUT DATA

PCA is nice for gaining graphical insight into the dominant patterns in a data table, without being burdened with redundancy (the same information repeated in different ways by different variables). It is therefore useful for initial inspection of raw data. Thereby, one can often find outliers and other unexpected oddities.

☐  COST-EFFECTIVE EXPERIMENTAL PLANNING: PRINCIPAL PROPERTIES DESIGN

PCA is a useful tool in experimental planning, because it enables us to make semi-controlled experimental designs based on large amounts of readily available data. Principal properties (PP) design is illustrated in Chapter 7, and explained in more detail in Chapter 11.

☐  SIMCA CLASSIFICATION

The ability to detect whether or not a sample is an outlier relative to a bi-linear model makes PCA useful also for classification purposes. This is briefly explained in Figure 3.4. In SIMCA classification different sets of samples (*classes*) are analysed separately by PCA. The different class models may have different numbers of PCs.

Each sample is then fitted to every class model, to check which samples fit to which classes. If two class models strongly overlap in the X-space, many samples will belong to both. If the classes are clearly distinct, most samples will only belong to one of the classes.

New samples may also be fitted to these classes; hitherto unseen sample types will fit to neither class. SIMCA classification is discussed in more detail in Chapter 9.

☐  PROCESS MONITORING

Since PCA is a bi-linear method, it may compress many input variables into a few PCs plus residuals. The output from PCA may be used for maintaining an overview of many measurements from a dynamic process, e.g. an industrial production process or a set of medical instruments monitoring a patient. Abnormalities in the scores and/or the residuals give automatic outlier warnings if unexpected sample qualities or instrument problems arise. This use of bi-linear modelling is even easier with PLSR as illustrated in Chapters 7 and 15.

### 5.4.2. How to Use PCA

For simple, homogeneous data tables, the PCA may be applied precisely as shown in this chapter. If the different samples constitute distinct groups, it may be helpful to model the groups separately by SIMCA analysis, as described above.

❒ BALANCING THE ERROR LEVELS IN THE INPUT VARIABLES: STANDARDISATION

If the input variables come in very different scales, and/or are suspected to have very different error levels, it is important to weight the input variables. The most common way to weight variables is to standardise them, forcing each variable to have a total standard deviation of 1 before they are submitted to the bi-linear analysis. This is attained by defining the weights as the inverse of the total initial standard deviations of the respective input variables (eqs. (8.1) and (8.2)). Examples and details on standardisation and other *a priori* weighting schemes are given in Chapter 8.2 and Appendix A8.

❒ CHOOSING THE OPTIMAL NUMBER OF PCS, $A_{OPT}$

Methods for statistical assessment and optimisation of bi-linear models will be discussed in Chapter 10. For now, we only mention that cross-validation is one way to assess how many PCs to include in a PCA model (Wold, 1978). Other criteria may also be included:

Interpretability is often a useful indicator of statistical validity. Lack of smoothness of loadings and scores, for data where smoothness is expected, is a good indicator of useless PCs. For instance, Figure 5.2j shows an unexpectedly non-smooth score vector for PC3, $t_3$. Subsequent PCs showed very non-smooth, noisy loadings as well (not shown here).

## 5.5. TEST QUESTIONS

1. What is a principal component?
2. Why is each variable usually mean-centred before bi-linear modelling?
3. What is the difference between the first and second PC?
4. What is the difference between scores and loadings?

## 5.6. ANSWERS

1. A principal component is a pattern of co-variation between a set of variables observed in a set of samples. Example: when you smile happily, we can see how your whole face wrinkles; when you are angry, your whole face may wrinkle the other way.
2. In order to see the systematic differences between samples clearly, we mean-centre each variable first.
3. The first PC picks up the most important pattern of co-variation. The second PC picks up the most important *remaining* pattern of co-variation, 90° to the first one, and so on.
4. *Score* is old Norse for *skara*, to cut or shear, and is a mark given to each

sample for each latent variable (PC). *Loading* is a type of weight (some would call it weighing) that each latent variable ascribes to each input variable, much like the loadings that artists give to various aspects in their paintings.

# Chapter 6

# Analysis of Two Data Tables X and Y: Partial Least Squares Regression (PLSR)

*This chapter brings in a second table of variables, called the Y-variables. The main bi-linear modelling tool, the partial least squares regression (PLSR), is introduced. We shall see how we can get reliable information from one data table, **X**, that is relevant to another data table, **Y**.*

*The basic principles behind PLSR, with special focus on multivariate calibration and prediction, will first be described. The method is shown also to work for non-smooth data. Thereafter a mini-example of how to predict two unknown variables is shown.*

## 6.1. MODELLING Y FROM THE ESSENCE OF X

Chapters 2 and 3 addressed the need to extract reliable and relevant information about quality. In Chapter 5 we learnt how this is done for one data table by Principal component analysis (PCA). But what if we want to explore relationships between two data tables, **X** and **Y**, or want to predict one from the other?

Figure 5.5 showed that the first few bi-linear score vectors from **X**, $t_a$, $a = 1, 2, ...$ are useful for reconstructing the X-variables. However, the score vectors from **X** are also powerful for modelling other variables (*Y-variables*). This is why multivariate calibration works, and this is why bi-linear modelling is useful in interdisciplinary scientific investigations.

Figure 6.1 illustrates how the two first latent *super-variables* $t_1$ and $t_2$ from **X** can be used for predicting one variable **y**, for the blue- and red-litmus data in the LITMUS-II data set. In this case the Y-variable *%Litmus* is the total relative concentration of litmus in the samples, irrespective of their pH-defined colour. Its value ranges from 0 to 100%.

With only one single Y-variable, we designate it as column vector **y** (lower case boldface, cf. Chapter 4.1). In contrast, when there are several Y-variables, they are denoted as matrix **Y**. But this formalistic distinction will sometimes be ignored, to ensure that the same notation may be used irrespective the number of Y-variables.

### 6.1.1. Seeing the Same Data From Three Different Angles

Figure 6.1 shows the 3D plot of **y** vs. $t_1$ and $t_2$. In Figure 6.1a it is seen from above, along the Y-axis. For some of the 19 samples, the value of **y** is given, with dashed lines between the blue and the red samples with the same value of **y**. The triangular shape from 100% blue litmus via clear water to 100% red litmus is evident.

Figure 6.1b shows the same figure, looked at more horizontally, from the angle of the clear water samples. The triangular shape is still evident. However, Figure 6.1c shows what happens when we rotate Figure 6.1b horizontally to another angle. In this data set all the samples lie along a more or less straight line! (the model of a straight line through the points may be written $\hat{y} = 58 + t_1 \times 45 + t_2 \times 50$).

Of course not all data sets behave as nicely as this. But the figure demonstrates how and why multivariate calibration compresses the X-variables into a few latent variables $t_1, t_2, \ldots$, which in turn are used for modelling the Y-variable(s).

### 6.2. BLM: FROM PCA TO PLSR

In Figure 6.1 the modelling of **y** from **X** was done by the so-called principal component regression (PCR), which consists of PCA of **X** into **T**, followed later by regression of **Y** on **T** (see e.g. Martens & Næs, 1989; Kramer, 1998). In the rest of this book PCR will be replaced by a similar, but more powerful BLM method: the PLSR, where the modelling of **X** and **Y** is done simultaneously in order to *ensure Y-relevant* PCs from **X**.

The difference between the two methods is not dramatic. But the reason why only PLSR is used in the rest of this book is that the PCs in PLSR become more Y-relevant than in PCR. Consequently, PLSR can deliver simpler, more compact regression models than PCR. This makes the models easier to interpret and statistically more reliable, and PLSR more versatile (cf. Figures 3.3, 3.4 and Appendix A3.1).

The difference between PCA and PLSR will now be briefly illustrated. The X-data used Figure 6.1 (OD spectra of $N = 19$ blue and red litmus samples at $K = 48$ wavelength channels) are shown again in Figure 6.2a in a 3D view.

The score and loading plots from the PLSR models are to be interpreted just like the ones from PCA for only one block of variables **X** (Chapter 5). This is

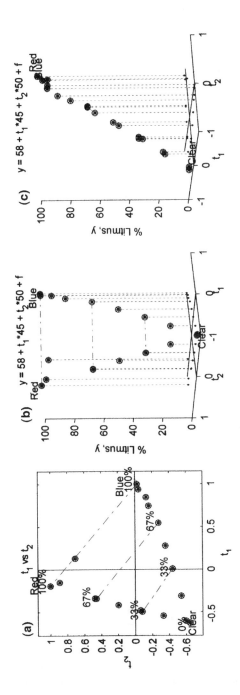

**Figure 6.1.** Relating a Y-variable to the first PCA components from **X**. A PCR solution seen from three angles. (a) Score plot seen from above, with Y-value (relative %litmus concentration) written for some samples. Some samples with the same Y-value (100, 67, and 33%) are connected by dashed lines. (b) The same plot seen from the side: **Y** vs. $t_1$, $t_2$. (c) The same plot seen from another side: **Y** vs. $t_1$, $t_2$.

why PCA is regarded as a special case of the general two-block bi-linear modelling, with no Y-variables (cf. Figure 3.4).

In the following, the Y-variable (*%Litmus*) is first related graphically and
statistically to the first two PCA score vectors from OD data **X**, $t_1$ and $t_2$.
Then it is demonstrated that the modelling works the same way, whether or
not the data have a smooth appearance.

### 6.2.1. Score Plot From PCA and PLSR

When again modelling **y** = litmus concentration from **X**, but now by PLSR instead
of PCA, the score plot in Figure 6.2b is obtained, instead of that in Figure 6.1a.

The value of **y** (*%Litmus*) has been used as the name for some of the samples
in the score plot. The sample configuration in the $t_1$ vs. $t_2$ plots appears to have
been slightly rotated clockwise from Figure 6.1a. But apart from a trivial difference in scaling of the axes, not much else has changed.

This rotation from the PCA to the PLSR solution is due to the fact that the first PC $t_1$
from PLSR picks up more of the Y-relevant X-variations than the first PC $t_1$ in PCA.

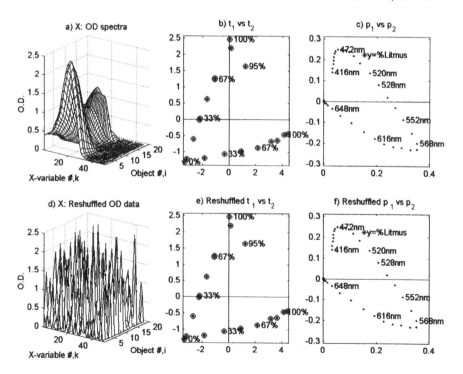

**Figure 6.2.**    Bi-linear modelling does not *need* smooth spectra or smooth time series. Upper:
PLSR modelling of the input OD spectra in terms of score and loading plots. Lower: PLSR
modelling of the same data, with *reshuffled* order of the variables and of the samples.

Since this set of X-data requires only two PCs to be sufficiently modelled, and since the PCs in both PCA and PLSR are orthogonal to each other ($t_1' t_2 = 0$), the second PC, $t_2$, is being rotated with $t_1$. The scaling of the score vectors will be discussed in Appendix A5.4.

## 6.2.2. Loading Plot From PLSR

Figure 6.2c shows the corresponding PLSR loadings for the X- and Y-variables, in analogy to the PCA loadings in Figure 5.4b. Here, the wavelengths have been used as names for some X-variables.

The variable $y = \%Litmus$ is also marked. The value of $y$ varies between 0 and 100%, while the X-variables only vary between 0 and max. 2.5 OD units. $y$ was therefore scaled *a priori* by a factor 1/75 in order to ensure that the Y-loadings $q_1$, $q_2$ become visible in the same loading plot as the X-loadings $p_1$, $p_2$ in Figure 6.2c. Better scaling methods will be shown later.

## 6.3. SMOOTH CURVES ARE NICE, BUT NOT NECESSARY

Few data give nice and smooth curves like the present OD spectra. But bi-linear modelling is equally applicable for smooth and for non-smooth data, as will now be demonstrated. This makes bi-linear modelling powerful as a data analytical method.

The X-data used in the present example (OD spectra of blue and red litmus samples) are smooth in two different respects. As the 3D view in Figure 6.2a shows, X-variables $x_{k-1}$, $x_k$, $x_{k+1}$ (adjacent column vectors in $X$) are similar, and samples $x_{i-1}$, $x_i$, $x_{i+1}$ (adjacent row vectors in $X$) are similar.

Figure 6.2d represents input X-data from non-smooth variables in a non-smooth series of samples. We can no longer interpret the structure just by visual inspection of these input data. However, the two input data sets in (a) and (d) are the same, except that $X$ has had the order of its columns (48 wavelength channels) randomly reshuffled. Likewise, the order of its rows (19 samples) has been randomly reshuffled; the Y-data were systematically reshuffled in the same sample order as in the X-data. The samples and variables in Figure 6.2e,f are still designated by their names, like for the un-shuffled data in Figure 6.2b,c.

The results show that the bi-linear model has not changed due to the loss of smoothness! The reason is that the two solutions use the same structure model, and neighbourhood information about what is adjacent to what in the input table is not used. Hence, the model structures to be illustrated in the next chapter are equally valid for both the smooth and the systematically reshuffled data, and for any other data.

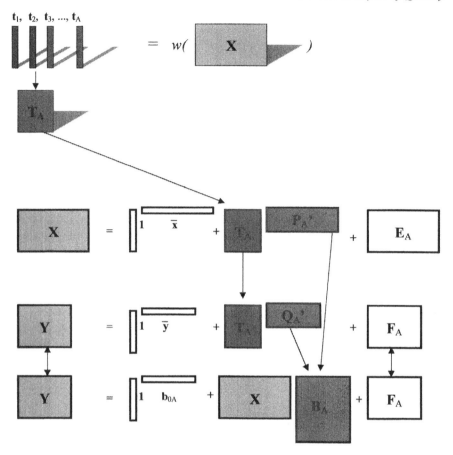

**Figure 6.3.** BLM: the matrix structure of the bi-linear structure model for **X** and **Y**. A set of X-variables, collected in matrix **X**, are compressed into $A$ PCs, called the score matrix $\mathbf{T}=[\mathbf{t}_a, a = 1, 2, ..., A]$. These are in turn used for modelling both **X** and **Y**, in terms of their loading matrices $\mathbf{P}'$ and $\mathbf{Q}'$ and their residuals **E** and **F**. **Y** may equivalently be expressed directly from **X** by the regression coefficients **B**. The subscripts indicate the rank of the model, A. The darkness indicates the intensity of information.

## 6.4. CALIBRATION AND PREDICTION MODELS

### 6.4.1. The Calibration Model

Figure 6.3 summarises two-block bi-linear modelling. Multivariate data analysis software employs sophisticated mathematical methodology, most of which is of no concern to the user. The actual steps that the user has to go through in bi-linear modelling will now be outlined.

The reader is recommended to try to understand the estimation steps conceptually, from the compact matrix *pictograms* in Figure 6.3; that would make it easier to read the rest of the book, and easier to do the multivariate data analysis of the reader's own data later.

Assume that we have input data from $K$ X-variables and $J$ Y-variables data for $N$ samples (rows); hence, matrix $\mathbf{X}$ has dimensions $(N \times K)$ and $\mathbf{Y}$ dimensions $(N \times J)$. (In the LITMUS data set above, we had $N = 19$, $K = 48$ and $J = 1$).

The statistical process of estimating a bi-linear $\mathbf{X}$–$\mathbf{Y}$ model then consists of the following steps (cf. Figure 6.3):

(1) A few (*more than enough*) latent variables (PCs), $a = 1, 2, \ldots$, are extracted as functions of the many $\mathbf{X}$ variables (not from the Y-variables!)

$$[\mathbf{t}_1, \mathbf{t}_2, \mathbf{t}_{3,\ldots}] = w(\mathbf{X}) \tag{6.1a}$$

where in PLSR, function $w(\mathbf{X})$ is defined so that the first few PCs are as Y-relevant as possible. The choice of $w(\ )$ depends on which BLM method is used. In this book we use the stabilised PLS regression method, due to its flexibility.

(2) When the optimal number of PCs, $A = A_{Opt}$ has been determined by some validation method (cf. Chapter 10), the $A$ score vectors are stored in one common score matrix $\mathbf{T} = [\mathbf{t}_1, \mathbf{t}_2, \mathbf{t}_3, \ldots, \mathbf{t}_A]$. In Figure 6.3, all the parameters are indexed by $A$; in the text this is dropped for convenience.

(3) The $\mathbf{X}$-variables are modelled by the X-scores $\mathbf{T}$, in terms of X-loadings $\mathbf{P}$ and X-residuals $\mathbf{E}$, just like in PCA (Chapter 5)

$$\mathbf{X} = \mathbf{1}\bar{\mathbf{x}} + \mathbf{T}\mathbf{P}' + \mathbf{E} \tag{6.1b}$$

The Y-variables are likewise modelled by X-scores $\mathbf{T}$, in terms of Y-loadings $\mathbf{Q}$ and Y-residuals $\mathbf{F}$

$$\mathbf{Y} = \mathbf{1}\bar{\mathbf{y}} + \mathbf{T}\mathbf{Q}' + \mathbf{F} \tag{6.1c}$$

Equivalently, the Y-variables may be modelled directly from the X-variables via the regression coefficient matrix $\mathbf{B}$ $(K \times J)$

$$\mathbf{Y} = \mathbf{1}\mathbf{b}_0 + \mathbf{X}\mathbf{B} + \mathbf{F} \tag{6.1d}$$

where $\mathbf{b}_0$ $(1 \times J)$ is the model offset, i.e. the Y-values expected if $\mathbf{X} = 0$.

The coefficients $\mathbf{B}$ are estimated as a function of the X- and Y-loadings $\mathbf{P}'$ and $\mathbf{Q}'$, as well as the function $w(\mathbf{X})$; see eq. (6.6a) on p. 384. The actual values in residuals $\mathbf{E}$ and $\mathbf{F}$ in eq. (6.1b,c) and in $\mathbf{B}$ and $\mathbf{b}_0$ in eq. (6.1d) will depend on the number of PCs $a = 1, 2, \ldots, A$ used in the model.

In Figure 6.3, the main information lies in the scores $\mathbf{T}$, the loadings $\mathbf{P}$ and $\mathbf{Q}$ and the estimates of regression coefficients $\mathbf{B}$. The other elements are drawn in white, because they are usually rather uninteresting. Either they reflect what does not vary between samples (the means $\bar{\mathbf{x}}$ and $\bar{\mathbf{y}}$ and the offset $\mathbf{b}_0$), or they contain mainly *noise* (residuals $\mathbf{E}$ and $\mathbf{F}$).

This means that the many X-variables are compressed into a few PCs $\mathbf{T}$, and both $\mathbf{X}$ and $\mathbf{Y}$ are then approximated from $\mathbf{T}$. Technical details on PLSR will be given in Appendix A6.

☐ SUMMARISING THE SIMPLIFIED REPRESENTATION OF THE BI-LINEAR MODEL

In the following, the formal vector $\mathbf{1}$ and the subscript $_A$ in Figure 6.3 have been dropped. The calibration model may then be summarised as shown below.

---

The calibration model summarised:

$\mathbf{T} = w(\mathbf{X})$

$\mathbf{X} = \bar{\mathbf{x}} + \mathbf{TP}' + \mathbf{E}$

$\mathbf{Y} = \bar{\mathbf{y}} + \mathbf{TQ}' + \mathbf{F} \equiv \mathbf{b}_0 + \mathbf{XB} + \mathbf{F}$                        (6.2)

---

If the structure in the X-data is nice, then the X-residuals $\mathbf{E}$ will be small. If the $\mathbf{X}-\mathbf{Y}$ relationship is nice, then the Y-residuals $\mathbf{F}$ will also be small. Many nicely behaved data sets give small residuals both in $\mathbf{E}$ and $\mathbf{F}$ after just a few PCs; that makes the data analysis easy.

☐ UNDER- AND OVERFITTING

If too few PCs are included in the bi-linear model, the description of $\mathbf{X}$ and $\mathbf{Y}$ will be unsatisfactory. This is called underfitting. If too many PCs are included, this is called overfitting.

Usually, overfitting is a more serious problem than underfitting, because we like to get good fit between model and data. The optimal number of PCs is **not** that which directly makes fitted residuals $\mathbf{E}$ and $\mathbf{F}$ as small as possible. For instance, since $\mathbf{T}$ comes from $\mathbf{X}$, we can make the X-residuals $\mathbf{E}$ as small as we want, just by including enough PCs in $\mathbf{T}$.

Instead, the optimal number of PCs is usually determined from estimates of the prediction error in $\mathbf{Y}$ (as if predicting $\mathbf{Y}$ from $\mathbf{X}$ in new samples via eq (6.2); this reaches a minimum after a certain number of PCs, and may then increase again). Estimating the optimal number of PCs by such cross-validation is explained in Chapter 10.

### 6.4.2. The Prediction Model

There are two distinct phases in data-driven predictive modelling.
   (1) The training or calibration phase
   Here the model for predicting Y-variables from the X-variables is determined, based on empirical data $(\mathbf{X}, \mathbf{Y})$ and prior knowledge (eq. 6.2).
   (2) The prediction phase

Here the model with "known" parameters $w(\ )$, $\mathbf{x}$, $\mathbf{P}$, $\bar{\mathbf{y}}$ and $\mathbf{Q}$, is applied to data from the X-variables in new samples, in order to predict the unknown values of the Y-variables, or for explorative purposes.

---

The prediction model for sample $i$ summarised:

$$\mathbf{t}_i = w(\mathbf{x}_i) \tag{6.3a}$$

$$\mathbf{x}_i = \bar{\mathbf{x}} + \mathbf{t}_i\mathbf{P}' + \mathbf{e}_i \tag{6.3b}$$

$$\mathbf{y}_i = \bar{\mathbf{y}} + \mathbf{t}_i\mathbf{Q}' + \mathbf{f}_i \equiv \mathbf{b}_0 + \mathbf{x}_i\mathbf{B} + \mathbf{f}_i \tag{6.3c}$$

---

In practice these steps are performed after eq. (6.3a):

$$\hat{\mathbf{x}}_i = \bar{\mathbf{x}} + \mathbf{t}_i\mathbf{P}'; \quad \mathbf{e}_i = \mathbf{x}_i - \hat{\mathbf{x}}_i \tag{6.4a}$$

$$\hat{\mathbf{y}}_i = \bar{\mathbf{y}} + \mathbf{t}_i\mathbf{Q}'_i \text{ or } \hat{\mathbf{b}}_0 + \mathbf{x}_i\hat{\mathbf{B}}; \tag{6.4b}$$

$$\mathbf{f}_i = \mathbf{y}_i - \hat{\mathbf{y}}_i \text{ (if data for } \mathbf{y}_i \text{ exist)} \tag{6.4c}$$

---

The next mini-example (Chapter 6.5) serves to illustrate a project with distinct calibration and prediction phases. Here we calibrate with extremely few calibration samples, but for more than one Y-variable at the same time.

## 6.5. A MINI-EXAMPLE: LEARNING HOW TO PREDICT TWO UNKNOWNS

In Figure 6.1 we showed that the total amount of litmus was known in the 19 samples, in terms of a relative measure, *%Litmus*. But the level of the two individual forms of litmus, (blue and red litmus) was unknown. The purpose of this illustration is therefore to learn how to predict, simultaneously, the relative concentration of two *analytes*, namely blue litmus $\mathbf{y}_1$ and red litmus $\mathbf{y}_2$ in the 19 samples in Figure 6.2a.

Both the calibration and the prediction phases of the example will be described in terms of the six *Steps I–VI* defined in Figure 3.2a,b in the Layman's guide.

### 6.5.1. Calibration Phase

*I. Purpose.* The overall purpose is to determine the relative concentration of each of the two molecular states of litmus, here termed blue litmus and red litmus, in a certain type of samples (represented by the 19 first samples in Table 4.1).

In the calibration phase, the goal is therefore to make a calibration model for predicting $y_1$ (=relative concentration of *blue litmus*) and $y_2$ (=relative concentration of *red litmus*), from OD data $X$ for this type of sample.

The calibration set should be as small as possible, in order to minimise the cost of generating reference data for $y_1$ and $y_2$. This way to use the BLM is similar to what is called *curve-fitting* or *direct unmixing* (Martens & Næs, 1989, p. 168).

## II. Experimental planning

❏   A CALIBRATION SET WITHOUT REDUNDANCY

If we know *a priori* that

(a) there are only a few, well known constituents present in this type of sample;

(b) there are no unexpected interfering phenomena (unknown constituents, changes in the known constituents or instruments effects);

(c) that the instrument response varies sufficiently linearly with constituent concentrations; and

(d) that the calibration data, which we are going to obtain, will be more or less error-free;

then we may rely on this theoretical knowledge, and *calibrate with very few samples*.

At the extreme, one only needs one sample for each phenomenon to be modelled: in this case one blue and one red litmus sample. To become independent of possible offsets in the input variables, we add one more sample – pure water – to represent the *origin* or *baseline*.

Here we choose three calibration samples for which we do know the concentration of both red *and* blue litmus without any experimental work at all. We know that sample 2 represents pure *blue* and sample 19 represents pure *red* litmus. In addition, sample 10 has no litmus at all.

| Name | Description | $Y = [y_1, y_2]$: | |
|------|-------------|-------------------|---|
| 2  | 100% blue litmus     | 1 | 0 |
| 19 | 100% red litmus      | 0 | 1 |
| 10 | Pure water, no litmus | 0 | 0 |

***III. Experimental work.*** The relative light intensity (transmittance, $T$) spectra of these three samples are measured (of course, in this demo example we have measured them already).

***IV. Pre-processing and QC of data.*** The transmittance data are linearised

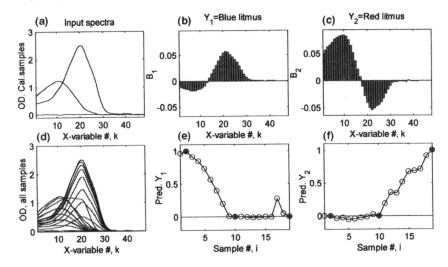

**Figure 6.4.** Multi-component modelling by PLSR: a very small example. (a) The $N = 3$ input spectra for calibration. $\mathbf{X} = [\mathbf{x}_1; \mathbf{x}_2; \mathbf{x}_3] = $ OD of [blue litmus, red litmus, clear water], with relative compositions $\mathbf{Y} = [\mathbf{y}_1; \mathbf{y}_2; \mathbf{y}_3] = [1, 0, 0; 0, 1, 0; 0, 0, 0]$. (b) Regression coefficient vector $\hat{\mathbf{b}}_1$ for $\mathbf{y}_1 = $ blue litmus, estimated by PLSR using $A = 2$ PCs. (c) Regression coefficient vector $\hat{\mathbf{b}}_2$ for $\mathbf{y}_2 = $ red litmus. (d) $N_{\text{new}} = 19$ input spectra $\mathbf{X} = [\mathbf{x}_i; i = 1, 2, ..., 19]$ for prediction of $\mathbf{y}_1$ and $\mathbf{y}_2$. (e) Predicted relative content $\hat{\mathbf{y}}_1$ for samples 1–19; ⊛ the $N = 3$ calibration samples. (f) Predicted relative content $\hat{\mathbf{y}}_2$ for samples 1–19.

according to Beer's law: OD $= \log(1/T)$, defined as row vectors $\mathbf{x}_i$, $i = 1, 2, 3$ and joined in a matrix $\mathbf{X}$.

The X-data for these three samples are shown in Figure 6.4a. The *blue litmus* sample has the highest, rightmost OD peak, the *red litmus* sample the intermediate, leftmost peak, and water shows only baseline signal near zero.

***V. Data analysis.*** The resulting mean and two first X-loading spectra are similar to those in Figure 5.2c,g,k. Therefore, the 2-PC calibration model is here only summarised by the regression coefficient vectors $\hat{\mathbf{b}}_1$ and $\hat{\mathbf{b}}_2$ (cf. Appendix A6).

Figure 6.4b shows the regression coefficients vector $\hat{\mathbf{b}}_1$ for $\mathbf{y}_1 = $ *blue litmus*. It shows that in order to predict *blue litmus*, the OD in the X-variables in range $k = 15$–30 are weighted positively. However, the X-variables in the range $k = 1$–15 are weighted negatively. Why? This is how the multivariate calibration automatically learns from the calibration data how to compensate for interferences: in this case the *red litmus*.

The opposite pattern is seen for $\mathbf{y}_2 = $ *red litmus* in Figure 6.4c, which shows $\hat{\mathbf{b}}_2$. Since *red litmus* gives a weaker colour signal than *blue litmus*, the former must be more strongly amplified than the latter. Consequently, the B-values for $\mathbf{y}_2$ are larger than for $\mathbf{y}_1$.

*VI. Conclusions*. Based on a training set of only three samples, we have been able to establish a calibration model, summarised by the regression coefficients $\hat{\mathbf{B}} = [\hat{\mathbf{b}}_1, \hat{\mathbf{b}}_2]$ and $\hat{\mathbf{b}}_0 = [\hat{b}_{0,1}, \hat{b}_{0,2}]$.

As long as our assumptions (a–d) hold, the model should be applicable for new samples of the same kind.

## 6.5.2. Prediction Phase

*I. Purpose*. From the obtained calibration model we can predict the relative concentrations of *blue* and *red litmus*, where $\hat{\mathbf{y}}_i = [\hat{y}_{i1}, \hat{y}_{i2}]$, from their OD spectra $\mathbf{x}_i$, in new samples.

*II. Experimental planning*. A set of 19 available red, blue and purple samples, including the three calibration samples, is chosen to be used as the prediction set.

*III. Experimental work*. The transmittance ($T$) spectra of the remaining new 16 samples is likewise *measured*.

*IV. Pre-processing and QC of data*. The light measurements are linearised by the same transformation, $OD = \log(1/T)$, and defined as spectra $\mathbf{x}_i$, $i = 1, 2, \dots, 19$. Figure 6.4d shows these X-data, (which of course are the same as in Figure 6.2a).

*V. Data analysis*. The predicted relative concentrations are given by the curves in Figure 6.4e,f for *blue* and *red litmus*, respectively: $\hat{y}_{i,1}$ and $\hat{y}_{i,2}$, $i = 1, 2, \dots, 19$. The three calibration samples, $i = 2, 19$ and $10$ are marked by asterisks.

Figure 6.4e shows that the predicted relative concentration of *blue litmus* falls smoothly from around $\hat{y}_{i1} = 1$ (100%) in samples $i = 1$ and $2$ to $\hat{y}_{i1} = 0$ in samples 9 and beyond. But at sample 17 it temporarily rises again.

For *red litmus*, Figure 6.4f shows that $\hat{y}_{i,2}$ remains near 0 until sample 10, thereafter it increases more or less smoothly.

Note that the predicted relative concentration of *red litmus* is slightly negative around sample $i = 5$. It is physically impossible to have negative amounts of a chemical compound. This illustrates that *predicted* values $\hat{\mathbf{y}}_i$ are not *true* values.

But the error is probably small, as indicated by the fact that the sum of predicted *blue* and *red litmus* (not shown here) is quite close to the known, total concentration of *%Litmus* (Figure 6.1c), as expected.

*VI. Conclusions*. In general, we have reached our goal. Sample 17 seems to have a higher level of *blue litmus* than its neighbours, indicating a higher pH. But it mostly contains *red litmus*. This corresponds to the fact that its colour was noted to be purple. But its OD spectrum did not fit as well to the 2-PC model (in $\mathbf{E}_2$ its residual was similar to that shown in Figure 5.2). So the quantitative results cannot fully be trusted for this sample.

There seemed to be a small deficiency in the calibration model, leading to small, but systematic errors during prediction: Some negative predictions were obtained in the concentration of *red litmus* (Figure 6.4f). The most probable explanation is that the process analytical instrument was used in such a way that its response was not as linear as we assumed in the calibration phase. With no intermediate litmus concentrations in the calibration set, the data analytical modelling could neither detect it, nor compensate for it.

The ensuing example in Chapter 7 modifies and expands this little demo of a multivariate project. It demonstrates a better way to design small, cost-effective calibration experiments: a way that is not so sensitive to errors in the theoretical assumptions.


## 6.6. WHEN AND HOW

PLSR of two data matrices **X** and **Y**, is a particularly versatile tool for data analysis. Therefore, the following list of potential applications is far from complete. But it may give some ideas about various ways to apply it.

### 6.6.1. When to Use PLSR

❑ PRELIMINARY INSPECTION OF RAW INPUT DATA

Look for outliers and unexpected patterns, in loadings, scores and residuals.

❑ PREDICTIVE MODELLING

Learn how to predict variables **Y** from variables **X**, based on a calibration set of data. Use the resulting model to predict **Y** from **X** in new samples. This is illustrated more or less in every example in the book, because, in principle, predictive ability is important even when the primary purpose of the data analysis is interpretation. This is the topic discussed in Chapter 7.

❑ INTERDISCIPLINARY STUDIES

Put different types of input variables in **X** and **Y**, and look at the loading plots to see how they relate to each other. Chapter 8 will discuss this more interpretative and explorative approach in soft modelling.

❑ CLASSIFICATION BY DISCRIMINANT ANALYSIS AND SIMCA

*Symmetrical classification by discriminant analysis*: Represent classes of samples by different class indicator variables in **Y**, and try to predict these from observed variables **X** in one single DPLSR model. This, together with SIMCA classification of heterogeneous sample sets are topics in Chapter 9.

❑ ANALYSIS OF EFFECTS IN DESIGNED EXPERIMENTS

See how response variables **Y** are modelled by design variables **X**. This is particularly useful when several response variables are to be analysed. Use cross-validation to assess statistical reliability, and use plots for understanding the observed effects. Examples of APLSR is given, e.g. in Chapters 11 and 16.

### 6.6.2. How to Use PLSR

❑ BALANCING THE ERROR LEVELS IN THE INPUT VARIABLES, IF NEEDED

Make sure that the error levels in the different variables in **X** are not outrageously different. If they are, then scale the X-variables differently. Similarly, scale different Y-variables differently, if they are expected to have outrageously different error levels. This *a priori* weighting is described in Chapter 8.2.3 (see also Appendix A3.2).

❑ CHOOSING THE OPTIMAL NUMBER OF PCS, $A_{Opt}$

Model interpretability and cross-validation (Chapter 10) are useful tools for that.

❑ MODEL VALIDATION AND INTERPRETATION

Check the predictive ability and the model stability by cross-validation with jack-knifing (Chapter 10).

Look at scores, loadings, regression coefficients and residuals, to detect outliers, to confirm expected patterns and to discover unexpected patterns.

### 6.7. TEST QUESTIONS

1. What is the similarity and difference between the two BLM techniques PCA and PLSR?
2. Are the PC scores in PLSR defined from the X-variables or from the Y-variables?
3. Do all projects have a prediction phase?
4. What is the difference between the bi-linear model and the linear model summary?

**6.8. ANSWERS**

1. PCA models one single table of input variables **X**. PLSR models two sets of input variables **X** and **Y**. In PCA, **X** is modelled by **T**, its most dominant PCs, leaving X-residuals **E**. In PLSR, **X** and **Y** are both modelled by **T**, the most Y-relevant PCs from **X**, leaving X-residuals **E** and Y-residuals **F**.
2. The PC scores **T** represent linear combinations of the X-variables.
3. No, not always in the sense that we really intend to predict **Y** from **X** in new samples of the same general kind. But conceptually, the ability to predict **Y** from **X** for *this kind of sample* is still important. Otherwise, what is the meaning of mathematical modelling? This internal predictive ability is the basis for the cross-validation.
4. The *bi-linear* model has *two* sets of parameters, conventionally called scores **T** for the samples, and loadings **P** and **Q** for the X- and Y-variables. The *linear* model summarises these loadings into *one* set of parameters, the estimated regression coefficients **B̂**.

# Chapter 7

# Example of a Multivariate Calibration Project

*This chapter explains more about predictive modelling. An example illustrates the use of principal properties (PP) design in cost-effective calibration projects. Non-selective, but cheap, X-data available in many samples, are analysed by PCA, and a small, but particularly informative subset of calibration samples is selected. These few samples are then submitted to expensive Y-measurements and used as a training set for multivariate calibration by PLSR modelling.*

*The resulting calibration model is then ready to be applied to new samples in order to predict their unknown, costly Y-values from these cheap X-measurements. Abnormal new samples are automatically tagged as outliers.*

## 7.1. PREDICTIVE MODELLING

Assume that we want to replace a traditional, expensive, slow or noxious way of measuring quality **Y**, by a cheaper, faster and/or safer multichannel measurement method **X**.

The present approach of multivariate predictive modelling is useful in many different types of applications, not only for *hard* physical measurements. The following are illustrated in this book (cf. Table 7.1):

**Table 7.1.** Examples of **X** and **Y** in multivariate calibration.

| X | Y | Typical application | Chapter |
|---|---|---|---|
| Sensory analysis | Consumer response | Marketing test | 3.5 |
| Visual light spectra | Chemical composition | Industrial process monitoring | 6,7 |
| NIR reflectance spectra | Chemical composition | Agricultural quality assessment | 12 |
| Questionnaire data | Job satisfaction | Management assessment | 13 |
| Quantum computations | Toxicity | Environmental assessment | 14 |
| Fluorescence spectra | Chemical composition | Industrial process monitoring | 15 |
| Chemical composition | Mould growth | Food safety assessment | 16 |

Data-driven predictive modelling requires good data for both **X** and **Y** in a training set of samples. Since it is usually expensive to measure the reference data **Y**, an important topic is to choose a training set of samples that is small enough to be affordable, and yet informative enough to give a calibration model with sufficient predictive ability. This will here be attained by choosing the training set by so-called *principal properties (PP) design* (Wold et al., 1986).

The present mini illustration employs the LITMUS-II data set previously seen, so that the reader can check that the mathematics work as expected.

But let us from now on assume that we have no previous knowledge of these data.

The example has a calibration phase and a prediction phase. In each the essential data analytical work *Steps I–VI* from Figure 3.2 are described.


## 7.2. CALIBRATION

*I. Purpose*. In the case at hand, there is one desired quality **y** (total *%Litmus*), which we would like to learn how to predict from *colour* spectra **X**, measured as OD at $K = 48$ wavelength channels by a process analytical instrument. We assume that the established reference method for measuring **y** is too slow to be used for our present purpose: on-line process monitoring and control. This is why we want to replace it by the high-speed OD measurements **X**.

Assume that we do not know enough about the possible selectivity and linearity problems in this type of measurements to make a traditional *hard* model based on first-principles of causality. Instead, we plan to use empirical X- and Y-data to learn how to predict **y** from **X** by multivariate calibration in this kind of sample.

*II. Experimental planning*. Assume further that we can only afford to measure **y** in a few samples, say, five. *Which five samples should we choose?*

### 7.2.1. Principal Properties Design

In the method of PP design the idea is to select a few calibration samples from a large set of relevant samples, by a preliminary data analysis of cheap X-measurements, thereby getting just a few informative samples for expensive Y-measurements. This data analysis is done by bi-linear PCA, explained in Chapter 5. More detail on the PP design will be given in Chapter 11.2.3.

Assume that the quality we want to determine, **y**, is the total relative *%Litmus* (the sum of blue and red litmus), and that this **y** is still *unknown*. However, cheap **X**-measurements (OD spectra) have been obtained for *many* (19) relevant samples. This available X-matrix is shown in Figure 7.1a, with some of the spectra marked explicitly for illustration.

From prior knowledge, we expect that these X-variables are relevant for the

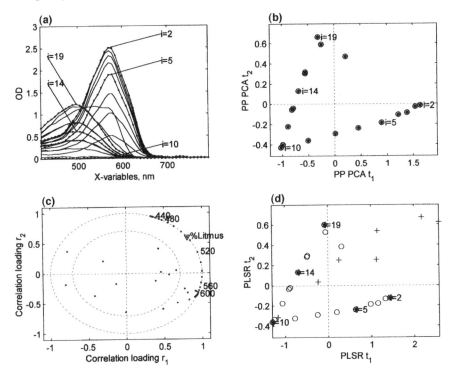

**Figure 7.1.** A calibration- and prediction-project: PCA-based PP design, followed by PLSR multivariate calibration, and finally, by prediction new samples. (a) Spectra **X**: OD at 48 wavelengths (fast, cheap measurement) available for 19 blue/red samples of litmus. Assumed situation: Y-measurements are expensive; no Y-data yet known. How few Y-measurements can we get away with? (b) PP design: score plot of two first PCs from PCA of **X**. ⊛, five of the 19 samples, spanning the X-space, chosen for slow, expensive Y-measurement (*%Litmus*). (c) Correlation loadings from PLSR: **y** vs. **X** in $N$ = five chosen calibration samples; first 2 PCs. (d) PLSR scores of the five calibration samples, with remaining 14 blue/red samples (o) and seven new samples (+) fitted by prediction afterwards (eq. (6.3a)).

prediction of **y**. Then it is reasonable to expect that a set of samples that span the systematic variations in **X**, will also represent the **X**–**y** relation well.

PCA is a good method for gaining insight into how the different samples span the X-data. Therefore the $19 \times 48$ matrix **X** is submitted to PCA. The PCA score plot in Figure 7.1b maps the configuration of all these (19) available samples with respect to the so-called PP, i.e. the scores of the latent variables $t_1$ and $t_2$.

The idea behind experimental design based on PP is now to choose a small subset of samples that span the PCA score space (Figure 7.1b) well. One such set of samples is $i$ = 2, 10 and 19, i.e. the ones used in the previous example (cf. Chapter 6.5). However, in case there is *non-linear instrument response*, and in order to be able to *check the predictive reliability* of the model, we add some *intermediate* samples, yielding a more *space-filling* design. So our present calibration set contains samples $i = 2, 5, 10, 14$ and 19. (We could have chosen to include the intermediate, purple sample 17, but since that was found to be a little abnormal in Figure 5.2, we skip it).

***III. Experimental work.*** Having chosen our small, but presumably representative and particularly informative set of samples from the PP scores in Figure 7.1b, the Y-variables are measured in these samples.

In this *as if* illustration, it corresponds to measuring the total relative *%Litmus* in samples $i = 2, 5, 10, 14$ and 19. Their y-values are found to be $\mathbf{y} = [100, 67, 0, 50, 100]'$.

***IV. Pre-processing and QC of data.*** The transmittance spectra had already been linearised by the OD conversion. Once chosen from the score plot, the OD spectra of these five calibration samples were marked explicitly in Figure 7.1a.

It may be argued that the span from sample 2 to 5 is too short, compared to the span from sample 5 to 10. The reason for this choice is that we expect possible non-linearities to manifest themselves most strongly for the blue samples in the range between samples 2 and 5. So the calibration data $\mathbf{X}$ and $\mathbf{y}$ are accepted for analysis.

***V. Data analysis***

### 7.2.2. Bi-linear Regression Modelling by PLSR

A regression model was then developed, based on the Y- and X-data for this small *calibration* sample set. After 2 PCs the residuals in $\mathbf{X}$ and $\mathbf{y}$ were very close to zero. Hence, it was decided to use 2 PCs in the calibration model.

### 7.2.3. Inspecting the Bi-linear Model

The asterisks in Figure 7.1d represent the scores $\mathbf{t}_1$ and $\mathbf{t}_2$ for the five calibration samples obtained by the PLSR. (The other symbols will be discussed later.) Compared with their PCA scores in Figure 7.1b, the configuration of the five calibration samples is virtually identical; it has just been rotated clockwise a little.

This similarity between the PCA and PLSR solutions is not surprising, considering that the dominant variations in $\mathbf{X}$ are so relevant to $\mathbf{y}$. If $\mathbf{X}$ had contained major systematic variations of no relevance to $\mathbf{y}$, the two solutions would have been more different.

Figure 7.1c displays the two first PLSR components, in terms of their correlation loadings, i.e. their correlation coefficients $\mathbf{r}_1$ and $\mathbf{r}_2$ (cf. Appendices A5.5 and A6.11) between the new latent variables $\mathbf{t}_1$ and $\mathbf{t}_2$ and the input variables in $\mathbf{X}$ and $\mathbf{y}$. The outer dotted ellipse represents completely (100%) explained variance, and the inner ellipse 50% explained variance. Some X-variables are named explicitly by their wavelengths (in nanometres).

The figure shows that **y** (triangle) is nearly completely modelled by the 2-PC regression model (i.e. high percent explained variance), and so were all the X-variables in the colour wavelength range from 440 to about 650 nm. Beyond 650 nm only spurious correlations are evident.

### 7.2.4. Summary of the Calibration Model: Regression Coefficients $\hat{b}$

❑ THE RELEVANT X-VARIABLES

The 2-PC calibration model is summarised by the regression coefficient vector $\hat{b}$ in Figure 7.2. It shows that the wavelength range from 440 to about 580 nm (X-variables $k = 1$–20) contributes to the prediction of **y**.

Beyond 580 nm the X-variables may well measure *blue* litmus (cf. Figure 7.1a), but they do not respond to *red* litmus, and therefore they are not used for determining the *total* litmus content with this 2-PC model. They may hence be seen as irrelevant for our present **y**.

❑ ESTIMATE OF RELIABILITY

The error bars in Figure 7.2 show estimates of the reliability range of $\hat{b}$, as represented by $\hat{b} \pm 2\hat{s}(\hat{b})$. It shows that for prediction of **y** = *%Litmus*, there is no relevant *and* reliable information above about 570 nm.

These standard uncertainties $\hat{s}(\hat{b})$ were estimated by *cross-validation* of the model parameters (*jack-knifing*). The whole calibration model was re-estimated five times, each time in turn keeping one of the five calibration samples out from the calibration set. The resulting perturbations in $\hat{b}$ were summarised to yield the standard uncertainty estimates $\hat{s}(\hat{b})$. The cross-validation (with jack-knifing) will be explained in more detail in Chapter 10.

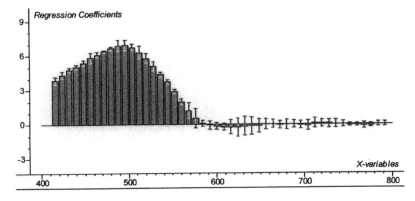

**Figure 7.2.** A calibration and prediction project, continued: prediction of **y** from **X**. Regression vector $\hat{b}$ (2 PCs), estimated from the chosen $N$ = five calibration samples. The estimated uncertainty of $\hat{b}$ is given by $\hat{b} \pm 2\hat{s}(\hat{b})$.

We could have chosen to down-weight or otherwise eliminate the X-variables above 570 nm from the modelling. But in this case the modelling is good enough anyway, so this is ignored here. Chapter 14 shows a more complex case where elimination of useless X-variables greatly improves the predictive ability for **y**.

*IV. Conclusions.* In this calibration phase of the example, we have now developed a multivariate calibration model, based on a very small, but empirically selected sample set. The model required 2 PCs, because two phenomena had to be modelled: presumably total litmus concentration (**y**) and a colour variation (blue/purple/red). Internal validation indicated that the model seems to be statistically reliable. We are now ready to apply it to new samples.

## 7.3. PREDICTION

*I. Purpose.* We would like to know *%Litmus* not only in the five calibration samples, but also in new samples.

*II–IV. Experimental planning and work, pre-processing and QC.* All the 19 samples for which X-data already exist are chosen for prediction of **y**. (Remember, in this demo example, we act *as if* we did not know it already.) Their X-data were shown in Figure 7.1a.

Later, X-data were similarly obtained for seven extra samples (file LITMUS-III). We know that these seven samples were made by more or less dilution of the same bottle of 100% litmus solution as the other samples, so we know that the true value of *%Litmus* must be between 0 and 100%. For simplicity, their spectra are not shown here, since they are presumably of the same general kind as samples 1–19.

*V. Data analysis*

## 7.3.1. Predicted Scores

The data for all 19 samples in Figure 7.1a were defined as **X** and submitted to the two-component bi-linear regression model. The open circles in Figure 7.1d show the scores for the prediction samples with unknown, but desired values for **y**. As already mentioned, the asterisks represent the five calibration samples.

The seven extra samples were also submitted to this bi-linear prediction modelling. Their scores (+) lie between those of *blue* ($i = 2$, 5) and of *red* ($i = 19$, 14) litmus. But some of them lie *beyond* the previous maximum range represented by $i = 2$ and 19. That is a little alarming. Should we have looked at their raw data, after all? Let us first try to predict the **y** from **X** in all the samples.

### 7.3.2. Predicted Y-values with Outlier-sensitive Uncertainty Estimate

The predicted values of *%Litmus*, $\hat{y}_i$ are shown in Figure 7.3, for the five calibration samples, for the rest of the 19 samples as well as for the seven extra samples ($i = 20$–26). The grey bars around the predicted values represent $\pm$ an outlier-sensitive estimate of the standard uncertainty of $\hat{y}_i$ (Chapter 10).

As expected, each of the samples in the calibration set and the first prediction set have predictions $\hat{y}_i$ between about 0 and 100% of the reference sample, and have small standard uncertainties. So the PP approach seems to have been successful.

But some of the seven extra samples (20, 22, 23 and 24) have predicted litmus concentrations $\hat{y}_i$ far above 100%, indicating that something is wrong with them. They also have abnormally large estimated uncertainties for their predictions $\hat{y}_i$.

Extra-samples 21, 25 and 26 have Y-predictions within the feasible range (0–100%), so from a conventional calibration point of view they seem OK. But the multivariate modelling shows that even samples 21 and 26 have abnormally large uncertainty ranges, which indicates that the calibration model does not necessarily apply for them, and their predicted Y-values must therefore be regarded with scepticism.

### 7.3.3. Inspecting the Detected Outliers

Whenever unexpected abnormalities are revealed during the data analysis, it is advisable to try to understand the reason for the problems. Sometimes that is easier said than done. But let us assume that we have later measured **y** by the traditional reference method in all the 26 samples: too late for any process

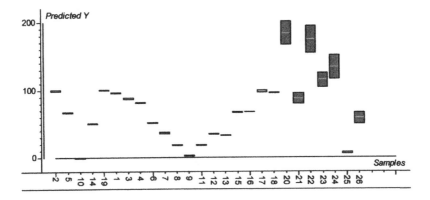

**Figure 7.3.** A calibration and prediction project, continued: prediction of **y** from **X**. Prediction of *%Litmus* from OD spectra: $\hat{y}_i$, with estimated individual uncertainty $\pm$ yDev$_i$ (eq. (10.6) in Appendix A10), for the calibration set ($i = 2, 5, 10, 14$ and 19), the 14 remaining initially available blue/red samples, and for seven new and totally unknown samples ($i = 20$–26).

control, but at least in time to learn to avoid similar problems in the future. Figure 7.4 shows some details for the three sets of samples.

◻ CALIBRATION SAMPLES

The first row shows how the variation in the input X-spectra of the five calibration samples (Figure 7.4a) is reduced to almost nothing after 2 PCs (Figure 7.4b). Figure 7.4c shows the predicted vs. the measured value of **y** in these calibration samples. As expected, the relationship seems nice and linear (samples 2 and 19 lie on top of each other at 100%).

It is not surprising that the 2-PC model gives a nice linear calibration fit for only five calibration samples. It may be a result of so-called *overfitting*. After all, if we had only *two* calibration samples, a straight line would fit them perfectly, whether or not the line is meaningful for other samples. Having five well-chosen calibration samples is a little better, but the risk for overfitting is still there. We

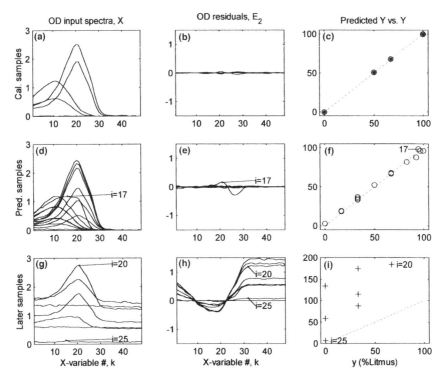

**Figure 7.4.** Example of calibration and prediction, continued: why outliers? Top: the data of the five calibration samples. Middle: the 14 remaining blue/red samples with unknown **y**. Bottom: the seven new samples with unknown **y**. Left: input OD spectra **X**; Middle: OD residuals after fit to the two-component calibration model, $E_2$. Right: predicted $\hat{y}$ (ordinate) vs. **y** (abscissa, obtained later).

checked for overfitting by cross-validation (Chapter 10), but even that is somewhat uncertain, with so few calibration samples.

◻ PREDICTED SAMPLES

Instead, in the second row of subplots, we use the remaining initial samples ($i = 1, 3, 4, 6–9, 11–13, 15–18$) as an *independent test set*. Figure 7.4d shows their X-data. Figure 7.4f demonstrates that the calibration model has very good ability to predict **y** also for unknown samples of this type.

◻ THE OUTLIERS

Sample 17 displays a small unmodelled spectral residual (Figure 7.4e), but this causes only a slight over-estimation of **y** (Figure 7.4f).

However, the third row shows why the seven extra samples ($i = 20–26$) were tagged as *outliers*. Their input spectra (Figure 7.4g) show abnormal vertical offsets. These unexpected features yielded some abnormal scores (Figure 7.1c). Their abnormality is even more clearly evident in $E_2$ (Figure 7.4h), as large unmodelled residuals after two PLSR components. Except for sample 25, they are obviously not of the sample type for which the calibration model applies!

Consequently, it is not surprising that their predicted values of *%Litmus* ($\hat{y}_i$) are grossly erroneous (Figure 7.4i). Based on spectroscopic background knowledge, we can conclude that the large spectral baseline offsets in Figure 7.4g indicate either an instrument failure or some type of optical problems in these samples.

(As some readers may have guessed, the third set actually represents the litmus samples 20–26 in Table 4.1, also shown in Figure 4.1c, with various amounts of white zinc oxide powder added to give uncontrolled turbidity variations.)

Note that for some of the extra samples, even predictions $\hat{y}_i$ that were in the feasible range of 0–100% (samples 22 and 26) are, in fact, grossly erroneous. But we already expected that, due to their abnormally high uncertainty range in Figure 7.3.

A sample's degree of abnormality *inside* the bi-linear model may be detected by comparing its scores (Figure 7.1c), to the distribution of scores in the calibration samples. Its degree of abnormality *outside* the bi-linear model may be detected by comparing its X-residual level to that of the calibration samples. Together, they cause the outlier-sensitive uncertainty bars for $\hat{y}_i$ (Figure 7.3). These outlier detectors are based on the so-called *leverage* and the *root mean square error of prediction* in **X**, $h_i$ and RMSEP(X)$_i$. More detail on these will be given in Chapter 10, and they will be used in practice in Chapters 12 and 15.

The automatic outlier detection is a feature lacking in conventional univariate calibration of instruments (Figure 4.1f).

## 7.3.4. Updating the Model

Now that we have taken the trouble to measure **y** in all these samples, we can recalibrate with all 26 samples instead of just five samples in the calibration set. The resulting updated calibration model (not shown here) needs $A_{Opt} = 3$ PCs to give adequate modelling. Its predictive ability for **y** was actually shown already in Figure 4.1i. Updating of models will be discussed in more detail in Chapter 15.3.

*VI. Conclusions.* The prediction phase of this example has shown that based on the 2-PC calibration model for clear water solutions of litmus, we could successfully predict **y** = *%Litmus* in new samples representing clear water solutions, from the measurements of **X** = OD at 48 wavelength channels.

But when turbid samples arrived, their predicted Y-values were grossly wrong. However, since they were automatically tagged as outliers in **X**, these errors were stopped from being used for decision making.

Later, the model could be updated, but now it required one more PC, in order to account for the new phenomenon (turbidity). (Upon closer scrutiny, is was even found that the fourth PC gave predictive improvement. The cause of this is probably some type of instrument non-linearity or the turbidity).

## 7.4. SUMMARY OF ALL THE LITMUS DATA SET ILLUSTRATIONS

### 7.4.1. Increasing complexity

Going back to Figure 4.1, we can recapitulate these method-chapters so far.

In Chapter 4 the input data were presented (Table 4.1), and the selectivity problem was defined.

In Chapter 5 we first analysed the simple one-phenomenon data (*blue litmus* only), with **X** = OD spectra (Figure 4.1a) by bi-linear principal component analysis (PCA). One single PC was sufficient to give good modelling of **X**.

Then the bi-linear PCA modelling was applied to a data set with two phenomena (*blue* and *red litmus*, Figure. 4.1b), requiring two PCs. One sample, $i = 17$, showed a small abnormality in residual $E_2$ in Figure 5.2.

In Chapter 6 we explained how bi-linear regression allows us to predict one or more Y-variable from the X-variables. The reason for using partial least squares regression (PLSR) instead of just extending PCA to PCR was explained.

A mini-example of calibration and prediction in Chapter 6.5, based on the data in Figure 4.1b, showed how the selectivity for both *blue* and *red litmus* ($y_1$ and $y_2$) could be attained, by naïve multivariate calibration on only 3 samples. The model required 2 PCs.

In Chapter 7 the two phenomenon data set (Figure 4.1b) was used for demonstrating how to make cost-effective experimental designs (only 5 samples) for multivariate calibration. In this case we only calibrated for one Y-variable, $y =$ *%Litmus*. Still, the model required 2 PCs.

When new samples arrived with unexpected, unmodelled qualities (turbid samples, Figure 7.1c), their predicted Y-values were dangerously wrong. But they were automatically tagged as outliers, so no harm was done. After even their Y-values had been measured by traditional means, the whole model was updated. But now the PLSR model required at least 3 PCs.

### 7.4.2. So, How was the Selectivity Enhancement in Figure 4.1 Attained?

Bi-linear modelling by PLSR was used for generating the selectivity enhancement in the bottom row in Figure 4.1, compared with the best univariate calibration (middle row).

For each of the three full data sets (columns of subplots) a multivariate calibration model was generated, based on $X =$ OD data (the top row of subplots) and $y =$ *%Litmus* (abscissa in the middle row of subplots).

For the one-constituent data (left column of subplots), a 2-PC solution was actually used. Why 2 PCs? The second PC implicitly modelled some of the slight, unexpected *curvature* seen in the Figure 4.1d)!

For the two-constituent data (middle column of subplots), a 3-PC solution was likewise chosen, and for the three-constituent data (right column of subplots) a 4-PC solution was used. The optimal number of PCs was chosen by graphical inspection and cross-validation (to be discussed in Chapter 10).

In summary, Chapters 4–7 have used the smooth litmus spectra as input data in order to explain the mathematics of the bi-linear structure modelling. But Figure 6.2 shows that smoothness was not at all required in BLM. Chapter 8 will illustrate bi-linear modelling of interdisciplinary and less well-behaved data, in order to fill some important method gaps still open.

### 7.5. TEST QUESTIONS

1. Think of an example of X- and Y-variables that would be interesting for you.
2. Why were the litmus-containing samples red, blue or white?
3. Can Y-predictions for outliers look normal?
4. Can Y-predictions for outliers be useful?

### 7.6. ANSWERS

1. You do the thinking. Some ideas are given in Table 7.1.
2. Red and blue colours were caused by the water solutions having been made acidic or alkaline by the addition of acid or lye, that caused the molecular structure of the anthocyanines in the litmus to give two different, equally

characteristic colours. The white colour was something quite different: a fine white powder added.

3. The really dangerous type of outlier is one where the predicted Y-value by chance looks normal, but is just nonsense! In traditional calibration of analytical instruments, this nonsense cannot be detected and will therefore be put to use like any other result. But they can be detected automatically by the multivariate modelling in **X**.

4. The outlier warnings alert us to all kinds of strange samples. Most will be garbage, some will be gold.

# Chapter 8

# Interpretation of Many Types of Data X ⇔ Y: Exploring Relationships in Interdisciplinary Data Sets

*This chapter demonstrates how the BLM method may be used for quantitative interpretation of many different types of input variables.*

*To make different variables in a data table compatible with each other, the technique of a priori standardisation is shown. The use of object indicator-variables is explained. The effect of swapping the variables in X and Y is demonstrated, and handling of non-linearities in the X–Y relationship is discussed. Finally, how to interpret BLM plots and choosing a few, but relevant variables as quality criteria is discussed.*

*The COCOA data set is to be used for demonstration taking the sensory, chemical and physical data in the mixture design into account.*

## 8.1. SENSORY AND CHEMICAL DATA

### 8.1.1. Finding Quality Criteria

As mentioned in Chapter 2, *quality* may be defined in different ways. What consumers define as *quality* is mostly related to what was defined as *excellence/goodness* (Figure 2.5). But what does *excellence* or *goodness* mean for a certain type of product? Which *quality criteria* should be measured in order to optimise and monitor the quality to satisfy consumers' needs? How to select as *few* quality criteria as possible without losing essential information in the data?

The present example will focus on interdisciplinary interpretation and data analytical principles necessary to find reliable and relevant quality criteria for a rather complex type of products, cocoa-milk drinks. When consumers push the *Cocoa* or *Hot Chocolate* button on instant dispensers of hot drinks, what they usually get is pre-mixed cocoa-milk powder, dissolved in hot water. Sometimes, some consumers dislike the product. So a producer company wanted to learn how to optimise the quality of hot cocoa-milk drink, made by mixing cocoa, sugar and milk powder.

## ☐ DIFFERENT APPROACHES TO DETERMINING QUALITY

*Production engineers* know how a product has been designed. But if asked to make it differently, they need to know in which way it should be different, and how this should be attained. In other words: how does the production design affect the properties of the product that the human senses detect, and that the consumers like/dislike?

*Marketing people* would probably propose that the production engineers make a few different cocoa-milk products, and then do a consumer study. That makes sense, since the consumer response was the problem in the first place. However, it is very expensive. Consumers may have quite different preferences, and are notoriously imprecise in their judgements. Therefore, many consumers are required before reliable results can be obtained. At the present stage of the project, only a small pilot consumer study with a few samples could be afforded (cf. Chapter 3.5).

*Physicists and chemists* would probably suggest that the production engineers produce many different cocoa-milk products, and then measure various physical and chemical properties. Information from modern analytical instruments is obtained faster and much more reliably than consumer studies! But what does all this precise information about qualities (*qualitas*, Table 2.2) tell us about quality as perceived by humans?

*Sensory scientists* would claim that the only relevant eating qualities are those that can be perceived by the human senses (assuming that no health factors are involved). The problem is to measure these perceived quality dimensions (appearance, taste, smell, texture) in a reliable way. But that has already being done routinely for a long time in many fields like food science, in the cosmetics industry etc., where sensory methods provide meaningful and reproducible measurements of quality. Sensory science is described more in Appendix A2.

## ☐ INTERDISCIPLINARY PROJECT: THE COCOA DATA SET

All of these types of product descriptors within the various disciplines are involved in the present project: for a consciously chosen set of products, design variables, physical/chemical measurements and sensory descriptions were obtained. For a few of the products, chosen on the basis of PP design, also consumer responses were acquired (Folkenberg et al., 1999). Figure 3.6 showed which sensory variations seemed relevant for predicting consumer preference (file COCOA-III). But what causes the sensory variations in physical and chemical terms?

By graphical BLM *data mapping*, the intention was to allow the different professions to relate their own perspective and measurements to that of the others. Thereby, a common understanding could hopefully be developed, about which production changes caused what physical and chemical changes, and how these changes affected the perceived quality.

The present small COCOA data set has been chosen to contain quite different types of input variables, but to have so obvious structure that the reader can follow the statistical method-descriptions intuitively, in the following project.

> The project description has been adapted didactically for this book, but the input data have been extracted as a subset of samples and variables from Folkenberg et al. (1999) without any modification.

To illustrate how multiresponse PLSR analysis works, we shall analyse this data set just like we would analyse any other multidisciplinary data set, as if we did not know their rather obvious structure.

### 8.1.2. Approaching Interdisciplinary Data

The purpose of the data analytical methodology to be presented here is to enable the reader to analyse a diversity of input variables.

❑ BLM: Bringing Various Types of Quality Measurements Together

For a given set of samples, different types of variables may be obtained to describe them. There may be valuable information in each of them, but to interpret such raw data is not trivial. The multivariate data analysis allows us to interpret interdisciplinary data.

❑ Being Aware of One's *A Priori* Expectations

Usually, we have an expectation about what the data should reveal. Since this expectation will float in the back of our mind and affect our interpretation, in any way, it is just as well to make it explicit and check it, as soon as the data are ready for analysis. That prevents us from being overwhelmed by the data analysis.

The subsequent multivariate data analysis can still reveal unexpected and interesting patterns, as long as we maintain a minimum of cognitive flexibility!

❑ Balancing Variables With Widely Different Input Ranges

If the X- and Y-variables are scaled differently from each other it poses no statistical problem. But if variables *within* **X** are scaled very differently, the data analyst must be careful, and rescale them to ensure that their levels of variability are compatible with each other, and at least not orders of magnitude different. The same is true if there are several differently scaled variables in **Y**. The technique of *standardisation* will be explained.

◻ CHOICE OF WHAT TO PUT IN **X** AND IN **Y**

If there are several different types of variables in a data set – e.g. chosen design variables and measured response variables of different kinds – the data analyst faces some decisions. Should they all be lumped together in a matrix **X** and analysed by PCA, or should they be split into **X** and **Y** and analysed by PLSR? In that case, what should be **X** and what should be **Y**? Does it matter?

◻ HANDLING NON-LINEARITIES

When relating many different types of variables to each other in a given set of samples, the data analyst must look for – and expect to find – some non-linear relationships. This chapter discusses how to detect that, and how to deal with it.

## 8.2. ANALYSING MANY TYPES OF X- AND Y-VARIABLES

*I. Purpose*. The purpose of the present study was to explore the sensory, physical and chemical quality variations in a designed set of cocoa-milk samples. What are the relationships between sensory and chemical/ physical data? Which quality criteria should be chosen to be used for predicting future samples?

*II. Experimental planning*. Seven different products are to be prepared, according to a reduced three-component mixture design (Figure 11.3, p. 220): seven mixtures were made of cocoa powder, skim milk powder and sugar. Each mixture was produced three times. Each designed powder mixture was mixed with a fixed amount of hot water, and the same samples were submitted to various laboratory analyses.

*III. Experimental work*. The seven cocoa-milk products, each produced in three replicates, were submitted to descriptive sensory analysis and to various physical and chemical analyses. One of the samples (mixture 5) was presented as twice as two independent products to the sensory panel, and here named *5a*, *5b*, i.e. to get sensory *repeatability* replicates.

The sensory analysis of the eight samples (three replicates) was performed by a panel of seven trained assessors. Only the mean of the seven assessors is reported here. The sensory analysis of the three produced replicates was separated in time, several days apart. The sensory variables are defined on a 0–15 scale.

The sensory variables are here named in lower case letters. For instance, the sensory colour darkness is here named *colour*. The physical and chemical variables are named in *CAPITAL* letters. There are five sensory descriptors, colour darkness (*colour*), *cocoa-odour*, smooth texture (abbreviated to *smooth-txtr*), *milk-taste* and sweet taste (*sweet*). This subset was selected to span all the important human sense modalities. In addition there are the three chemical design variables, *%COCOA, %SUGAR* and *%MILK*, which define how the seven cocoa-milk products were mixed (representing percentage of dry matter).

The samples were measured instrumentally with respect to various rheological properties, in two repeatability replicates. Only the mean *VISCOSITY* is used here. Instrumental colour was measured by a reflectance spectrophotometer, and expressed in the conventional *L* (lightness), *a* (redness/greenness), *b* (blueness/yellowness) units. Only the measured colour lightness is here used, and referred to as *COLOUR* lightness: it is given in a unit where 0 = black and 100 = white.

Chemical measurements were also performed, but for simplicity they are here ignored. Chemical information about the samples is instead represented by the three design-variables *%COCOA, %SUGAR* and *%MILK*.

## IV. Pre-processing and QC of data

### 8.2.1. Preliminary Inspection of the Raw Data

Graphical inspection of the raw data did not reveal any gross errors. For simplicity, therefore, only the mean of the replicates will be used in this chapter; (individual data for the three full replicates will be used in Chapter 10.5).

The input data have already been given in Chapter 1 (Table 1.1). Figure 8.1 shows the means and the total standard deviations of the variables. We note, for

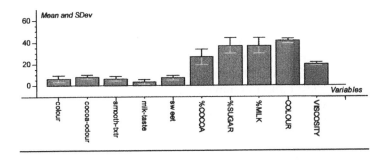

**Figure 8.1.** Interdisciplinary input data set COCOA: summary statistics. Graphical overview of the mean $\pm$ total initial standard deviation **s**, for the sensory and chemical/physical variables in Table 1.1.

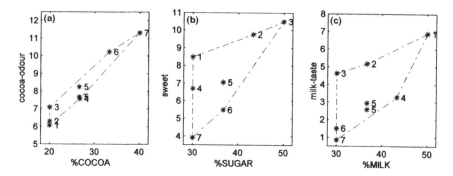

**Figure 8.2.** A preliminary look for causal relationships in the raw data. (a) Sensory *cocoa-odour* vs. *%COCOA*. (b) Sensory *sweet* taste vs. *%SUGAR*. (c) Sensory milk-taste vs. *%MILK*. The mixture design is outlined by the dashed curves; numbers represent cocoa-milk mixture name. The two replicates for mixture *5* are virtually on top of each other in (b).

now, that the variables have quite different means (eq. (4.5), p. 371) and different total initial standard deviations (eq. (4.7), p. 372).

## 8.2.2. Checking *A Priori* Expectations

In the present example we expect e.g. the *cocoa-odour* to increase with *%COCOA*, *sweet* taste is expected to increase with *%SUGAR*, and *milk-taste* is expected to increase with *%MILK*.

While we are inspecting the raw data for gross errors, let us look for these causal relations. Figure 8.2 checks these expected sensory-chemical relationships. The dashed curves outline the triangular experimental mixture design: with samples *1*, *2* and *3* along one side, samples *1*, *4* and *7* along another and samples *1*, *6* and *7* along the third. Sample *5* is the centre point with its two sensory repeatability replicates; in Figure 8.2b they are virtually identical.

*Cocoa-odour* is indeed found to increase more or less linearly with *%COCOA*. *Sweet* taste and *milk-taste* are also seen to increase with their respective causal ingredients. But, surprisingly, their responses are far less selective: for instance, at a constant *%SUGAR* of 30%, the perceived *sweet* taste decreases strongly from sample *1* (most milk) to sample *7* (most cocoa). At a constant *%MILK* of 30%, the perceived *milk-taste* is seen to decrease strongly from sample *3* (most sugar) to sample *7* (most cocoa). Some curvatures can also be seen in the expected causal relations.

Thus, the input data seem to make sense, since we can see the general patterns expected. But the data obviously contain some unexpected effects! What more do the input data have in store for us?

**V. Data analysis**

### 8.2.3. Standardising the Input Variables to the Same Initial Variability

The input variables, in Figure 8.1, were found to have quite different means and standard deviations. These differences are of no interest to us *per se*, but they can destroy the multivariate modelling.

Before submitting such *mice and elephants* to multivariate modelling, we make the input variables more compatible, by applying some sort of *a priori* pre-processing. The simplest approach is to *standardise* all of the input variables.

❑ Mean Centring to Remove Level Differences

Since we are interested in the *between-sample variation* of the variables, not in their *absolute levels*, we mean centre each of them. After the mean centring, every variable has a mean of 0. This is the same operation shown in row 1, Figure 5.2 for the Litmus data.

❑ Dividing by Standard Deviation to Remove Range Difference

In order to avoid that valid variation in some variables is swamped by noise in other variables in the same data table, the variables may be *scaled* prior to bilinear modelling. The simplest way of scaling them is to divide them by their total initial standard deviation.

That makes sense if we can assume that all the variables have approximately the same *relative error level,* i.e. that the uncertainty standard deviation $s_{err}$ is proportional to the total initial standard deviation, $s$ for each variable. After division by their total standard deviation $s$, we have forced each variable to have a total initial standard deviation of 1.

❑ Standardisation

Together, the two operations of mean centring and division by the total standard deviation, is called *standardisation.*

Mathematically, the standardisation of an input variable number $k$, e.g. $\mathbf{x}_{k,\,input}$, is described by

$$\mathbf{x}_k = \left(\mathbf{x}_{k,input} - \bar{\mathbf{x}}_{k,input}\right) \times v_k \tag{8.1}$$

where $v_k$ represents *a priori scaling weights*. In conventional standardisation, these scaling weights are simply defined as the inverse of the total initial standard deviations (eq. (4.7), p. 372)

$$v_k = 1/s_k \tag{8.2}$$

It should be noted that default standardisation should *not* be applied, if some of the variables contain only baseline information (cf. the LITMUS data in Figure 4.1a–c), because this makes the modelling more noisy. Then the input variables may be left unscaled, or the standardisation may then be modified, as described in Appendices A8.2 and A8.3.

### 8.2.4. One Matrix BLM: PCA as a Preliminary Inspection Tool

A quick look at the PCA of the whole data matrix can also help understanding the data. That was, in fact, already done in Chapter 1. All the variables in Table 1.1 were put into the X-block, standardised and submitted to the PCA process explained in Chapter 5.

The score plot for the first 2 PCs in Figure 1.4a (p. 20) showed the seven cocoa mixtures to fall in a triangular pattern, as expected, with the two sensory repeatability replicates for mixture 5 close to each other in the centre of the mixture design. The corresponding loading plot in Figure 1.4b showed the main patterns of intercorrelations between the variables in Table 1.1.

The obtained PCA solution does not look wholly unexpected. But assume that we want to assess to what degree people's sensory perception provide information about the chemical composition and physical properties of the samples, or vice versa. The one-matrix PCA (**X**) analysis does not provide explicit answers to that, but the two-matrix PLSR (**X**, **Y**) analysis does.

### 8.2.5. Two Matrices BLM: PLS Regression

❑ STANDARDISING THE INPUT VARIABLES

PLSR allows us to define some variables as **X** and other variables as **Y**, thereby learning how to predict **Y** from **X**. But, like in PCA, scaling differences must first be considered.

Chapters 6 and 7 showed that the input variables in **Y** do not have to be given in the same unit as those in **X**: the scaling differences *between* **X** and **Y** is accounted for by the Y- and X-loadings in the BLM. But if the input variables *within* **X** or *within* **Y** are of different types, and hence given in different units, then it is important to *rescale* them prior to the PLSR modelling.

For simplicity, all the variables in Table 1.1 were standardised (eqs. (8.1) and (8.2)) prior to the PLSR modelling.

❑ TWO ALTERNATIVE CHOICES OF **X** AND **Y**

Figure 8.3 shows two different ways of relating the sensory variables to the chemical and physical variables. Figure 8.3a shows the results when **X** = sensory variables and **Y** = chemical and physical variables. Figure 8.3b shows the result after the definitions of **X** and **Y** have been swapped. In both cases, 2 PCs were required.

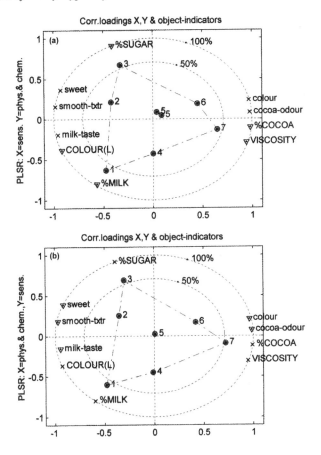

**Figure 8.3.** Choice of **X** and **Y** in PLSR. Correlation loadings. × = X-variables; ∇ = Y-variables; dotted ellipsoids = 50 and 100% explained variance for X- and Y-variables; 1–7 = sample indicator variables; dashed lines = outline of the mixture design. a) **X**, sensory variables; **Y**, chemical design variables and physical measurements. b) **X**, chemical design variables and physical measurements; **Y**, sensory variables.

The large dotted triangles in Figure 8.3 outline the sample set, by the use of *object indicator-variables*.

*Object indicator-variables*: The cocoa mixture samples, which we could have plotted in a score plot like Figure 1.4a, are instead represented directly in the correlation loading plot (⊛ in Figure 8.3). This is attained by the use of *object indicator-variables*. For each row number $i = 1, 2,..., N$ in **X**, a new *indicator* variable $\mathbf{d}_i$ has been defined. This indicator-variable column contains *zeroes* except for row number $i$, where it is 1. Together the indicator variables form a so-called *identity matrix* of product indicator-variables with $N$ rows and $N$ columns, here called by its traditional name,

$$\mathbf{I} = [\mathbf{d}_i, \ i = 1, \ 2, ..., N] = \tag{8.3}$$

```
1 0 0 0...0
0 1 0 0...0
0 0 1 0...0
0 0 0 1...0
    ....
0 0 0 0...1
```

With $N = 8$ samples, there are eight such object indicator-variables in the present data set. Since this set of object indicator-variables in eq. (8.3) provides no new structure information, it may be handled in several equivalent ways, without affecting the resulting BLM solution. It may either be put into $\mathbf{X}$ or $\mathbf{Y}$ as extra variables (Figure 3.4), or they may (like here), be treated as *passive* variables (Appendix A10.2): After the PLSR model had been established with latent variables $\mathbf{t}_a \# a = 1,2,...$, the correlation coefficients $r_{ia}$ between each object indicator-variable $\mathbf{d}_i$ and each latent variable $\mathbf{t}_a$ were computed and plotted with the other correlation loadings (eqs. (5.14) in Appendix A5 and eq. (6.7)) in Appendix A6.

## 8.2.6. Predicting Chemistry and Physics (Y) From People's Perception (X)

Each of the two subplots in Figure 8.3 summarises a complete PLSR model, in terms of its correlation loadings for the first 2 PCs. The X-variables are represented by $\times$ and the Y-variables by $\nabla$. The configuration of samples is shown by their object indicators ($\circledast$).

Figure 8.3a shows the human assessor's ability to *look* at these cocoa samples, *smell* them, *taste* them and feel their *texture* in the mouth ($\mathbf{X}$), and how this relates to the chemical and the physical properties ($\mathbf{Y}$).

❒  THE BLM PROCESS

As outlined in eq. (6.2) p. 118, this BLM modelling was made by:

1. Finding the main systematic patterns of variation (latent variables or PCs) as functions of the X-data (here: sensory inputs) $[\mathbf{t}_1, \mathbf{t}_2,....] = w(\mathbf{X})$;
2. These are collectively referred to as score matrix $\mathbf{T} = [\mathbf{t}_1, \mathbf{t}_2,...]$;
3. Checking how well these PCs are able to model the X-data:
   $\mathbf{X} = p(\mathbf{T}) + \mathbf{E}$;
4. Learning how the PCs model the samples' Y-data (here chemical composition and physical instruments) $\mathbf{Y} = q(\mathbf{T}) + \mathbf{F}$;
5. Optionally, making the equivalent linear model summary: $\mathbf{Y} = b(\mathbf{X}) + \mathbf{F}$.

❐ OPTIMAL MODEL RANK: HOW MANY PCs, $A = 1,2,..,A_{Opt}$?

By some kind of validation (Chapter 10), we decide the optimal model rank. In the present case, this is rather easy: 2 PCs appear to be both necessary and sufficient.

It is *sufficient*, because with $A_{Opt} = 2$, all of the X- and Y-variables fall quite close to the circle of $r^2 = 1$ (100% explained variance). Hence, there is little or no need for a third PC. It is *necessary*, because if only 1 PC had been used, some variables like *%SUGAR* and *%MILK*, *sweet* and *COLOUR* would not have been sufficiently explained.

❐ SYSTEMATIC VARIATION PATTERNS?

By observing a model plot, as in Figure 8.3a, the main systematic variation patterns gradually emerge in the mind. This process is helped by ensuring that the input data tables have clearly interpretable *names* of the individual variables and samples.

❐ INTERPRETATION OF A PLOT

This is an illustration of a common way to browse through a model plot:

*A pair of correlated variables:* Figure 8.3a shows the X-variable *cocoa-odour*, and the Y-variable *%COCOA* to fall close to each other, to the far right in the plot. This means that they have the same type of correlation to the first 2 PCs. Since they fall near the 100% variance circle, this correlation to the bi-linear model is more or less complete and hence they are also nearly 100% correlated to each other. This can be confirmed by going back to the raw data (cf. Figure 8.2a), where $r_{cocoa\text{-}odour, \%COCOA} = 0.97$).

*A group of correlated variables:* Two other variables, sensory *colour* and instrumental *VISCOSITY* show similar near-100% relations to the model, near *cocoa-odour* and *%COCOA*. From background knowledge it would be tempting to interpret this causally: when *%COCOA* increases, so does the sensory *colour* darkness, *cocoa-odour* and *VISCOSITY*. In support of this preliminary interpretation, cocoa mixture 7, which has the highest cocoa content, is most strongly associated with this group of cocoa-related variables.

*Anti-correlated variables:* The BLM plot in Figure 8.3a shows that these four cocoa-related variables are strongly *anti-correlated* to other variables. For instance, *colour* darkness is the almost perfect *antipode* to *COLOUR* lightness. This is as expected, since a high-cocoa sample like 7 would have high value of sensory *colour* darkness and low value of instrumental *COLOUR* lightness. Looking at the raw data (Figure 1.2c, p. 15) indeed confirms this; $r_{colour,COLOUR(L)} = -0.98$. Likewise, *smooth-txtr* and *VISCOSITY* are seen to be almost perfectly anti-correlated.

*The object indicators:*   The product indicator-variables 1–7 have slightly different properties than ordinary variables: they do not come close to the 100% explained ellipse except for very dominant samples. But they show how the patterns of variables relate to the patterns of samples. For instance, cocoa mixtures *1*, *2* and *3* (containing low cocoa percentages) are the most negatively associated with the variable *%COCOA*. The two repeatability replicates of mixture *5* fall near the origin, which corresponds to the mean sample quality, and hence has very little correlation to the PCs in the model.

*A systematic pattern:*   In summary, Figure 8.3a shows that the data indicate a strong, underlying variation-pattern associated with *%COCOA*, in this case seen horizontally, along the *East/West* direction in the plot. When *%COCOA increases* (e.g. from mixture *2* to *7*), then there is an *increase* in *colour* darkness, *cocoa-odour* and *VISCOSITY*, and simultaneously a *decrease* in *sweet*, *smooth-txtr* and to some extent *milk-taste* and *COLOUR* lightness.

The remaining variables, *%MILK* and *%SUGAR* are to some degree anti-correlated with *%COCOA* along the horizontal *West/East* direction, as expected, since their sum is 100%, so when one goes up, at least one of the others has to go down. Correspondingly, mixtures *1* (mostly milk) and *3* (mostly sugar), like *2* (their mixture), lie *to the West* of the mean sample quality (*5*).

*Another systematic variation pattern, different than expected:*   A second systematic type of variation is also seen in Figure 8.3a, in the vertical direction, orthogonal to PC1. Upon closer scrutiny it seems reasonable to interpret this pattern as a systematic *%MILK* vs. *%SUGAR* variation, spanning from high-milk mixture *1* (and to some extent *4*) via mixtures *2*, *5* and *7*, with equal amounts of milk and sugar, to mixture *3* (and to some extent *6*).

Not unexpectedly, sensory *sweet* taste is positively correlated to *%SUGAR*, and sensory *milk-taste* and *COLOUR* lightness positively correlated to *%MILK*. But *milk-taste* is not *uniquely* correlated to *%MILK*, and *sweet* is not *uniquely* correlated to *%SUGAR*. In addition, *milk-taste* and *sweet* are also strongly anti-correlated to *%COCOA,* more so than *%MILK* and *%SUGAR* are anti-correlated to *%COCOA*. Why? That cannot be elucidated from these data alone. Three hypotheses were listed in Chapter 1.4.3.

Together, these two systematic patterns attain close to 100% explained variance, even for these last variables. As usual, when unexpected patterns are found, we check the raw data. Figure 8.2b showed, e.g. that even at a *constant %SUGAR* of 30%, *sweet* taste decreases with cocoa content, from mixture *1* via mixture *4* to mixture *7*. Likewise, Figure 8.2c showed that at a constant *%MILK* of 30%, the *milk-taste* decreases with cocoa content from mixture *3* via mixture *6* to mixture *7*. So the effect discovered in the BLM plot is real enough.

In the present case the two valid PCs happen to coincide more or less with two possible causal patterns: PC1 = increasing *%COCOA* (and concomitant decrease in *%MILK* and *%SUGAR*), PC2 = increasing *%SUGAR* and decreasing *%MILK* at constant cocoa levels.

This simple structure primarily arose because so many of the variables are dominated by one of the causal phenomena (varying cocoa percentage) and therefore defined PC1 uniquely. Since the mixture design was quite balanced, the remaining sugar/milk-variation was picked up by PC2.

## ❏ Summarising the BLM Findings

Figure 8.3a showed an apparently successful description of the five sensory X-

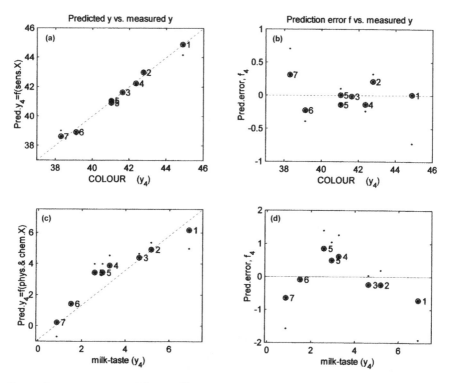

**Figure 8.4.** Prediction of **Y** from **X**. Top: Five sensory X-variables used for predicting a physical Y-variable, *COLOUR* lightness. Bottom: Five chemical/physical X-variables used for predicting a sensory Y-variable, *milk-taste*. Left: Predicted vs. measured **Y**. Right: Residual $\hat{\mathbf{F}}$ ( $= \mathbf{Y} - \hat{\mathbf{Y}}$) vs. **Y** for given Y-variable. (Dots represent predictions from leave-one-out cross-validation: to be explained in Chapter 10).

variables into a two-dimensional map, with a good potential for predicting the five chemical and physical variables (the Y-variables).

Consequently, Figure 8.4a shows an example of such a prediction of **Y** from **X**, for the value of the fourth Y-variable, *COLOUR* lightness. The values predicted ($\hat{y}_4$, ordinate) from the five sensory descriptors (**X**) is plotted against its measured values ($y_4$, abscissa), within the set of eight calibration samples (⊗).

Figure 8.4b shows the corresponding lack-of-fit residual, i.e. the difference (ordinate) between predicted and measured and , $f_4 = \hat{y}_4 - y_4$, plotted (amplified) against its measured values, $y_4$ (abscissa). These calibration residuals look more or less randomly scattered, indicating a satisfactory modelling of this variable.

### 8.2.7. Swapping X and Y

❐ PREDICTING PEOPLE'S PERCEPTION (**Y**) FROM CHEMISTRY AND PHYSICS (**X**)

For the sake of completion, Figure 8.3b shows the PLSR correlation loading plot after having interchanged **X** and **Y**. **X** now contains the chemical design variables *(%COCOA, %SUGAR, %MILK)* plus the two physical measurements *VISCOSITY* and *COLOUR* lightness, and **Y** contains the sensory variables.

The reader is encouraged to try to perform the interpretation of Figure 8.3b alone. Remember that mixtures *1, 3* and *7* contained the maximum percentages of milk, sugar and cocoa, respectively. The results in Figure 8.3b are seen to be very similar to those in Figure 8.3a.

❐ WHEN **X** AND **Y** AGREE

Figure 8.3a,b shows two-block BLM patterns very similar to the previous one-block BLM patterns (Figure 1.4). This similarity illustrates the close resemblance in how PCA and PLSR perform, when the data are well-behaved, and all the X-data are *relevant* to **Y**.

❐ WHEN **X** AND **Y** DISAGREE

In other data sets, the swapping of **X** and **Y** may appear to give different results. Usually the difference is just a trivial flipping or rotation of axes; that has no consequence for the interpretation or the predictive ability.

But if the two sets of variables *disagree* strongly about the main pattern of samples, then the two BLM solutions will differ beyond trivialities. This situation only arises when the data show poor model fit and bad predictive ability. That is a warning that there is trouble in the data, or maybe something very interesting: look more closely at the raw data!

☐ WHAT SHOULD BE **X** AND **Y**?

More discussion of how to choose what to put in **X** and what to put in **Y**, was given in Chapter 3.3.2. Briefly, variables to be predicted in the future are in this book defined as **Y**. If prediction is not a central concern (beyond predictive model validation), the block of variables that spans the most interesting variation and/or are measured with the best precision should be used as **X**.

But the choice is usually not very critical. If in doubt, the user should try to swap **X** and **Y**.

Remember that the latent variables are obtained from **X**, not from **Y**. So for instance, if there is only *one* Y-variable and many X-variables in a model, and we swap them, then there will only be *one* X-variable, **X** = [$x_1$], and hence only *one* PC. For more details on PLS regression modelling of mixtures, see Kettaneh-Wold (1992).

☐ CAUSALITY MODELLED IN THE COMPUTER OR IN THE MIND

Figure 8.3a,b shows that the causality in the data does not have to be reflected in how the data are modelled mathematically, **Y** = $f$(**X**) as briefly discussed in Chapter 1.4. Obviously, in this case the chemistry and physics of the samples may be seen as causal for the sensory perception, and not the other way around.

But the non-causal BLM in Figure 8.3a, where **Y** ⇒ **X**, in this case gave the same mental interpretation as the causal BLM in Figure 8.3b, where **Y** ⇐ **X**, since **X** and **Y** agree well on what is the important pattern of the samples.

### 8.2.8. Non-Linearities

☐ SEEING AN UNMODELLED CURVATURE

Based on the model in Figure 8.3b we should now be able to predict the sensory Y-variables from the chemical and physical X-variables, via the two latent variables $t_1$ and $t_2$. This is illustrated for one of the sensory variables (*milk-taste*) in Figure 8.4c.

However, upon closer scrutiny, *milk-taste* in these samples ⊛ shows a more or less well defined pattern of unmodelled structure: the predictions (ordinate) are too high in the middle, and too low at either end of the milk-taste scale, compared to the measured values (abscissa). The curvature error becomes even more evident when inspecting the residuals (Figure 8.4d).

This is the characteristic sign of unmodelled *curvature*: a certain type of non-linear relationship between variables. Going back to the raw data, the curvature in Figure 8.2c indicates that the sensory assessors perceive the milk-related increase in *milk-taste* most strongly, when the *%MILK* is above about 40%.

❏ Handling Curvatures

Such non-linearities are often encountered in explorative data analysis, and they have to be handled, one way or the other.

In the present case we choose to handle it just by *"eye-balling"*: the effect is small and easy to understand, so there is no need to model it mathematically. We just leave it as it is!

In some cases, with many precise variables in **X**, non-linear **X**–**y** relationships may automatically be *implicitly modelled*, but with the price of one or more extra PCs (Chapters 7.4 and 12). This is usually not a problem.

In other cases, non-linear **X**–**y** relationships can really destroy the modelling. In such cases we should try to improve the modelling. This can be done in various ways:

1. Many types of response non-linearities may be *removed*, or at least *reduced*, by improving the data pre-processing (Chapters 4 and 12).
2. The bi-linear BLM method may be extended to *polynomial* structure, by extending the set of X-variables with their squares and pair-wise interactions defined as new X-variables. This is described in more detail in Appendix A8.1 and is used also in Chapters 14.2 and 16.3.

## 8.2.9. Quality Criteria

❏ Choosing a Few Quality Criteria

In general, the following selection rules are useful for choosing a few quality criteria from a BLM plot, without loosing essential information. Variables should:

1. discriminate between the samples, i.e. show high loadings in the valid PCs;
2. be non-redundant, i.e. since variables along the same PC reflect the same underlying phenomenon, only one or two from each PC are needed;
3. be relevant, i.e. give meaningful interpretation w.r.t. project purpose;
4. be cognitive clear (with respect to sensory descriptive analysis), i.e. show reliable results reflecting intersubjectivity among the assessors.

The present study suggests that the quality variations in this kind of three-component mixtures may be adequately monitored by the combination of *two* quality criteria.

One possible set of quality criteria is of course the production variables defining the chemical composition, e.g. *%COCOA* and *%SUGAR*, representing PC1 and PC2, respectively (with *%MILK* defined by difference, if so desired).

An alternative set of quality criteria could be based on sensory analysis, e.g. sensory *colour* and *sweet* taste. But due to the unexpectedly strong impact of *%COCOA*, one should probably add another descriptor, e.g. the *milk-taste*.

A third alternative is to rely on the physical measurements in the study. *VISCOSITY* and *COLOUR* lightness could in principle also be used. But like the sensory variables, they had problems spanning the milk/sugar type of variability, PC2 alone.

☐ OPTIMISING THE QUALITY

When comparing the sensory results with the small consumer pilot study (Chapter 3.5.2) in terms of these quality criteria, it was found that the tested consumer group preferred the product to be sweeter and with less pronounced cocoa character than expected.

This is an indication in which direction *%SUGAR* and *%COCOA* should be optimised, and what that would mean from a sensory and a physical perspective. If the consumer study had been larger, with more samples tested, it could have helped us to find the optimal product quality for the chosen type of customers, based on e.g. BLM-based *response surface* analysis (Chapter 16.2).

*VI. Conclusions.* An interdisciplinary data set was analysed by PCA and PLSR. The example shows how variables of widely different types may be brought together and interpreted.

The importance of standardising the diverse set of input variables was first stressed. The similarity between PCA and PLSR was shown. The technique of representing the samples in the loading plot by indicator-variables was demonstrated. The choice of **X** and **Y** was discussed, and the effect of swapping **X** and **Y** in a well-behaved data set was demonstrated. Furthermore, an unexpected curvature was detected, and the handling of such non-linearities was discussed. Finally, the choice of a few quality criteria was demonstrated, for use e.g. in quality optimisation.

From an application point of view, this analysis showed the producers of the cocoa-milk powder-mix that *sweet* taste and *milk-odour* was more strongly affected by *%COCOA* than expected. The relationship between sensory and instrumental quality measurements was good, but the sensory response for *milk-taste* was somewhat non-linearly related to the latter. Moreover, it showed that the *VISCOSITY* in these mixtures was primarily controlled (+) by *%COCOA*, but to some degree also (−) by *%SUGAR*. Various alternative sets of criteria for measuring and monitoring the quality of the cocoa-milk product were proposed.

This small data set will later be expanded, in terms of more samples and variables (Chapter 9) and of replicates (Chapter 10).

## 8.3. TEST QUESTIONS

1. How are sensory data generated?
2. Why is it sometimes useful to standardise input variables?
3. When to use PCA and when to use PLSR?
4. What should be **X** and what should be **Y** in PLSR?

## 8.4. ANSWERS

1. Sensory data are generated by a trained panel of people, who report what they sense with their eyes, hands, nose and mouth. They report by using a set of agreed-upon descriptors (words) with well-defined meanings, and with well-defined quantitative scales anchored by reference samples (cf. Appendix A2).
2. It is useful to scale input variables within **X** or **Y**, if they have very different error levels. Standardisation secures the right of each one of them to have equal say-so in the modelling. Beware: that may give undue say-so to variables with nothing to say.
3. For well behaved data, PCA and PLSR give more or less identical interpretations. But PLSR is usually easier to validate, e.g. for finding the optimal number of PCs. Also, if you want to be able to interpret or predict some variables **Y** from other variables **X**, then you should use PLSR. For ill-behaved data with low percentage explained variance in the modelling, PLSR makes it easier to reveal what the problems are.
4. Relax! Follow the guidelines Chapters 3.3.2 and 8.2.7.

# Chapter 9

# Classification and Discrimination $X_1$, $X_2$, $X_3$, ...: Handling Heterogeneous Sample Sets

*This subchapter explains how a multivariate data set, which contains two or more distinct classes of samples may be studied by BLM. The resulting models may then be used for classifying individual samples. Two classification techniques, soft independent modelling of class analogy (SIMCA) classification and discriminant PLSR (DPLSR), will be explained, and related to outlier detection.*

## 9.1. HETEROGENEOUS SAMPLE SET

The previous COCOA data set (COCAO-I) of eight samples is now expanded with seven more samples, based on another cocoa raw material, and six more variables will be included in the data set. The new data set, named COCOA-II, is here analysed as if the previous COCOA data set were unknown to us.

Two BLM-based techniques, *asymmetric* classification by SIMCA modelling and *symmetric* classification by DPLSR will be explained and applied. Again, we shall follow the data analytical *Steps I–VI* from Figure 3.2.

## 9.2. SIMCA AND DPLSR

*I. Purpose*. The purpose is now to study how the sensory and physical qualities of the cocoa-milk samples change with cocoa raw material and with the percentage of cocoa, sugar and milk. How can different *classes* of samples be revealed? How are they analysed?

***II. Experimental planning***. A set of 15 samples of cocoa, sugar and milk powder is to be produced, according to a combined factorial/reduced-mixture design: the previous eight samples (*minus* mixtures): *1−*, *2−*, *3−*, *4−*, *5−* (*a*), *5−* (*b*), *6−* and *7−* (Chapter 8) are included. The added seven new samples are named *plus* mixtures: *1+*, *2+*, *3+*, *4+*, *5+*, *6+* and *7+*. These are based on a different cocoa raw material, but otherwise prepared according to the same reduced three-component mixture design as the *minus* mixtures.

The 15 mixtures are to be produced in three replicates. These 45 samples are then to be analysed with respect to ten sensory descriptors: *colour* darkness, *cocoa-odour, milk-odour,* thick texture (*thick-txtr*), *mouthfeel,* smooth texture (*smooth-txtr*), creamy texture (*creamy-txtr*), *cocoa-taste, milk-taste* and *sweet* taste. They are also to be characterised by their chemical composition (*%COCOA, %SUGAR* and *%MILK,* sum = 100%), and by physical measurements of the amount of *SEDIMENT,* of *COLOUR* lightness and of *VISCOSITY.*

***III. Experimental work***. The mixtures were produced and analysed in random order, as described in Chapter 8.

***IV. Pre-processing and QC of the data***. The different measurements were collected in tables, inspected for gross mistakes and averaged over the replicates.

Table 9.1 shows the results. A marked increase in *VISCOSITY* between the *minus* and *plus* samples is immediately noted. Figure 9.1 shows the distribution of each input variable, in terms of their 0, 25, 75 and 100% percentiles (Chapter 3.5.1) before and after standardisation.

Figure 9.1a shows the distributions of the input data. Large variations in level and range are evident. Figure 9.1b shows the same after standardisation of each variable (eq. (8.1)) to a mean of 0 and a standard deviation of 1. The variables are now much more compatible. Some variables show skewed distributions: in particular the sensory term *milk-odour* and the three chemical design variables and the physical measurements *SEDIMENT* and *VISCOSITY.* But a quick graphical inspection of the input data of the variables indicated no obvious errors, and the skewness is not strong enough to warrant any pretreatment. Hence, we accept these input data and proceed to the multivariate modelling.

***V. Data analysis***

### 9.2.1. Initial Analysis: Discovering Classes of Samples

The standardised variables in Figure 9.1b were split into two groups so that the ten sensory variables were defined as **X** and the six chemical and physical variables as **Y**.

This choice of **X** and **Y** is based on the fact that we have a higher number, and a more diverse selection of sensory variables than of chemical and physical variables. To be sure that we do not lose essential dimensionality in the BLM, we define the sensory variables as **X**. This choice corresponds to a study of how the sensory assessors might predict the chemical and physical properties of the samples, based on what they sense (cf. Chapter 8.2).

The X- and Y-data were submitted to PLSR. $A_{Opt} = 3$ PCs were now found to be necessary and sufficient in order to obtain adequate modelling of all the variables.

◻  LOADING AND SCORE-PLOTS

The model shown in Figure 9.2 was obtained. The correlation loadings of the ten X- and six Y-variables are shown in Figure 9.2a for PC1 (abscissa) vs. PC2 (ordinate), and in Figure 9.2b for PC1 and PC3 (ordinate). The corresponding scores of the $N = 15$ samples are shown in Figure 9.2c,d.

*PC1 vs. PC2:*   An obvious splitting of the 15 samples into two groups is evident in the scoreplot of $t_1$ vs. $t_2$ (Figure 9.2c): the *minus* samples, *1−* to *7−* lie in an elongated cluster, above and clearly distinct from the cluster of *plus* samples, *1+* to *7+*.

*The difference between the two sample sets: stabiliser added.* The cocoa raw material in the *plus* samples was essentially the same as in the *minus* samples, except that a so-called *stabiliser* had been added. The purpose of this *stabiliser* is to thicken the cocoa-milk drinks, by making them more viscous and thereby to stop cocoa particles from sedimenting in the liquids.

Without having been given any information about this stabiliser (except as text for sample names), the BLM has revealed the heterogeneity of the sample set.

The effect of the stabiliser splits the sample set mainly in the PC2 direction, and this effect correlates (Figure 9.2a) with *SEDIMENT* and oppositely with *VISCOSITY* and with sensory *mouthfeel, thick-txtr, creamy-txtr* and *smooth-txtr.*

*PC1 vs. PC3:*   Looking at the score plot of $t_1$ vs. $t_3$ in Figure 9.2d, we recognise the triangular design pattern which we also saw in Figure 8.3. By also keeping an eye on the loadings in Figure 9.2b, we see that PC1 mainly spans the effect of *%COCOA*, and PC3 spans the contrast between *%SUGAR* and *%MILK*.

> The main difference from Figure 8.3 is that the direction of PC1 has flipped; the variables *%COCOA, cocoa-taste, cocoa-odour* (overlapping) and *colour* are now on the left side in Figure 9.2b, and so is cocoa mixture *7*, while in Figure 8.3 they were on the right side. In addition, since the number of samples and variables has increased: there are more samples or variables printed on top of each other. In real, interactive data

**Table 9.1.** Extended set of cocoa mixtures, COCOA-II: input data from six chemical/physical variables and ten sensory variables, measured in $N = 15$ cocoa products.

| | | %COCOA 1′ | %SUGAR 2′ | %MILK 3′ | SEDIMENT 4′ | COLOUR 5′ | VISCOSITY 6′ |
|---|---|---|---|---|---|---|---|
| 1− | 1 | 20.00 | 30.00 | 50.00 | 2.60 | 44.89 | 18.62 |
| 2− | 2 | 20.00 | 43.30 | 36.70 | 2.65 | 42.77 | 18.00 |
| 3− | 3 | 20.00 | 50.00 | 30.00 | 2.40 | 41.64 | 17.84 |
| 4− | 4 | 26.70 | 30.00 | 43.30 | 3.10 | 42.37 | 20.63 |
| 5− | 5 | 26.60 | 36.70 | 36.70 | 3.55 | 41.04 | 19.66 |
| 5− | 6 | 26.60 | 36.70 | 36.70 | 3.55 | 41.04 | 19.66 |
| 6− | 7 | 33.30 | 36.70 | 30.00 | 4.30 | 39.14 | 21.33 |
| 7− | 8 | 40.00 | 30.00 | 30.00 | 4.70 | 38.31 | 22.57 |
| 1+ | 9 | 20.00 | 30.00 | 50.00 | 0.12 | 44.25 | 486.00 |
| 2+ | 10 | 20.00 | 43.30 | 36.70 | 0.09 | 41.98 | 441.00 |
| 3+ | 11 | 20.00 | 50.00 | 30.00 | 0.10 | 41.18 | 436.00 |
| 4+ | 12 | 26.70 | 30.00 | 43.30 | 0.10 | 41.13 | 478.00 |
| 5+ | 13 | 26.60 | 36.70 | 36.70 | 0.10 | 40.39 | 503.00 |
| 6+ | 14 | 33.30 | 36.70 | 30.00 | 0.10 | 38.85 | 514.00 |
| 7+ | 15 | 40.00 | 30.00 | 30.00 | 0.09 | 37.91 | 548.00 |

| | | Colour 1' | Cocoa-odour 2' | Milk-odour 3' | Thick-txtr 4' | Mouthfeel 5' | Smooth-txtr 6' | Creamy-txtr 7' | Cocoa-taste 8' | Milk-taste 9' | Sweet 10' |
|---|---|---|---|---|---|---|---|---|---|---|---|
| 1- | 1 | 1.67 | 6.06 | 7.37 | 5.94 | 7.80 | 8.59 | 6.51 | 6.24 | 6.89 | 8.48 |
| 2- | 2 | 3.22 | 6.30 | 5.10 | 6.34 | 8.40 | 9.09 | 7.14 | 7.04 | 5.17 | 9.76 |
| 3- | 3 | 4.82 | 7.09 | 4.11 | 6.68 | 8.29 | 8.61 | 6.76 | 7.26 | 4.62 | 10.50 |
| 4- | 4 | 4.90 | 7.57 | 3.86 | 6.79 | 8.58 | 5.96 | 5.46 | 8.77 | 3.26 | 6.69 |
| 5- | 5 | 7.20 | 8.25 | 3.03 | 6.58 | 8.77 | 6.09 | 5.52 | 9.07 | 2.94 | 7.05 |
| 5- | 6 | 6.86 | 7.66 | 2.96 | 7.26 | 8.65 | 6.74 | 5.65 | 8.78 | 2.58 | 7.04 |
| 6- | 7 | 10.60 | 10.24 | 1.57 | 6.51 | 9.70 | 4.55 | 4.62 | 11.44 | 1.51 | 5.48 |
| 7- | 8 | 11.11 | 11.31 | 1.25 | 7.04 | 9.72 | 3.42 | 4.11 | 12.43 | 0.86 | 3.91 |
| 1+ | 9 | 3.06 | 6.97 | 5.40 | 9.84 | 9.99 | 10.67 | 9.11 | 7.66 | 5.71 | 8.24 |
| 2+ | 10 | 6.02 | 8.61 | 3.75 | 10.01 | 9.92 | 10.86 | 8.64 | 7.66 | 4.86 | 8.71 |
| 3+ | 11 | 7.94 | 8.40 | 2.95 | 9.61 | 9.92 | 10.84 | 8.26 | 8.32 | 4.09 | 9.67 |
| 4+ | 12 | 9.17 | 9.30 | 2.86 | 10.68 | 11.05 | 10.48 | 8.20 | 10.40 | 2.22 | 6.43 |
| 5+ | 13 | 10.46 | 10.14 | 1.90 | 10.71 | 10.64 | 9.60 | 7.84 | 11.05 | 2.01 | 7.02 |
| 6+ | 14 | 12.40 | 11.30 | 1.18 | 10.64 | 11.09 | 7.24 | 7.23 | 11.78 | 1.65 | 5.59 |
| 7+ | 15 | 13.46 | 11.49 | 1.56 | 11.31 | 11.36 | 7.22 | 6.86 | 12.60 | 1.06 | 4.34 |

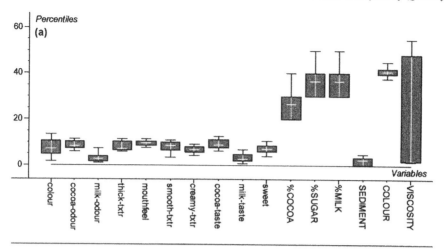

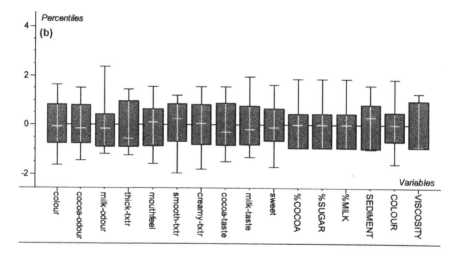

**Figure 9.1.** Extended set of cocoa mixtures: statistics for input data. Distribution summaries for each of the six chemical/physical variables and ten sensory variables over 15 cocoa mixture samples, in terms of their 0, 25, 50, 75 and 100% percentiles. (a) Input data. (b) Input data after standardisation.

analysis this is not a problem, because we can use colours and interactively enlarge cluttered areas of a plot.

❏ DEALING WITH A HETEROGENEOUS SAMPLE SET

We might have settled with the present bi-linear model for all 15 samples, since

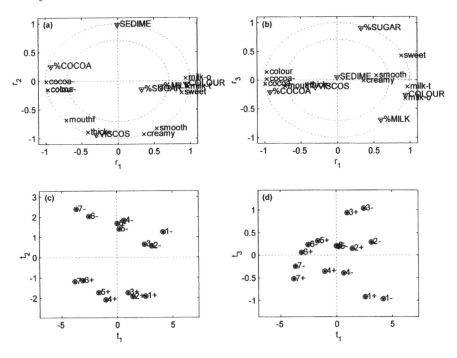

**Figure 9.2.** Heterogeneous sample set: clusters discovered in the score plot. (a,b) Correlation loadings for **X** = sensory variables (×) and **Y** = chemical/physical variables (∇). (c,d) Score plots for the $N = 15$ cocoa samples. Numbering represents the cocoa mixture name. + and − mark samples with and without texture *stabiliser* added. (a,c) PC1 vs. PC2. (b,d) PC1 vs. PC3. Due to the high number of variables, only the six first characters are printed. Plot overlaps: (a) *colour* and *cocoa-odour*; *thick-txtr* and *VISCOSITY*; *COLOUR*, *milk-taste* and *%MILK*. (b) *VISCOSITY*, *thick-txtr* and *mouthfeel*.

the resulting three-component model is more or less complete for all the variables, and since the model is easy to understand.

However, if obvious groups of samples are found, it is usually a good idea to try to handle the heterogeneity explicitly. In the present case we have found an apparent sample grouping. But as the sample names indicate, we did already expect the *minus* and *plus* classes to be distinct, or at least interesting.

Such class information may be used explicitly in the modelling, in order to simplify the models, in order to study what distinguishes the classes from each other, and to learn how to classify new samples.

First, we shall use the asymmetrical classification method, the SIMCA analysis. Then we shall use a simpler alternative, the symmetrical DPLSR.

### 9.2.2. Classification by Separate Modelling of Each Sample Group: SIMCA

One way to use the information about sample grouping is to split the samples

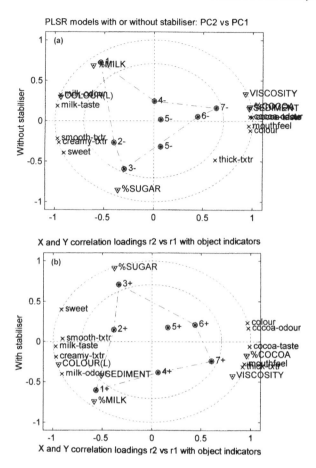

**Figure 9.3.**    Splitting a heterogeneous sample set into two data sets for separate modelling. Modelling six chemical/physical variables (**Y**) from ten sensory variables (**X**) by PLSR: Correlation loading plot with sample indicators. (a) The WITHOUT STABILISER model for the eight cocoa samples without stabiliser (seven mixtures and one sensory replicate for mixture *5−*). Overlaps: *COLOUR* and *milk-odour*; *SEDIMENT* and *%COCOA*; *cocoa-odour* and *cocoa-taste*. (b) The WITH STABILISER model for the seven cocoa mixtures with stabiliser. Overlaps: *mouthfeel* and *thick-txtr*.

into classes, and model each of them separately by BLM. We would expect each class to yield a simpler (lower-dimensional) model than the joint model in Figure 9.2.

If we had kept all the variables as one table, **X**, then we might have modelled the data of each class by PCA (Figure 3.4). In the present case we have chosen to split the variables into **X** and **Y** in order to study the predictive sensory-chemical relationship, so we use the two-block version of BLM, PLSR, in the SIMCA analysis.

☐　SEPARATE BI-LINEAR MODELS FOR EACH CLASS OF SAMPLES

Within each class the X- and Y-variables were standardised and submitted to PLSR. Both in the WITHOUT STABILISER and the WITH STABILISER class, a two-component model ($A = 2$) was found to describe most of the non-random variation in the X- and Y-data. Figure 9.3 shows the results. For choosing $A_{Opt}$, the optimal number of PCs in PLSR, the modelling of the Y-variables is usually considered as being most important. But in classification, the modelling of the X-variables is also important, since that is where most of the classification takes place.

*WITHOUT STABILISER:*　The *minus* samples (without stabiliser) yield a model (Figure 9.3a) similar to that obtained previously for the same samples, with fewer variables (Figure 8.3). The first PC reflects the effect of increasing *%COCOA*, and PC2 the effect of replacing *%MILK* with *%SUGAR*.

The product indicator-variables, now included in **Y**, still outline the design triangle. But the triangle is more distorted than in Figure 8.3, so the new variables seem to have introduced more non-linearity or noise. This is further evidenced by the increased distance between the two replicates of mixture $5-$. The sensory variable *thick-txtr* is rather badly described by the model of these samples without stabiliser.

Inspection of the raw data of *thick-txtr* indicates that it has very little variation within this sample set. The standardisation has probably amplified measurement noise to the extent that it contaminates the model.

*WITH STABILISER:*　The *plus* samples (with stabiliser) (Figure 9.3b) similarly reflects the design triangle (dotted outline), more or less as expected. But note how PC2 has flipped, compared to Figure 9.3a! This is clearly evident e.g. for the variables *%MILK* and *%SUGAR* and samples *1* and *3*. In itself, such a flipping of a PC is trivial, as explained in more detail in Appendix A5, eq. (5.11) since

$$t_2 \times \mathbf{p}'_2 = t_2 \times 1 \times \mathbf{p}'_2 = t_2 \times (-1 \times -1) \times \mathbf{p}'_2 = (-t_2) \times (-\mathbf{p}_2)'$$

But although it has no consequence for the modelling, this ambiguity may require some training to learn to see and accept.

In this class, the variable *SEDIMENT* is badly described by the model of these samples. Again, this is apparently due to lack of variation in the input data. Since the purpose of the stabiliser was to stop sedimentation of cocoa, there is very little sedimentation in these viscous samples.

*Re-modelling without standardisation of X:*　Considering that all the X-data come from the same source (sensory analysis), there is reason to expect that all the input variables in **X** have about the same error levels. Therefore, maybe the *default* standardisation of all the variables, even in **X**, was not such a good idea, after all?

The two class models were therefore re-computed, with unscaled X-variables (scaling weights $v_k = 1$, $k = 1$, $2,...,K$ in eq. (8.1)), but with standardised Y-variables (eq. (8.2)), as before (since **Y** does contain *mice and elephants*).

The two class models were called class 1: WITHOUT STABILISER and class 2: WITH STABILISER. Again, each class model required 2 PCs. When plotted as correlation loadings, the results were more regular, as expected, but rather similar to those in Figure 9.3, and will therefore not be shown here.

◻ CLASSIFICATION OF EACH SAMPLE IN EACH OF THE CLASS MODELS

*The SIMCA method:* This *asymmetrical* classification method was developed by Svante Wold (1983b), who named it SIMCA (Soft Independent Modelling of Class Analogy). It consists of fitting the X-data of each sample *i* to each class model *c*, to see how well the classes are separated from each other (cf. Figure 3.4).

New, unknown samples may also be classified in this way. For interpretation, one may study how the different X-variables contribute to distinguishing between the classes. The technical details are described in Appendix A9.

First, the row vector of X-data of sample *i*, $x_i$, is fitted to class model *c* to obtain its score vector, $t_{i,c}$.

Then the residual X-vector, $e_{i,c}$ is computed, and distance between sample *i* and class *c* is estimated. This distance is here called root mean square error in **X**, RMSE(X)$_{i,c}$ (eq. (9.4), p. 396).

We have thus split the present data set, and have made two class models. Figure 9.4 shows the results of this SIMCA classification.

*Cooman's plot:* Figure 9.4a shows the distances RMSE(X)$_{i,c}$ of each sample $i = 1$, $2,...,15$ (named *1−*, *2−*, *...,7−*, *1+*, *2+,...*, *7+*) to class model $c = 1$ (WITHOUT STABILISER, abscissa) and to class model $c = 2$ (WITH STABILISER, ordinate).

The vertical line represents the statistical classification limit of the model $c = 1$; the horizontal line marks the corresponding classification limit for the model $c = 2$. This so-called *Cooman's plot* splits the possible outcome of the *two-class asymmetric* classification into four rectangular regions:

1. Class 1: inside class 1 but outside class 2;
2. Class 2: inside class 2 but outside class 1;
3. Classes 1 and 2: inside both class 1 and 2;
4. Outliers: outside both class 1 and 2.

Figure 9.4a shows that all the *minus* samples fall inside the class WITHOUT STABILISER and outside the other class (section 1), and all the *plus* samples fall inside the class WITH STABILISER and outside the *other* class (section 2). Hence, the separation of the sample set into two distinct classes was highly successful.

In practice, such a complete separation of classes is rare. Usually, some of the samples will belong to both classes (section 3), because the classes are somewhat overlapping. Sometimes, there are samples that do not fit to any of the classes (section 4), because they are outliers.

Other classification tools may also be used, to check how each sample fits to each class, based on the leverage $h_{i,c}$ and various Y-measures. This will be discussed in Chapter 9.3, and demonstrated in, e.g. Figures 7.3, 7.4, 14.4 and 15.9.

*Discriminative ability:* Figure 9.4b shows how the individual X-variables contribute to the classification of the 15 samples. The discriminative ability of a variable is the ratio between the between-classes and the within-classes squared distances. Variables with discriminative ability much higher than, e.g. four, contribute strongly to the SIMCA classification.

S. Wold used *discrimination power* for this expression in his first papers (Wold, 1983b). We use *discriminative ability* in this book, since *power* has a special meaning in this book (in *design power estimation*).

In this case the texture variables *creamy-txtr* and *thick-txtr* show the highest discriminative ability. The two other texture variables (*mouthfeel* and *smooth-txtr*) are also important, as expected. But what about the variable *sweet*?

*Checking an unexpected effect:* The contribution by *sweet* taste to the *stabiliser* classification is a surprise, and calls for closer scrutiny of the input data. Figure 9.4c shows the *sweet* taste (ordinate) plotted against *%SUGAR* for all 15 samples. In order to avoid cluttering the plot, only the − and + part of the sample names is printed.

The figure shows small sweetness effects of the stabiliser addition, but it is rather complex: for most of the mixtures, e.g. at high *%SUGAR*, the addition of stabiliser decreased the sweetness of the samples (+ lies below −). But for the least *sweet* samples (the cocoa-rich mixtures 6 and 7), the addition of stabiliser increased the sweetness. Inspection of the other input variables revealed the same small phenomenon for *milk-taste*, but not for any of the other sensory variables. Why this complex response? Well, that is for new experiments to elucidate. But there is a hypothesis: could it be that the increased viscosity in the mouth reduces the accessibility of the taste-giving compounds at the receptors in the taste buds of the tongue? This would cause fewer sugar molecules to be sensed in the *plus* samples than in the *minus* samples, and the *sweet* taste would *decrease* with addition of the texture stabiliser. But at high cocoa levels (samples 6 and 7), this viscosity effect could be even more pronounced for the

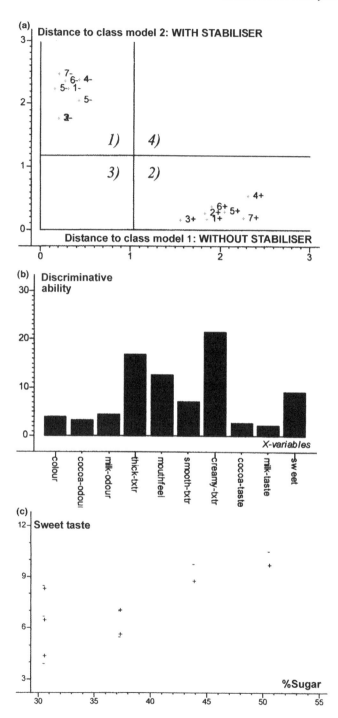

bitter cocoa compounds, which appears to mask some of the sweetness; for these samples the *sweet* taste could *increase* with addition of the stabiliser. The same would apply for the molecules responsible for *milk-taste*.

In order to make SIMCA classification reject extreme samples, the distance from sample $i$ to class $c$, $\mathrm{RMSE(X)}_{i,c}$ (eq. 9.4) may be modified by sample $i$'s leverage inside class $c$, $h_{i,c}$, which may be computed as a summary of the $A$ score values (cf. eq. (9.3) in Appendix A9). For simplicity that is ignored here.

The SIMCA classification method has been termed *asymmetrical* since it works even if there had been only one single class. In that case, the classification amounts to a conventional outlier test. The next section explains an alternative, so-called *symmetrical* classification method, DPLSR. In order to be able to establish a classification model, two or more classes of samples must then be seen already in the input data.

### 9.2.3. Classification by DPLSR

In the SIMCA modelling we first developed *separate* class models, and then checked how the individual samples fitted to each of these models. In DPLSR all the samples are kept inside one *common* model, and used for learning how to discriminate between the different classes. Within the BLM framework this is here called DPLSR. The PLSR modelling is performed just like any other PLSR modelling. The term DPLSR just indicates that discrimination information is used as Y-variables.

> Even this method was originally published by Wold et al. (1983b), who prefers to call it PLS-discriminant analysis.

#### ❑ The DPLSR Method

The DPLSR consists of building a conventional PLSR model, but with class indicator variable(s) as **Y**, and with whatever other variables available as **X**, as illustrated in Figure 3.4. The class differences are then seen, e.g. in the loading plot.

Once established, this model allows us to classify even new samples, by predicting the degree to which each new sample seems to belong to each class (**Y**) from their X-data.

---

**Figure 9.4.** SIMCA classification. (a) Distances to the two PLSR models *WITH STABI-LISER* and *WITHOUT STABILISER*, for each of the cocoa-milk samples with (+) or without (−) stabiliser. (b) Discriminative ability of the X-variables between classes *WITH STABILISER* and *WITHOUT STABILISER*. (c) Confirmation of a modelling discovery in the input data: systematic differences in the relationship between *sweet* taste and *%SUGAR*, between samples with(+) and without (−) texture stabiliser.

In the present case there are only two classes: cocoa-milk samples with and without stabiliser (+ and −). We can distinguish between two classes by the help of one single *indicator variable*. Thus we define two new variables *STABILISER* and *NO-STABILISER*, to represent the samples with or without stabiliser. The binary Y-variable has value of 1 for samples with stabiliser (e.g. *1+* and *6+*) and some lower value (e.g. 0, or as chosen here, −1) for samples without stabiliser (e.g. *1−* and *7−*).

We could have chosen to ignore one of these two perfectly anti-correlated Y-variable; that would not affect the model. If there had been more that two classes of samples, we would just add more such class indicator variables in **Y**.

*Calibration and prediction:*    Moreover, only eight of the 15 samples (*1−*, *3−*, *4−*, *7−*, *1+*, *2+*, *5+*, *6+*) will now be used in the training set, in order to simulate the situation in which we have a limited training set from which we shall learn how to classify also future unknown samples.

The remaining seven samples (*2−*, *5−(a)*, *5−(b)*, *6−*, *3+*, *4+*, *7+*) will then be regarded as *future, unknown samples*, to be fitted to the discriminant model afterwards.

□   JOINT DISCRIMINANT MODEL

All the sensory and chemical/physical variables were standardised and used as X-variables, with the twin indicator variable *STABILISER* and *NO-STABILISER* used as **Y** in the PLSR.

2 PCs were found to provide most of the modelling of the Y-variables.

Leave-one-sample-out cross-validation (Chapter 10) showed that the two- and three-PC solutions predicted 93 and 98% of the Y-variance, respectively. This is unusually high predicted variances for DPLSR. In other data sets where the classes are less sharply defined, lower estimates of the predictive ability must be expected. The predictive ability may then in fact be higher than it immediately appears. The reasons is that while **Y** has only binary values, e.g. 1 and −1, the Y-predictions $\hat{\mathbf{Y}}$ can take any value, and this creates unnecessarily high differences $\mathbf{Y} - \hat{\mathbf{Y}}$. Y-predictions of e.g. $\hat{y}_{i,1} = 0.8$ or 1.6 for $y_{i,1} = 1$, or $\hat{y}_{i,1} = -0.7$ or $-2.1$ for $y_{i,1} = -1$, will generate *apparent* errors in the Y-predictions, even though the predicted classification is OK. This may be corrected by *optimal scaling* (Gifi, 1990) of the Y-predictions, e.g.

$\hat{y}_{i,1 \text{ corrected}} = 1 \; if \; \hat{y}_{i,1} > 0.5$

$\hat{y}_{i,1 \text{ corrected}} = -1 \; if \; \hat{y}_{i,1} < -0.5$                                        (9.1)

*Correlation loading plot, with object indicator-variables:*    Figure 9.5a shows the main information in the resulting model, in terms of the correlation loadings and sample indicators for the first 2 PCs. The eight samples used as training set are indicated by the symbol ⊛. After the model had been established, the X-data of the remaining seven *unknown samples* were also fitted to the model, eq. (6.3a); these are symbolised by the open circles.

Figure 9.5a shows that the variation in the Y-variables *STABILISER* and *NO-STABILISER* (▽) are close to 100% explained by the X-variables. They correlate primarily with the first component, but to some extent also with the second component.

The correlation pattern of $y_1 = STABILISER$ is almost identical to that of X-variables *VISCOSITY* and *thick-txtr*, and almost completely anti-correlated with that of *SEDIMENT*. The other texture variables *smooth-txtr, creamy-txtr* and *mouthfeel* also pull more or less in the same direction, but they show some differences, in that *smooth* and *creamy* texture appear also to correlate to *%MILK*, and *mouthfeel* instead to correlate somewhat with *%COCOA* and the corresponding flavour variables.

The discriminating Y-variable *STABILISER* lies in the West direction in the correlation loading plot. Correspondingly, the configuration of calibration samples in this direction groups the samples with stabiliser (+) distinctly from those without (−). The latter lie in the direction of $y_2 = NO\text{-}STABILISER$ and *SEDIMENT*, as expected.

Even the unknown samples (○) place themselves into these two groups, which shows that the sample scores themselves can be used for *visual* classification. However, more explicit statistical classification can also be attained, based on the prediction of $y_1$:

*The regression coefficient vector (the discriminant function):*     Figure 9.5b shows the regression coefficients $\hat{\mathbf{b}}_1$ for the discriminant prediction: $\hat{y}_{i,1} = \hat{b}_{0,1} + \mathbf{x}_i \hat{\mathbf{b}}_1$ (eq. (6.6a), p. 384). Here, the regression coefficients $\hat{\mathbf{b}}_1$ for the standardised variables are given for PLSR models with both $A = 1, 2$ and 3 PCs (increasing darkness in the plot). The coefficients are seen to be rather insensitive to the actual model rank.

These standardised regression coefficients show that higher-than-average *VISCOSITY, thick-txtr, mouthfeel, smooth-txtr* and *creamy-txtr* and lower-than-average *SEDIMENT* values contribute to a discrimination between samples with and without stabiliser. This corresponds well to the variables' distance from the origin in Figure 9.5a in the direction of *STABILISER*, and also to what we would expect, causally.

The third PC gives a slight further increase in the discriminative ability, apparently by bringing in small contributions of *sweet* taste (and its associated variables *%SUGAR* and *%MILK*).

*Classifying old and new samples by using the discriminant function:*     When the $\hat{\mathbf{b}}_1$-vector with 3 PCs was applied to the X-vector of each of the 15 samples in eq. (6.4b, p. 119), the predicted $y_1$ memberships in the two classes were obtained:

Figure 9.5c shows that the eight leftmost samples constitute the *old* samples

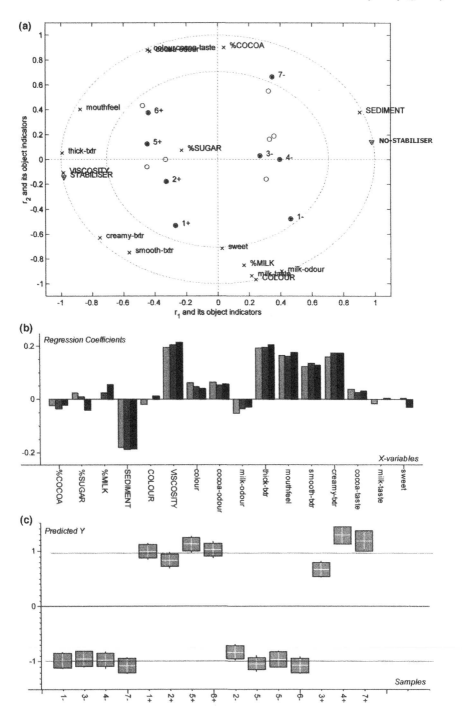

from the training set, and are of course well classified. More importantly, the seven rightmost *new, unknown* samples are *also* seen to be very well classified.

The bars, representing $\pm 1$ estimated outlier-sensitive standard uncertainty of the predictions, $\hat{s}(\hat{y}_{i,1})$ ( $=$ YDev$_i$ for the first y-variable, STABILISER, eq. (10.6), p. 407), is seen to be rather small. The correct classification ($+1$, $-1$) lies within $\hat{y}_{i,1} \pm 2\hat{s}(\hat{y}_{i,1})$ for all of the samples. Hence, the DPLSR gave successful classification with respect to the presence or absence of stabiliser in the cocoa-milk samples.

*VI. Conclusions.* Two classification techniques have been described, SIMCA and DPLSR. They were applied to a data set, COCOA-II, where they successfully distinguished cocoa-milk samples from two different raw material sources, with and without stabiliser. The interpretation here showed more or less the same sensory difference between the $+$ and $-$ class samples for both techniques.

## 9.3. OUTLIER DETECTION

### 9.3.1. The Importance of Outliers

Outliers are samples that are *abnormal* in some way. They may be the most valuable and informative samples of all. But they may also represent gross errors that just destroy a model. In either case, it is important to be able to discover them (cf. Chapter 7).

There is a close connection between sample *classification* and the detection of *outliers*. In both cases, a definition of what is *normal* is first established, by bilinear modelling of one or more training sets of samples. Then, individual samples are compared to this *standard of normality*, not unlike we humans judge each other culturally.

SIMCA analysis is essentially nothing more than outlier analysis with respect to one or more models. DPLSR relies on Y-prediction to discriminate between *known* classes, but can discover samples belonging to *unexpected classes* as outliers.

---

**Figure 9.5.** Discriminant PLSR. (a) Overview of main model structure. Correlation loading plot with sample indicators, for first 2 PCs from PLSR model of $\mathbf{Y}$ = indicator variables ($\nabla$) *STABILISER* ($+1$ = stabiliser, $-1$ = no stabiliser) and *NO-STABILISER* ($-1$ = stabiliser, $+1$ = no stabiliser) vs. $\mathbf{X}$ = six chemical/physical and ten sensory input variables ($\times$). Filled circles: eight cocoa samples used for calibration. Open circles: seven *unknown* samples for demonstrating predictive classification. (b) Discriminant regression coefficient $\hat{\mathbf{b}}_1$ for classification with respect to the presence of stabiliser. The $\hat{\mathbf{b}}_1$ for PLSR models with increasing number of latent variables, $A = 1$ (light grey), $A = 2$ (grey) and $A = 3$ (black) PCs, for the 16 standardised X-variables. (c) DPLSR classification. The values $\hat{\mathbf{y}}_1$ are predicted from $\mathbf{X}$ by the model with rank $A = 3$, in the eight calibration samples and the seven *unknown* samples. Grey error bars represent the outlier-sensitive uncertainty measure $+/-$ YDev$_i$ (eq. (10.6)).

## 9.3.2. Techniques for Detecting Outliers

Outliers may of course be detected just by looking graphically at input data. The *normal probability plot* (Figure 3.5h) is a useful tool in that context, but with a high number of input variables, this is impractical. Then the BLM method helps, by identifying the outliers automatically.

For a BLM with a given number of factors (*A*), outliers may be detected in **X**, **Y** or both. The basic idea is to compute various statistical measures for each sample, and compare them to few correspondingly *average* measures from few calibration sample set. Here are some useful outlier detection criteria:

❒ OUTLIER DETECTION IN **X**

*Leverage in X, $h_i$:*    Eq. (9.3) in Appendix A9 shows how extreme the position of a sample is INSIDE the model. Example: A new, hypothetical cocoa-milk sample with e.g. 90%*COCOA*, i.e. outside the normal range of 20–40 %*COCOA* calibrated for in the previous example, arrives and is analysed, and its X-data are fitted to the model with $A = 3$ PCs, obtained from $N = 15$ calibration samples. It contains no new phenomenon, but abnormal levels of a known phenomenon, so it would be tagged as an abnormal outlier *inside* the model, (*an extremist*).

*Residual in X, RMSE(X)$_i$:*    Eq. (9.4) shows the position of a sample OUTSIDE the model, i.e. the distance from its data $x_i$ to the model, measured by its X-residuals $e_{i,A}$. Example: Another new, hypothetical cocoa-milk sample, produced by mistake when someone diluted a normal powder mixture with *coffee* instead of water. Since it contains something quite new and very different from the calibration samples, its X-data do not fit the model: its RMSE(X)$_i$ is much higher than the average distance in the calibration set, RMSE(X). Hence it would be tagged as outlier *outside* the model, (*an alien*).

❒ OUTLIER DETECTION IN **Y**

*Outlier-sensitive standard prediction uncertainty, YDev$_i$:*    This pragmatic measure (eq. (10.6), Appendix A10) catches both *extremists* and *aliens*, since it responds to both $h_i$ and RMSE(X)$_i$. Example: Both of the above outliers would probably have been detected by this test, even without input Y-data.

*Prediction errors RMSE(Y)$_i$:*    The predictions do not fit with the input data: RMSE(Y)$_i$ (eq. (9.5b)) is high (only for samples with known Y-data). Example: Both of the above outliers are likely to be detected because they give strange and incorrect predictions of **Y**, although for different reasons. The *extremist* challenges the model to an extreme degree with its extreme scores values, and any non-linearity or estimation error in the Y-loadings **Q** will cause errors in $\hat{y}_i$. The *alien* obviously does not belong to our class, so we cannot expect it to behave like the normal individuals in its **Y**–**X** relations, can we?

❏ An Abnormal Inlier:

If a new sample had only *half* as much stabiliser as the *plus* samples in the previous example, this intermediate sample would NOT be seen as an outlier in the DPLSR model made for both the *minus* and *plus* samples (Figure 9.5):

The new sample would obtain intermediate sample scores, halfway between the *plus* and *minus* samples in Figure 9.5a, with low leverage $h_i$ and low X-error, RMSE(X)$_i$. The outlier-sensitive prediction uncertainty, YDev$_i$ would also be normal and small. The predicted classification value would be of $\hat{y}_i \approx 0$ in Figure 9.5c, close to true value $y_i = 0$ (halfway between $-1$ and $+1$). So the prediction error RMSE(Y)$_i$ would be low, even if we estimated it by proper cross-validation.

But how would this new, intermediate sample fare in the two SIMCA classes in Figure 9.4a? It would probably be an outlier both with respect to class 1: WITHOUT STABILISER and to class 2: WITH STABILISER! Details on warning limits for outliers are given in Appendix A9.3.

## 9.4. TEST QUESTIONS

1. How did we detect that the sample set was heterogeneous?
2. What is SIMCA classification?
3. What is discriminant PLSR?
4. What is the difference between SIMCA classification and DPLSR?

## 9.5. ANSWERS

1. We detected that the sample set was heterogeneous, by discovering clusters of samples in the PLSR score plot of the full sample set.
2. SIMCA classification consists of developing two or more separate BLM-based class models, and then checking how each sample fits into each of these class models.
3. Discriminant PLSR is a classification method in which the class information is used as one or more indicator variables in **Y**, for the full sample set.
4. The two methods usually give similar results. DPLSR is, operationally, the simpler of the two. SIMCA involves making several different class models. But if three or more classes lie like pearls on a string in the X-space, then the DPLSR will have problems with the middle class pearls. SIMCA classification handles even that situation: each class forms a local volume in the

X-space, etc., e.g. a round ball ($A = 0$ PCs), a long cigar ($A = 1$), a cigar box ($A = 2$). These classes are independent of each other, and may or may not be overlapping.

# Chapter 10

## Validation X? $\hat{Y}$??

*This chapter starts out by illustrating the importance of validating models and the need for sufficient redundancy in the input data. Then it explains the cross-validation method, which includes jack-knifing. This is used for choosing the right number of PCs, for estimating the predictive ability of the model and for assessing its parameter stability. The use of cross-validation in defining limits for outlier warning is described. Cross-validation at different validity levels is demonstrated.*

### 10.1. THE VALIDITY OF INFORMATION FROM BLM

#### 10.1.1. The Need for Validation Tools

The data-driven BLM method in this book is good at extracting patterns from input data. But how valid are these patterns? What is the quality of our obtained information?

The results from data-driven modelling cannot be trusted before they have been checked for external and internal validity. *External validity*, i.e. the generalisability of the conclusions, is to some degree assessed by interpretation of the obtained model structures in relation to the samples and variables used, and to external, *a priori* information. *Internal validity* is assessed, by checking the relevance and statistical performance of the modelling results, relative to the project purpose. The method of cross-validation (including jack-knifing) is used here, since it requires little or no statistical theory and delivers visually accessible graphical results.

The importance of ensuring satisfactory quality of the information from data analysis was pointed out in Chapter 2.2, and the two most important quality criteria for information, the *reliability* and *relevance*, were discussed.

#### 10.1.2. Reliability of BLM Results

The BLM method (Chapters 5 and 6) has been designed to extract systematic information structures that are *relevant* for both the X- and Y-data. But it is important to *optimise* this relationship, with respect to the number of PCs to be included in the final model.

It is also important to assess how *reliable* this optimal model really is, both with respect to how precise predictions it can give of **Y** from **X**, and how precisely its model parameters have been estimated.

In the validation of a model, two main measures of information reliability are estimated, the *predictive ability* and *parameter stability*. The predictive ability concerns the *model's ability to reconstruct the input variables* **Y** and **X** in independent samples, while the parameter stability concerns the *precision of the model's parameter values* ($\hat{\mathbf{B}}$, **T**, **P**, **Q**, etc.).

Small, unreliable PCs should be left unmodelled in the residuals **E** and **F**, and summarised into measures of what is the *usual* noise level in this kind of data. This definition of *normality* may in turn be used for detecting abnormal samples, i.e. outliers, as described in Chapter 9.3 and Appendix A9.3.

All of these needs for validation calls for a validation *tool* in the *soft modelling*. Chapter 3.4 outlined several approaches to the validation of models. The validation method chosen in this book has been designed to have good statistical properties but with no need for the user to learn abstract statistical theory. The method builds on contemporary statistical re-sampling theory, adapted to bi-linear modelling.

It combines two related re-sampling techniques, *cross-validation* and *jack-knifing*. In fact, the latter is part of the former, so in this book they are both treated as parts of the cross-validation.

### 10.1.3. A Visual Approach to Validation

The results from the present cross-validation method are directly visually accessible, as perturbations of the ordinary BLM output. In software this is implemented as an option that can be turned on and off, to ensure that uncertainty issues do not overshadow the discovery process during data analysis.

The uncertainties are also summarised in terms of well-defined error measures, which in this book are treated rather pragmatically, but which in turn may be used for significance estimation in traditional hypothesis testing, if so desired.

Traditional statistical testing is often considered too difficult and too trivial at the same time, by applied data analysts who often ignor it. It requires difficult, abstract statistical theory. At the same time, it is often perceived as a waste of time, at least in the mind of self-assured researchers who like to discover, not just to confirm. Unfortunately, validation is important. With the present, visually-oriented and data-driven approach, the model validation has hopefully been simplified and made more informative and agreeable.

### 10.1.4. How to Summarise the X- and Y-residuals

Figure 6.3 outlined how the X- and Y-data were modelled in terms of means $\bar{\mathbf{x}}$, $\bar{\mathbf{y}}$ plus the bi-linear contributions from $A$ (usually just a few) latent variables or PCs, $\mathbf{T} \times \mathbf{P}'$ and $\mathbf{T} \times \mathbf{Q}'$ plus residuals $\mathbf{E}_A$ and $\mathbf{F}_A$.

Figure 10.1 shows how these residual matrices $\mathbf{E}_A$ and $\mathbf{F}_A$ may be summarised in terms of so-called root mean square error, RMSE, in $\mathbf{X}$ and $\mathbf{Y}$ separately (cf. Appendix A9). For the samples, the RMSE summaries are given the index $i$: RMSE(X)$_i$, RMSE(Y)$_i$. For the X-variables it has index $k$, and for the Y-variables it has index $j$: RMSE(X)$_k$, RMSE(Y)$_j$. The total summaries are written without index: RMSE(X), RMSE(Y). When needed, the error estimates are given an index for the number of PCs used, $A$, as in RMSE(Y)$_{A=2}$.

As will be explained below, such summaries are made both for the so-called *unvalidated calibration residuals*, called RMSEC, and for the estimated prediction residuals, called RMSEP. The RMSEP results are of primary interest, since they can give realistic estimates of what prediction errors to expect in the long run. The RMSEC results are primarily used for technical diagnostic purposes in difficult data sets, since they are very simple, but give unrealistically low values.

### 10.1.5. Cross-validation (CV)

❑ Checking the Model's Predictive Ability and Parameter Stability

In data-driven modelling, it is important to have several informative samples for each type of variation to be modelled. Otherwise, the estimated model-para-

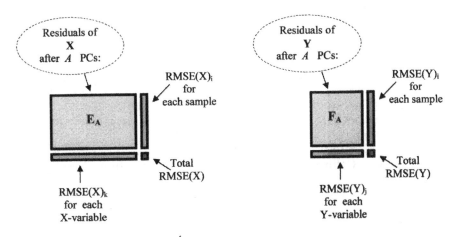

**Figure 10.1.** Summarising the X- and Y-residuals after $A$ PCs, $\mathbf{E}_A$ and $\mathbf{F}_A$. The meaning of the root mean square error (RMSE) summaries for samples $i$, variables $k$ and $j$, and their total summaries.

meters may be strongly contaminated by errors in the calibration data. Even though the model's fit to these calibration data (RMSEC) may look good, the actual predictive ability for new samples will be bad. This is what RMSEP from cross-validation is intended to reveal.

Cross-validation checks a model by repeatedly taking out different sub-sets of calibration samples from the model estimation, and instead using them as temporary, local sets of *secret* test samples. If the model parameter estimates are stable against these repeated perturbations, this indicates that the model is reliable. If the *secret* test samples are well-predicted from the other samples, this indicates that the model will have good predictive ability for new samples of the same general kind.

In the simplest case, each subset contains only one sample: this is called full leave-one-out cross-validation and is used in many of the illustrations here. Other ways of segmenting the samples during cross-validation are discussed towards the end of this chapter, and are used in the examples in Chapters 12–16.

## 10.2. CROSS-VALIDATION PRINCIPLES

Empirical input data have errors (sampling and measurement noise etc). In order to reduce the effect of these input errors on the final model, it is important to have *many* samples, and sufficiently *informative* (i.e. *representative* and *different*) samples. The least squares estimation method (Chapter 4.3) then ensures that the valid difference-information is picked up, while the random input errors to a large extent cancel each other out. But we would like to know *how well* this mechanism has worked, otherwise we cannot know if the results are reliable.

In cross-validation, that is seen by repeatedly making the data set *a little smaller*, and checking if the model then deteriorates a little (OK!) or a lot (bad!). If the calibration sample set has sufficient redundancy (repeated information), the model parameters remain stable even if various samples are left out from the calibration set.

If a sample set has too little redundancy, e.g. too *few* samples or too noisy data, the repeated sub-models in cross-validation will differ strongly from the full model, and not have the ability to predict the samples temporarily left out as *secret and unknown*.

### 10.2.1. Enough Calibration Samples to Guard Against Input Errors?

❐   A TOO SMALL CALIBRATION SET

Assume that we have only two variables, **x** and **y**, measured in only $N = 2$ samples, named *1* and *2*, and we use these as *calibration data*. They are represented by the symbol ⊕ in Figure 10.2a. A straight line can of course be drawn *perfectly* through the two points. However, that does not mean that the obtained

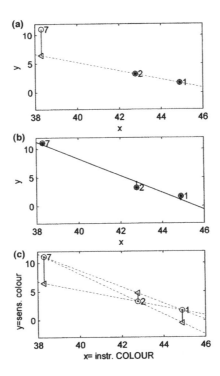

**Figure 10.2.** Cross-validation illustrated. Leave-one-sample-out cross-validation for univariate regression. Model: $\mathbf{y} = b_0 + \mathbf{x}b + \mathbf{f}$ *where* $\mathbf{y}$ = sensory *colour* darkness and $\mathbf{x}$ = physical *COLOUR* lightness for $N = 3$ samples, named *1*, *2* and *7*. (a) Checking a model by a new sample. Estimating a model-line from samples *1* and *2*, and checking how well it predicts sample *7*. (b) Estimating the full model: the least squares regression line for slope $\hat{b}$ and offset $\hat{b}_0$, estimated for all $N = 3$ samples. (c) Cross-validating by leave-one-sample-out: circles = input data; dotted lines = Sub-models $[\hat{b}_{-i}, \hat{b}_{0,-i}]$, $i = 1,2,3$, each estimated without sample $i$; triangles = predicted value of the input Y-value, $\hat{y}_{i,-i}$

line is a *perfect* statistical estimate! Since these data are real measurements, they are imperfect. So the apparently *perfect* line may be quite erroneous: it describes the calibration data all too well.

That becomes evident, when we apply the obtained *calibration model* line in Figure 10.2a to the data of the new sample 7. Presumably, this sample is of the same general kind as the other two. Yet, the line does not at all fit well for the new sample!

The data used here were taken from Table 1.1.

## ❑ A LITTLE LARGER AND MORE INFORMATIVE CALIBRATION SET

We now update our model, using all $N = 3$ available samples as calibration set.

By linear least squares regression (Chapter 4.3) the line in Figure 10.2b was estimated. Its formula (informally written $\hat{\mathbf{y}} = \hat{b}\mathbf{x} + \hat{b}_0$) is

$$\hat{\mathbf{y}} = -1.5\mathbf{x} + 68$$

The improved line fits sample 7 better than the old one, at the price that the fit to the previous two samples is no longer perfect. That is acceptable, because there is no point in being *precisely wrong*.

But how reliable is this new model line? And how good will our predictions of unknown y-values be, for still new samples of the same general kind, based on their known values of $x_i$ and this model line?

*Checking the model by leave-one-out cross-validation/jack-knifing:* Figure 10.2c shows the results of re-computing the linear model three times, each time with one of the three samples kept *secret*. This is the leave-one-out cross-validation. The circles represent the input data, the dotted lines represent three local sub-models $i = 1, 2, 3$, and the triangles represent the predicted y-values, obtained from these sub-model lines.

The line in Figure 10.2b, being based on all the input samples, is called the *full model*. It is our best least-squares estimate of a line through these points. So this is the model that we take with us from this analysis. The three lines in Figure 10.4c are called *sub-models*. Each of them is statistically less reliable, since it has fewer calibration samples. But together, the sub-models' variability sheds light on the reliability of the full model.

Based on how well each of the three sub-model lines in Figure 10.2c is able to predict the Y-value of its *secret* sample, we can make an estimate of the *predictive ability* of the full-model line in Figure 10.2b. This estimation of predictive ability for the input-variables is the essence of cross-validation.

Based on how the slopes of the three lines in Figure 10.2c deviate from the full-model line in Figure 10.2b, we can assess the reliability of the slope of the full-model line. This estimation of model parameter stability is the essence of what traditionally is called *jack-knifing*. It is here considered a part of the cross-validation.

The cross-validation estimate of predictive ability in **Y** is called RMSEP(Y). The estimate of the standard uncertainty (the uncertainty standard deviation) of the estimated slope parameter $\hat{b}$ is called $\hat{s}(\hat{b})$. These concepts will be explained in the next sections.

*The input data of the first mini-example:* The mini-example was made by selecting only three of the available samples from Table 1.1.

The variables in this data set represent two different ways to measure colour in cocoa-milk samples, the variables instrumental *COLOUR* lightness and the sensory *colour* darkness. They have previously been shown to be strongly intercorrelated (Figure 1.2a, p. 15).

Except for their opposite direction, we should in principle be able to predict one from the other by simple, univariate regression modelling.

*Conclusions from this first mini-data set:* Figure 10.2b indicated that instrumental *COLOUR* lightness (**x**) and sensory *colour* (**y**) indeed are nicely related to each other, as expected. But Figure 10.2c has told us that, based on this small data set alone, we cannot know if this relationship will hold in the long run. Given the apparent noise-level in these data, a higher number of samples is needed for assessing the validity of the results.

## 10.2.2. Every Phenomenon to be Modelled, Must be Seen at Least Twice

❑ A NEW DATA SET, WITH TWO PHENOMENA TO BE MODELLED

Assume that we want to model one Y-variable **y** from two X-variables, **X** = [**x**$_1$, **x**$_2$], and that we have only four or five calibration samples. This is illustrated in Figure 10.3.

The left side of Figure 10.3 has $N = 4$ calibration samples, named *1, 2, 3* and *7*, and spells cross-validation disaster, while the right side, which has $N = 5$ samples is satisfactory. The difference is one single sample, *6*. It accompanies the otherwise lone hero, sample *7*, and thereby stops it from being treated as a dangerous outlier during cross-validation.

Figure 10.3a shows **y** (○) plotted in 3D against **x**$_1$ and **x**$_2$ for the four samples *1, 2, 3* and *7*. The three first samples are seen to lie along a straight line *1–2–3* in the horizontal **x**$_1$ vs. **x**$_2$ plane. Thus they span only one dimension in **X**, and form a *1D subspace in our present 2D X-space*. The *lone hero*, sample *7*, has a higher level of **x**$_2$, and thus calls for a second dimension (PC) in the Y-model, at least if **x**$_2$ has any effect on **y** that needs to be modelled.

In Figure 10.3b the same data set has been supplemented with one more sample, *6*. All five samples are seen to lie in the same plane in the 3D space of [**x**$_1$, **x**$_2$, **y**]. Hence, this new sample provides one more point spanning the same second PC as sample *7*, off the linear *1–2–3* subset.

*The input data of the second mini-example:* The present mini-example concerns how to predict **y** = *sweet* taste from **x**$_1$ = *%SUGAR* and **x**$_2$ = *%COCOA*. The data are taken from Table 1.1 and concerns the same cocoa-milk samples as before. Figure 8.2 showed that sensory *sweet* taste was not uniquely defined by the sugar content in the cocoa-milk samples. Another ingredient, cocoa, reduced it. Shall we see that in the model?

❑ REGRESSION MODELLING: THE FULL MODEL

To what extent can we find a calibration model that predicts **y** well from **X** = [**x**$_1$, **x**$_2$] in these four or five samples? In both cases the *full* bi-linear regression model

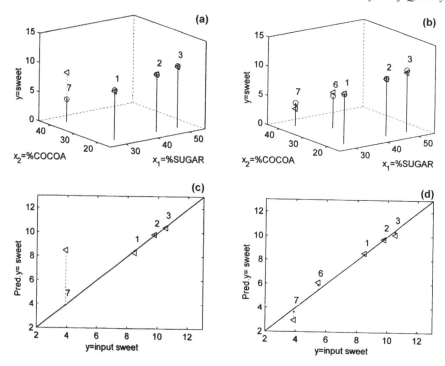

**Figure 10.3.** Cross-validation requires every modelled phenomenon to be observed at least twice. Leave-one-sample-out cross-validation for bi-variate regression, for $y$ = sensory *sweet* taste, $x_1$ = *%COCOA* and $x_2$ = *%SUGAR*. Model: $y = b_0 + x_1 b_1 + x_2 b_2 + f$. (a,c) Data from only $N = 4$ available samples (cocoa-milk mixtures named *1*, *2*, *3* and *7*); (b,d) data from $N = 5$ available samples (cocoa mixtures named *1, 2, 3, 6* and *7*); (a,b) input data (circles) and cross-validated predictions (triangles); (c,d) abscissa: measured values $y_i$, $i = 1,2,...,N$; ordinate: predicted values $\hat{y}_i$; diagonal line, $\hat{y}_i = y_i$; dots, full-model estimates of $\hat{y}_i$, using a 2-PC model; triangles, cross-validated predictions $\hat{y}_{i,-i}$, where $i = 1,2,..., N$.

needs to extract 2 PCs from $\mathbf{X}$, $\mathbf{t}_1 = w_1(\mathbf{X})$, $\mathbf{t}_2 = w_2(\mathbf{X})$, in order to span the X-space sufficiently to model $\mathbf{y}$.

*Calibration fit:* The Y-variable may now be reconstructed from $\mathbf{X}$ in each sample via the bi-linear model $\hat{\mathbf{y}} = \bar{y} + \mathbf{t}_1 q_1 + \mathbf{t}_2 q_2$ or its equivalent, $\hat{\mathbf{y}} = \hat{b}_0 + \mathbf{x}_1 \hat{b}_1 + \mathbf{x}_2 \hat{b}_2$. These reconstructed Y-values $\hat{\mathbf{y}}$, based on the *full* models using on all four or five calibration samples, are shown ($\cdot$) in Figure 10.3a,b. They are also given in Figure 10.3c,d, where the abscissa is the measured value of $\mathbf{y}$: the straight diagonal line is the target line, where perfect reconstructions would fall, $\mathbf{y} = \hat{\mathbf{y}}$.

The results show that in both cases, a very good full-model *calibration fit* can indeed be made to these calibration data. The dots show an almost *perfect* fit in the $N = 4$ case (Figure 10.3c) and a *good*, but not perfect fit in the $N = 5$ case (Figure 10.3d). But how can we know if these full-model results be trusted? Maybe these full models are overfitted and strongly noise-contaminated?

*Cross-validation: predictive ability and model stability:*   For the case of $N = 4$, the triangles ($\nabla$) in Figure 10.3a,c show the cross-validated Y-values, $\hat{y}_{i,-i}$, $i = 1$, 2,... $N$, each predicted from its local model $-i$, estimated with sample $i$ omitted.

The figures show that with sample 7 being all alone in requiring a second PC in the X-space, it is considered an *outlier* during the cross-validation: when temporarily omitting sample 7, the remaining calibration samples *1–2–3* can only give rise to one single PC, in this case pointing in the direction of $x_1$. The $x_2$ direction is not spanned at all! Consequently, sample 7 *stands out as a sore thumb*: the effect of $x_2$ on y is ignored in this sub-model. Since this effect is important in the present data set, the prediction of the Y-value for sample 7 is poor ($\hat{y}_{i,-i}$ equals that of sample 1, because they have the same value of $x_1$).

In contrast, with sample 6 and 7 both present in the calibration data set, ($N = 5$, Figure 10.3b,d) the cross-validation ($\nabla$) can afford to leave each of them out in turn. The other one will span the second direction in the X-space, anyhow, and thereby ensure proper 2-PC modelling of y in every sub-model.

*Predictive ability in Y:*   The errors in the Y-variable are summarised by the two root mean square error expressions in Table 10.1.

The calibration fit, RMSEC(Y), summarises the vertical errors (dots – circles) in Figure 10.3a,b, i.e. the vertical errors between ($\cdot$) and the diagonal line in Figure 10.3c,d. It shows that when $N$ increased from 4 to 5 with the addition of sample 6, the calibration fit RMSEC(Y) increased from its perversely low value (0.0006), to a slightly more realistic value (0.05), but this is still far too low.

On the other hand, the standard predictive uncertainty, RMSEP(Y) is very high (5.21) for $N = 4$, due to the lone hero, sample 7. But it decreases to a satisfactory level (0.29) with the inclusion of sample 6.

*Model parameter stability for the slope coefficient b:*   The stability of the obtained full model may also be estimated from the results in Figure 10.3. Table 10.1

**Table 10.1.** Validation of Figure 10.3: calibration fit RMSE(Y), cross-validated standard prediction uncertainty RMSEP(Y) and linear slope estimates, with reliability range estimated by jack-knifing.

| Samples $N$ | Calibration fit RMSEC(Y) | Predictive ability RMSEP(Y) | Model estimates $\hat{b}_1 \pm 2\hat{s}(\hat{b}_1)$ | $\hat{b}_2 \pm 2\hat{s}(\hat{b}_2)$ |
|---|---|---|---|---|
| 4 | 0.0006 | 5.21 | $0.10 \pm 0.01$ | $-0.22 \pm 0.23$ |
| 5 | 0.05 | 0.29 | $0.09 \pm 0.02$ | $-0.24 \pm 0.04$ |

shows, for instance, that while the full-model regression coefficient estimates $\hat{\mathbf{b}} = [\hat{b}_1, \hat{b}_2]'$ did not change much, the reliability range for the second X-variable, $\pm 2\hat{s}(\hat{b}_2)$, improved drastically, from 0.23 to 0.04, when including more than one sample for each type of phenomenon to be modelled. Details on how to estimate RMSEP(Y) and $\hat{s}(\hat{b})$ are given later, in particular in Appendix A10.3.

Repetition: we have previously seen (eq. (6.2)) that the full bi-linear predictive model

$$\hat{y} = \bar{y} + \mathbf{T}\mathbf{q}' \qquad \equiv \bar{y} + \sum \mathbf{t}_a q_a \qquad \equiv \bar{y} + \mathbf{t}_1 q_1 + \mathbf{t}_2 q_2$$

may be equivalently summarised by the full linear model

$$\hat{y} = \hat{b}_0 + \mathbf{X}\hat{\mathbf{b}} \qquad \equiv \hat{b}_0 + \sum \mathbf{x}_k \hat{b}_k \qquad \equiv \hat{b}_0 + \mathbf{x}_1 \hat{b}_1 + \mathbf{x}_2 \hat{b}_2 + ... + \mathbf{x}_K \hat{b}_K$$

Likewise, the cross-validated prediction of sample $i$ from the local, temporary sub-model $-i$ estimated without sample $i$, may for $i = 1, 2,..., N$ be written, both as bi-linear sub-model (eqs. (6.3) and (6.4))

$$\hat{y}_{i,-i} = \bar{y}_{-i} + \mathbf{t}_i \mathbf{q}'_{-i} \qquad \equiv \bar{y}_{-i} + \sum t_{a,i} q_{a,-i} \qquad \equiv \bar{y}_{-i} + t_{1,i} q_{1,-i} + t_{2,i} q_{2,-i}$$

and as linear sub-model

$$\hat{y}_{i,-i} = \hat{b}_{0,-i} + \mathbf{x}_i \mathbf{b}_{-i} \equiv \hat{b}_{0,-i} + \sum x_{k,i} \hat{b}_{k,-i} \equiv \hat{b}_{0,-i} + x_{1,i} \hat{b}_{1,-i} + x_{2,i} \hat{b}_{2,-i} + ... x_{K,i} \hat{b}_{K,-i}$$

The stability of both the bi-linear and the linear model parameters may now be assessed. For the linear parameters, Table 10.1 shows the full-model regression coefficients $\hat{b}_k$ for X-variables $k = 1,2$ (eq. (6.6a)). Their reliability ranges are based on standard uncertainties $\hat{s}(\hat{b}_k)$, which summarise the deviations $(\hat{b}_k - \hat{b}_{k,-i})$ (cf. eq. (10.10), p. 409). RMSEP(Y) (eq. (10.5g)) is estimated from $y_i - \hat{y}_{i,-i}$.

*Validity criterion: seen at least twice:*   In summary, the central concept of cross-validation is that the input sample set must have sufficient redundancy (repeated information). As long as each phenomenon to be modelled is seen at least twice with sufficient clarity, the cross-validation is able to distinguish it from input noise and outlier trouble.

Of course, it may happen that the same random noise pattern by chance is repeated two or more times in a data set. Therefore there is a small risk that the cross-validation may pick up a final, apparently valid PC that is based on noise, or at least strongly noise-contaminated. But the probability for that is smaller, the more X- and Y-variables are being modelled. This is one more reason for using multivariate data analysis.

## 10.2.3. For Readers Who Want a Short-cut

While the re-sampling concept behind cross-validation is simple, it is more difficult to understand the full *flexibility* of the cross-validation. Readers who find the topic too technical or tedious:

If you have understood Chapter 10 thus far, then you have understood the main points about cross-validation, and you may well leave this chapter here. But before you do, please read the following summary and admonition before you skip the rest.

## ❐ WHAT CROSS-VALIDATION IS USED FOR

The primary purposes of cross-validation are:

1. To optimise the model: determine the optimal number of PCs to be used in the final model. This is termed $A_{Opt}$.
2. To know if the model can be used in practice: Estimate the *predictive ability* of the final model, i.e. how good the model is for predicting **Y** from **X** in new samples of the same kind. The cross-validated prediction error estimate is termed the root mean square error of prediction of Y-variables, RMSEP(Y).
3. To assess the reliability of the modelling of various X- and Y-variables, and in particular to identify useless variables: estimate the *reliability of the estimated model parameters*, both the linear regression coefficients **B** and the bi-linear parameters **T**, **P** and **Q** (Figure 6.3).
4. To prepare for detection of *abnormal* outliers in the present data set and/or in the future: Estimate the X-error level RMSEP(X), etc. in the calibration data, thereby defining what is to be considered *normal*.

## ❐ A SHORT SUMMARY OF REST OF THE FIGURES

Figure 10.4a,b, in short, gives you a feeling for how RMSEP(Y) is estimated for a given Y-variable, and how the optimal number of PCs, $A_{Opt}$ is estimated. Figure 10.4c,d show how the *reliability of regression coefficients* **b** of five X-variables is estimated by *partial perturbations* during the cross-validation, and displayed in terms of reliability ranges $\hat{\mathbf{b}} \pm 2\hat{s}(\hat{\mathbf{b}})$, where $\hat{s}(\hat{\mathbf{b}})$ is called the estimated standard uncertainty of regression coefficients $\hat{\mathbf{b}}$.

Figure 10.5 shows how the reliability of a whole bi-linear model may be viewed at once, in terms of (a) *% Explained cross-validation-variance* in **X** and **Y**, (b) correlation loadings for complete model overview, (c) model stability of the samples' scores and (d) model stability of the variables' loadings. The model stability is visualised in terms of the *partial perturbation* lines. An example of summarising these into estimates of pragmatic reliability range rectangles is shown.

Figure 10.6 demonstrates that *mean-data* input tables give models with better apparent predictive ability (lower RMSEP(Y)) than raw-data tables that include *replicates*. It also shows the difference between *interpolation ability* and *repeatability/local reproducibility*.

Figure 10.7 follows up on the latter, by demonstrating how the estimated stability of a bi-linear model (a), depends on which validity level we choose to check: between-treatments (b), between-samples (c) or between-replicates (d). It also shows that in general, *many small perturbations amount to the same as a few larger perturbations.*

## ❐ WHAT TO REMEMBER WHEN YOU START ANALYSING YOUR OWN DATA

*Independent sampling at the desired validity level:*    When starting to do your own data analyses later, remember that the samples, or groups of samples (*segments*) to be taken out in cross-validation, must represent *independent* information. If you have repeatedly measured each sample several times, one immediately after the other, these repeatability replicates are *not* independent samples.

If you just want to check the short-term *repeatability* of your *measurement* device, a leave-one-out cross-validation with all the replicates included as samples in your data table is satisfactory. But if you want to assess your *reproducibility*, or your ability to draw *more general conclusions* or make reliable *predictions of Y in new samples*, then you cannot just run leave-one-out cross-validation on the whole input table, if it includes the individual replicates. Instead, you should first average the replicates, thereby reducing the number of samples.

*Internal validity is not the same as external validity:*    It should be noted that internal validity of a model, estimated from a given data set by cross-validation, does not automatically ensure external validity.

To ensure proper scientific reproducibility, with conclusions of more general value, the results must be supported and verified with new and completely independent evidence, as discussed in Chapter 3.4. To ensure stable predictions over a longer time span, older models should be checked from time to time, and updated if the system under observation or the measurement process has proven to be non-stationary (Chapter 15).

## ❐ FINAL ADMONITIONS

Remember that BLM is a powerful tool for finding patterns in data, but has no intelligence of its own. Don't fool yourself with it!

BLM will treat systematic mistakes in the data just like it treats valid patterns in the data. There is even a 5% risk that it finds small, but *"statistically significant"*, patterns in random measurement noise. Be careful, and make sure that you check your modelling conclusions, first in the input data and then by new experiments, if necessary.

With these admonitions, the eager or faint-hearted may now leave this chapter.

## 10.3. CROSS-VALIDATION, STEP BY STEP

The previous two mini-examples have illustrated how cross-validation works, what it requires, and what it can reveal about uncertainty in the data analysis. Further details will now be given.

### 10.3.1. A Small Data Set With One y, many X-variables

In Table 10.2, the five first columns are defined as five X-variables $\mathbf{X} = [\mathbf{x}_1\mathbf{x}_2,..,\mathbf{x}_5]$ and the last one is taken as **y**.

❒ INPUT DATA OF THIS THIRD MINI-EXAMPLE

In this third mini-example, the input data in Table 10.2 are shown after *standardisation*. They represent a subset of Table 1.1: in order to make the discussion of validity simpler, only the first replicate of mixture 5 is used, so there is only one sample per mixture, named *1, 2, 3, 4, 5, 6* and *7*. Only one of the sensory variables is used here, *sweet* taste (**y**). Later examples will explain model validation for many Y-variables.

❒ MAXIMUM NUMBER OF PCs

In the present data set used for illustration, there are five X-variables, so the maximum possible PCs to extract is five. But since $\mathbf{x}_1 + \mathbf{x}_2 + \mathbf{x}_3 = 100\%$, there are maximally four *different* PCs; the last one will have length $\mathbf{t}_5'\mathbf{t}_5 = 0$. So there is no sense in estimating it. Some software even becomes unstable when trying to extract such *null-space dimensions* (cf. Appendix A6.8).

Let us act as if we had not looked at these data previously. We do not yet know the optimal number of PCs to use in this data set. But we expect to find at least 2 PCs. To be on the safe side, we estimate *more than enough* PCs, say $A_{Max} = 4$.

Note that *a* is the index for an individual PC, $a = 1, 2, 3,..., A_{Max}$, while $A$ is the number of PCs used in a model, $A = 0, 1, 2,..., A_{Max}$. So when a model uses $A$ PCs, that means that PCs $a = 1, 2,..., A$ are being used; if $A = 0$, the input data are just mean-centred.

**Table 10.2.** Input data for validation example: some chemical, physical and sensory data from Table 1.1 after standardisation.

| | | %COCOA 1' | %SUGAR 2' | %MILK 3' | COLOUR 4' | VISCOSITY 5' | Sweet 6' |
|---|---|---|---|---|---|---|---|
| *1* | 1 | −0.87 | −0.87 | 1.73 | 1.54 | −0.66 | 0.46 |
| 2 | 2 | −0.87 | 0.86 | 0.00 | 0.59 | −1.01 | 1.01 |
| 3 | 3 | −0.87 | 1.73 | −0.87 | 0.08 | −1.10 | 1.32 |
| 4 | 4 | 0.00 | −0.87 | 0.86 | 0.41 | 0.46 | −0.31 |
| 5 | 5 | −0.00 | 0.00 | 0.00 | −0.18 | −0.08 | −0.15 |
| 6 | 6 | 0.86 | 0.00 | −0.87 | −1.04 | 0.85 | −0.83 |
| 7 | 7 | 1.73 | −0.87 | −0.87 | −1.41 | 1.54 | −1.50 |

❏ ESTIMATING THE FULL MODEL

The *full* model, including $\hat{\mathbf{B}}$, is estimated by BLM, *using all N samples*, and saved until we have decided the optimal number of PCs to be used.

❏ ESTIMATING *M* LOCAL SUB-MODELS

In cross-validation, the set of $N$ calibration samples is split into $M$ sub-sets. The BLM is re-estimated $M$ times, each time keeping out the samples allocated to subset number $m$ when estimating sub-model number $m$, for $m = 1, 2, .., M$. This is symbolised by a subscript $(_{-m})$, as in $\hat{\mathbf{B}}_{-m}$, $w_{-m}(\cdot)$, $\mathbf{P}_{-m}$ and $\mathbf{Q}_{-m}$. We switch to index $m$ (instead of $i$), because each segment $m$ may contain more than one sample $i$.

❏ APPLYING THE SUB-MODELS TO THE SAMPLES LEFT OUT AS SECRET

The data of the sample(s) in segment $m$ are then submitted to prediction modelling, using the corresponding sub-model with $A = 0, 1, 2,..., A_{\text{Max}}$. PCs. Their Y-predictions are denoted $\hat{\mathbf{Y}}_{m,-m}$, cf. Table 10.1 and eq. (10.4d).

Each of the $M$ local sub-models and their Y- and X-predictions are stored until we have decided how many PCs should be used.

❏ RESULTS FOR THE PRESENT DATA SET

*Leave-one-out cross-validation:* During the leave-one-sample-out cross-validation, one sample is taken out individually at a time. Hence, the number of cross-validation segments equals the number of samples, $M = N$.

Figure 10.4 shows the results from the leave-one-sample-out cross-validation with $M = N = 7$ segments, each with one alternative sample left out.

Each curve in Figure 10.4a shows, for the denoted sample $i = m$, the size of the prediction error in **y** as a function of the number of PCs, $A$. It is called the root mean square error of prediction for the Y-data in segment $m$, RMSEP(Y)$_m$, and has one value for each model rank, $A = 0, 1, 2,..., A_{\text{Max}}$.

❏ ESTIMATION OF THE MEAN ERROR CURVE RMSEP(Y) AND RMSEC(Y)

The root mean squared error of prediction, RMSEP(Y) (eq. (10.5e)) is calculated as the RMS average of the $M$ prediction-error curves in Figure 10.4a. This is given by curve 1 (black) in Figure 10.4b.

The RMSEP(Y) curve is our best estimate of how large error between predicted and measured Y-values we may expect in the long run, for new samples of the same general kind, when using our full model with the various numbers of PCs.

Curve 2 (grey) in Figure 10.4b represents the corresponding over-optimistic

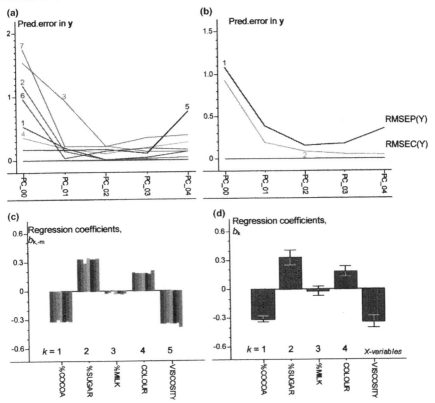

**Figure 10.4.** Estimation of optimal number of PCs, the predictive ability in **Y** and the reliability of the linear model, by leave-one-out cross-validation. PLSR of **y** = *sweet* taste on **X** = [*%COCOA, %SUGAR, %MILK, COLOUR* lightness, *VISCOSITY*] in *N* = 7 calibration samples (cocoa-milk mixtures 1–7, Table 10.2). (a) Cross-validated prediction errors RMSEP(Y)$_m$ for individual cross-validation segments *m* (samples *i* = 1,2,...,7),plotted against the number of PLSR components, *A* = 0,1,2,...,4. (b) Cross-validated error, RMSEP(Y), and the calibration fit, RMSEC(Y). (c) Jack-knifed sub-models **b̂**$_{-m}$ with 2 PCs, for segments *m* = 1,2,...,7, each based on only *N*-1 = 6 samples. (d) Full model **b̂** with *A* = 2 PCs, with jack-knifed estimated of its reliability range ± 2ŝ(**b̂**).

calibration model fit, RMSEC(Y) (eq. (10.3e)). It is shown to fall continuously with increasing number of PCs, and to give a serious under-estimate of the prediction error, compared to the RMSEP(Y) curve. This illustrates why RMSEC(Y) can be used neither for assessing the optimal number of PCs nor for estimating the predictive ability.

❐ Estimate the Optimal Number of PCs to Use, $A_{Opt}$

Based on the cross-validated estimate of Y-prediction error curve RMSEP(Y) (Figure 10.4b), the optimal number of PCs, $A_{Opt}$ may now be chosen.

There is a clear minimum RMSEP(Y) at $A = 2$. For $A < 2$ PCs we see that it is *underfitted* and too high, because valid structure in the data is still not modelled. For $A > 2$ it is *overfitted* and too high, because *errors* (noise, outlier structures, etc.) have now been pulled into the model, contaminating the model parameters and reducing its predictive ability.

In the present case the sample set was designed as a three-component mixture design, and the X-data should span a two-dimensional space. So *a priori* we expect that 2 PCs should be optimal. And indeed this is what the cross-validation indicated here.

In general it is advisable to choose an $A_{Opt}$ as low as possible. Chapter 12 shows that if the RMSEP(Y)-curve more or less flattens out, it is advisable to *stop before it reaches its minimum.*

$A_{Opt}$ is in some data sets found to be a little lower or higher than the expected number of independent types of variations in **X**, $A_{Expected}$. This is as it should be: For instance, if the data are few and noisy, we may find $A_{Opt} < A_{Expected}$: the model is hit by noise before it has been able to observe and pick up all the expected structures, or some X-structures are irrelevant for **Y**. On the other hand, if there are plenty of precise input data, but strong non-linearities within the X-variables and between the X- and Y-variables, then we may find $A_{Opt} > A_{Expected}$, because some extra PCs are needed to ensure implicit modelling of these systematic non-linearities (Chapters 7.4 and 12.2).

❐ Assessing the Predictive Ability RMSEP(Y)

The RMSEP(Y), the value of the curve at $A_{Opt} = 2$, is $RMSEP(Y)_{A=2} = 0.15$. Of course, this was based on the *M* perturbed models, not on the full model itself. But $RMSEP(Y)_{A=2}$ is our best estimate of how large the error between future Y-data and their Y-predictions from the full model will tend to be, in this general kind of samples. The smaller this value is, the better.

> We cannot assess a model without perturbing it unless we make additional theoretical assumptions, which this book seeks to avoid.

Since RMSEP(Y) is itself an estimate, it has its own estimation uncertainty. This may be assessed from the variations between the *M* cross-validation segments (Figure 10.4a), as illustrated in eq. (10.5h) and Figure 12.2b.

*Percentage Explained Variance in the Cross-Validation:*   Another useful measure of predictive ability is the percent explained cross-validation-variance, which expresses how much of the total initial variance ($RMSEP^2_{A=0}$) is removed by the model. For instance, in the present case, with initial $\text{RMSEP(Y)}_{A=0} = 1.08$ and final $\text{RMSEP(Y)}_{A=2} = 0.15$, the percent explained variance is:

$$\text{Percentage explained Y} - \text{variance} = 100\left(1.08^2 - 0.15^2\right)/1.08^2\% = 98\%$$

While the RMSEP may be the professional data analyst's choice, because it shows what kind of uncertainty to expect, the percentage explained variance may be the application researcher's choice, since it focuses on what *can* be done, not on what *cannot* be done. However, since the two measures reflect the same information, and since the application researcher ideally should be the data analyst, it is good to have both measures easily accessible in the software.

After a variable has been standardised, it has a total initial standard deviation (eq. (4.7)) of $s = 1.0$. Its initial ($A=0$) RMSEC (eq. (10.3b,c)) and RMSEP (eq. (10.5d,e)) (with leave-one-out cross-validation is then):

$$\text{RMSEC}_{A=0} = s\sqrt{[(N-1)/N]} \tag{10.1a}$$

$$\text{RMSEP}_{A=0} = s\sqrt{[N/(N-1)]} \tag{10.1b}$$

where $N$ is the number of samples in the set. The same proportional difference between $\text{RMSEC}_{A=0}$ and $\text{RMSEP}_{A=0}$ is retained if the variables have not been standardised.

This difference may of course be corrected for. However, it is useful to be able to distinguish RMSEC and RMSEP curves from each other, and this is easy when retaining their difference in $\text{RMSEC}_{A=0}$ and $\text{RMSEP}_{A=0}$. This small difference usually has no practical importance on the conclusions from a data analysis. In the present case, with $N = 7$, this means that after standardisation of the input variables, we find $\text{RMSEC}_{A=0} = 0.93$ and $\text{RMSEP}_{A=0} = 1.08$, instead of $s = 1.0$.

☐ INSPECTING THE FULL MODEL

We can now look at the full model with $A_{\text{Opt}}$ PCs. In this mini-example, we shall only assess the linear model-summary parameters; the stability of the bi-linear parameters will be discussed in the next example.

The bars in Figure 10.4d summarise the 2-PC model, $\hat{\mathbf{b}}_{A=2}$. The figure shows that four out of the five X-variables show large regression coefficients: The variables $\mathbf{x}_2$ and $\mathbf{x}_4$ give positive contributions to $\mathbf{y}$, while $\mathbf{x}_1$ and $\mathbf{x}_5$ give negative contributions; $\mathbf{x}_3$ does not seem to contribute much to the prediction of $\mathbf{y}$.

☐ ASSESSING THE STABILITY OF THE FULL MODEL: JACK-KNIFING

We may now assess the statistical stability of the obtained full model. This is done by studying the variation between the full model and the different sub-models computed with $A_{\text{Opt}}$ PCs during cross-validation. This is a BLM modification (Martens and Martens, 2000a,b) of the established *jack-knifing* technique (Efron and Tibshirani, 1993).

In Figure 10.4c the seven individual local sub-models, each using 2 PCs, are shown for each X-variable $k = 1, 2, ..., K (= 5)$: $\hat{b}_{k,-m}$, $m = 1, 2, ..., M (= N = 7)$. These individual regression coefficients constitute sub-model vector, $\hat{\mathbf{b}}_{-m}$, which already were used for predicting $\hat{y}_{m,-m}$ in Figure 10.4a.

The error bars in Figure 10.4d represent the estimated reliability range of the regression coefficients in the full model. This range is defined as $\pm 2\hat{s}(\hat{\mathbf{b}})$. The standard uncertainty $\hat{s}(\hat{\mathbf{b}})$ was obtained by summarising the partial perturbations between the full model (Figure 10.4d) and the cross-validation segments in Figure 10.4c, $\hat{\mathbf{b}} - \hat{\mathbf{b}}_{-m}$, over the $M$ segments (eq. (10.10)).

*Significance testing from the reliability range:* For each X-variable, $k$, the range $\hat{b}_k \pm 2\hat{s}(\hat{b}_k)$ may, under idealised conditions, be regarded as an approximate 95% confidence interval (ISO, 1995a). Therefore it is useful for checking against Type I error: Being fooled into believing effects caused only by input data errors.

If the range $\hat{b}_k \pm 2\hat{s}(\hat{b}_k)$ does *not* contain the value $b_k = 0$, then there is, ideally, only about 5% or less risk of committing Type I error. But if the uncertainty range does contain the value $b_k = 0$, then the estimated effect $\hat{b}_k$ cannot be trusted, because it may have been caused by random noise only: Even if the unknown, but *"true"* value, $b_{k,true}$ in fact had been equal to zero, there would be more than 5% risk of getting the result $\hat{b}_k$.

The figure shows that the range $\hat{b}_k \pm 2\hat{s}(\hat{b}_k)$ contains $b_k = 0$ only for the third X-variable. All the other X-variables appear to give reliable, non-zero contributions to the prediction of **y**. But of course, that does not automatically mean that these parameter estimates are *error free* and reflect *causal* contributions to **y**!

*In jack-knifing, the partial parameter perturbations are summed, not averaged:* Since the $M$ sub-models share many samples, they are not independent of each other and of the full model. Therefore, the deviations between the full model (Figure 10.4d) and the individual, local sub-models (Figure 10.4c) are called *partial perturbations*. They cannot be treated as if they were independent replicates of each other and just have their observed error contributions *averaged* (in the RMS sense).

Instead, their error contributions have to be *summed*. Therefore, the estimated standard deviation $\hat{s}(\hat{b}_k)$ is higher than the general level of the individual partial perturbations. That is evident from the wide reliability range $\pm 2\hat{s}(\hat{b}_k)$ in Figure 10.4d, compared to the apparent level of the partial perturbation in Figure 10.4c.

*Interpretation of this data set:* In Figure 10.4a samples *7* and *3* had the highest initial total Y-variation ($A = 0$). These samples are in fact the least and most *sweet* samples. Sample *7* and most of the other samples are well predicted by one-factor sub-models, but for, e.g. sample *3*, even a second PC is required. With more than 2 PCs in the local sub-models the prediction error goes up in several samples. Hence, 2 PCs appears to be the optimal compromise.

Figure 10.4c shows that sample *7* has the largest partial perturbation in the regression coefficients for *COLOUR* and *VISCOSITY*. This illustrates how the individual perturbation results in cross-validation also provides outlier-detection capability.

## 10.4. THE STABILITY OF BI-LINEAR MODELS

The previous example showed how to assess the stability of the *linear* model parameters, the B-coefficients. The next example demonstrates how to assess the stability of *bi-linear* model parameters, **T, P** and **Q**.

### 10.4.1. Model Reliability at a Glance

Figure 10.5 illustrates an efficient way to gain overview of a bi-linear model with many Y- and X-variables. In this mini-example there are $N = 7$ samples. Matrix **X** now has six sensory variables and matrix **Y** has four chemical and physical variables.

With more samples and variables, the plots get somewhat cluttered when not presented in colours and with interactive amplification.

◻ INPUT DATA

Even this data set concerns the cocoa-milk samples (Table 1.1, i.e. produced without texture stabiliser). To make the data analysis a little more challenging, the additional sensory variable thick texture (*thick-txtr*) has been added, with data taken from Table 9.1 These unstabilised cocoa-milk data are now analysed with **X** = sensory variables and **Y** = chemical and physical variables, like in Figure 8.3a.

As X-variables we use the previous sensory variables *(colour darkness, cocoa-odour, milk-taste, smooth-txtr* and *sweet* taste, plus the new sensory variable, *thick-txtr*). As Y-variables we use *%COCOA, %SUGAR, %MILK,* and *VISCOSITY*. The well-modelled but redundant variable *COLOUR* lightness has been taken out, in order to avoid cluttering the plots.

As in the previous mini-example, we have seven samples (each averaged over replicates), named *1–7*. The object indicator variable named *1–7* represent seven additional indicator variables. All the variables were standardised prior to the PLSR analysis.

◻ PASSIVE VARIABLES

One of the X-variables (*thick-txtr*) was found, in preliminary cross-validation, to give very uncertain regression coefficients $\hat{b}_{kj}$ for all the Y-variables. The whole model was therefore re-estimated, with *thick-txtr* treated as a *passive* variable (Appendix A10.2). Likewise, the object indicator variables 1–7 were passive.

Thereby we have ensured that these variables do not affect the obtained model, but we can still see how they fit to the model.

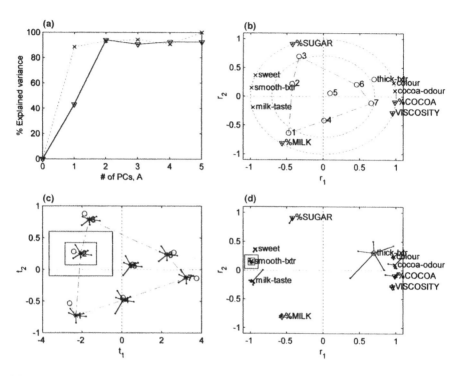

**Figure 10.5.** Three validation principles: prediction ability, bi-linear model interpretability and parameter stability. $J = 4$ chemical/physical variables ($\mathbf{Y} = \%COCOA$, $\%SUGAR$, $\%MILK$, $VISCOSITY$) and a passive variable (*thick-txtr*) modelled from $K = 5$ sensory variables ($\mathbf{X} = colour$, *cocoa-odour*, *smooth-texture*, *milk-taste* and *sweet*) in $N = 7$ cocoa-milk samples by PLSR. (a) Cross-validation: predictive ability in $\mathbf{Y}$ and $\mathbf{X}$, expressed as percent explained cross-validation-variance vs. number of PCs. (b) Graphical interpretation: correlation loading vectors $\mathbf{r}_1$ (abscissa) and $\mathbf{r}_2$ (ordinate), for the Y-variables ($\triangledown$), the X-variables ($\times$) and the passive variable ($\bigcirc$). Dotted circles = 50 and 100% explained variance (unvalidated). Numbers 1–7 = passive sample indicators (c) Jack-knife stability estimate of the samples: scores $\mathbf{t}_i = [t_{i1}, t_{i2}]$ (*) for samples $i = 1, 2, ..., N$. $\bigcirc$, the perturbation found when a sample *itself* was kept *secret*; other protruding line segments point to its partial perturbations caused by keeping one of the other $N$-1 samples *secret*. The rectangles around sample number $i = 2$ demonstrate approximate reliability ranges ($\pm 1$ and $\pm 2$ estimated standard uncertainty of its scores, $\hat{s}(t_i)$). (d) Jack-knife stability estimate of the variables: correlation loadings for the Y-variables ($\triangledown$), the X-variables ($\times$) and the passive variable ($\bigcirc$). Protruding line segments: partial perturbations by keeping one of the seven samples *secret* during crossvalidation. The rectangles around the variable *smooth-txtr* demonstrate approximate confidence regions ($\pm 1$ and $\pm 2$ uncertainty standard deviation of its correlation loadings, $\hat{s}(r_{smooth\text{-}txtr})$).

◻ ESTIMATE OF THE OPTIMAL NUMBER OF PCS, $A_{Opt}$

The X-and Y-tables were submitted to bi-linear modelling by PLSR, with leave-one-out cross-validation, i.e. with $M = N = 7$ segments, with $A_{Max} = 5$ PCs.

Figure 10.5a shows the percentage explained cross-validation-variance, summarised over all the samples for the four Y-variables (solid line) and for the five active X-variables (dotted line). It shows that **Y** requires 2 PCs to be described, although **X** was quite well described already by 1 PC. The 2-PC solution is very good in both **X** and **Y**. The prediction error in **Y** is reduced from $RMSEP(Y)_0 = 1.08$ to $RMSEP(Y)_2 = 0.26$ (94% explained cross-validation variance in **Y**) and the prediction error in **X** from $RMSEP(X)_0 = 1.08$ to $RMSEP(X)_2 = 0.28$ (93% explained cross-validation variance in **X**).

Note that with only five X-variables remaining active after the loss of *thick-txtr*, the maximum possible number of PCs is five. At five PCs the X-residuals $E_5$ are per definition all zeroes, and since $RMSEP(X)_5$ (eqs. (10.5d,f)) is corrected by the factor $5/(5-5) = 5/0$, it is %, which cannot be computed, formally. The software has just set it to 100%, by default.

◻ MODEL OVERVIEW: CORRELATION LOADING PLOT WITH OBJECT INDICATORS

Figure 10.5b summarises the 2-PC solution, in terms of the correlation loadings of the X-variables ( ×, eq. (5.14)) and the Y-variables (triangles, eq. (6.7)). The passive variable *thick-txtr* and the object indicators are marked with circles. The dashed curve outlines the mixture design of the samples.

In analogy to Figure 8.3a the figure shows the expected pattern: All the active X-and Y-variables lie close to the circle of 100% explained unvalidated variance, which means that their RMSEC(X) or RMSEC(Y) is near 0. But the passive variable *thick-txtr* is seen to have only a little more than 50% of its variance explained. Its data do not seem to agree with the other sensory variables. Why? Is it because it contains unique, but useful information, or because it has more noise? That will be seen from the cross-validation results below.

◻ THE STABILITY PLOT FOR THE VARIABLE LOADINGS

Figure 10.5d shows the correlation loadings for the X- and Y-variables again, this time connected by $M$ line segments to their correlation loadings obtained in the cross-validation sub-models $m = 1, 2,..., M ( = 7)$.

These partial perturbations immediately show that the model parameters for *thick-txtr* are very unstable, compared to the other variables. Inspection of the raw data reveals that this variable has very little variation between the seven samples; it was accidentally blown up by the *a priori* standardisation.

The X-variable *milk-taste* also appears to have a somewhat strange value in one of the segments. Closer inspection reveals that it occurs when sample *1*, which has highest *%MILK*, is left out. Again this shows how outliers may be detected as part of the jack-knifed stability assessment.

*The meaning of the partial perturbations:*   The stability plot shows the stability of the loading parameters. Line segment number *m* points from the full model to the position which a variable obtains in that sub-model. The length and direction of the segment shows how much the parameters for this variable change when cross-validation segments *m* is *left out*. If the line segment is long, it means that the sample(s) in cross-validation segment number *m* have unique information not represented by the other samples, making the model unstable.

The perturbations are again called *partial*, because the various cross-validation segments share most of the samples, so their data sets are not independent. Hence, the different sub-models are not independently estimated. Therefore, unless $M = 2$, (*split-half* cross-validation), the perturbation lines should *not* be interpreted as pointing to *independently* observed replicates. Instead, they show how stable each loading parameter is against *losing* segments $m = 1, 2,...,$ *M* of samples from the calibration set. In the present case of leave-one-out cross-validation, with $M = N$, they show the effect of removing, in turn, each individual mixture or treatment from the calibration set.

☐   APPROXIMATE RELIABILITY RANGE RECTANGLES AROUND THE VARIABLES

Based on the *M* perturbations between the cross-validation, we may estimate the uncertainty standard deviation of the individual loadings (eq. (10.9b,c)).

Since the perturbations in such jack-knifing are only *partial*, their uncertainty contributions must be *summed*, not *averaged*. In order to avoid cluttering the plot, this is only shown for one of the variables, *smooth-txtr*. Its reliability is indicated by approximate reliability ranges of $\pm 1$ and $\pm 2$ standard uncertainties along PC1 and PC2, symbolised by pragmatic little rectangles.

> More *advanced* ellipsoid displays of estimated confidence intervals may be estimated and drawn. But that would here be presumptuous and unwarranted, considering that the perturbations are probably largely due to small but non-random errors in the model specification, which leaves response curvature unmodelled in the residuals. Hence, the errors do not primarily reflect *random noise*, as assumed by classical statistical distribution theory. Before these partial perturbations can be studied, each of the *M* sub-models has been rotated orthogonally towards the full model, as described in Appendix A10, p. 409.

For *thick-txtr*, the reliability range (not shown here) extends far beyond the origin (0,0) in the plot, indicating that if the errors had been random, then the loading value for this variable could have been caused by these errors alone.

☐ THE STABILITY PLOT FOR THE SAMPLE SCORES

Figure 10.5c shows the corresponding partial perturbations of the *M* sub-models for each sample's scores. Again, the dashed line outlines the mixture design of the samples.

The circle for each sample marks the cross-validation segment *m* in which it was left out from the estimation of means and loadings, and instead modelled as an unknown sample. As exemplified for sample 2, the two rectangles represent the estimated reliability ranges of $\pm 1$ and $\pm 2$ approximate standard uncertainties along PC1 and PC2.

If a sample had been a particularly dangerous outlier of the influential type with both high leverage and high X-residuals, it would be seen here as a particularly long perturbation line segment to its circle-symbol.

That is not observed presently, which indicates that there are no gross outliers. But the perturbations are relatively large, which reflect the fact that the sample set is small: there are, on the average, only about three or four samples that clearly span each direction in the experimental design.

### 10.4.2. Summarising the Validity Assessment of this Model

If the present samples had been obtained just by random sampling, having a sample set with only $N = 7$ samples would be considered as too small, at least for more than a 1-PC solution. But since

1. this is a consciously designed sample set, produced under controlled conditions,
2. the cross-validation shows such good multivariate predictive ability in both **X** and **Y** and
3. the solution is easy to interpret, and corresponds to what was theoretically expected,

we accept the solution in Figure 10.5 as reliable. More detail on the design of sample sets will be given in Chapter 11.

### 10.5. DIFFERENT LEVELS OF VALIDITY

### 10.5.1. What Kind of Validity?

Figures 10.6 and 10.7 show how the model stability assessment changes, depending on what level of validity we try to assess.

Under normal circumstances it is recommended (Figure 3.2, *Step IV*) to work on means over replicates, and simply cross-validate between independent units of interest (here: *treatments*). The data set of mean samples is smaller and more precise, and therefore easier and more fun to work with. What usually interests a researcher is the ability to depict the **X–Y** relationship so well that the model allows good *interpolation between treatments* or whatever independent sample types we have, and that is what the leave-one-treatment-out cross-validation reveals.

But the cross-validation may also be used for checking other levels of validity, e.g. technical repeatability or general project reproducibility. As discussed in Chapter 2, the *repeatability* shows the short-term stability of the measurement procedures, and is of interest to the person responsible for the measurement instrument, but in principle, of no interest to anyone else. The more difficult and ambitious *reproducibility*, on the other hand, is what the scientific community asks for: To what extent can the conclusions be trusted and extrapolated?

## 10.5.2. Including the Replicates

The cocoa data set analysed in Figure 10.6 is the same as in Figure 10.5, with six sensory X-variables (one passive) and four chemical and physical Y-variables. But now we shall use the *three individual sensory replicates*, instead of their mean. So, instead of having just seven samples (seven rows in **X** and **Y**), we have $N = 21$ samples in the sensory data table **X**. For simplicity, the Y-data have just been copied three times for each sample, to make data table **Y** compatible with **X**.

❒   WHAT KIND OF REPLICATES?

In the present COCOA data set (COCOA-III), the three sensory replicates were performed at separate points in time, and on samples independently produced. This means that these production replicates are more independent of each other than ordinary repeatability replicates for the measurements (*pushing the same button again and again*).

On the other hand, the sensory panelists were the same in all three replicates, and so were the raw materials (cocoa, sugar and milk). Therefore the replicates do not fulfil the requirement of completely independent evidence. If some panelists had repeated some misunderstanding, or the raw material had been abnormal, the same systematic error would have been repeated each time!

Consequently, the between-replicates cross-validation is here expected to provide a validation assessment somewhere between trivial repeatability and fundamental reproducibility.

Sampling problems should not be forgotten (Gy, 1998). For instance, powders are notorious for agglomerating and segregating during transportation, and therefore represent major sampling problems. Of the present raw materials, only the sugar may be expected (Chapter 16) to be so pure and homogeneous that sampling problems were small. The dry cocoa and dry milk may have given some powder-sampling problems. Moreover, milk from different cows or different points in time may be quite different, and variations in cocoa bean production are known to be real. These sampling problems were sought minimised by the way the raw material was selected and handled. Moreover, the present results are so clear, that there is good reason to expect other samples or other batches of these raw materials to behave not too differently. But we cannot be sure.

## 10.5.3. Three Different Ways to Segment the Replicated Samples in the Cross-validation

With seven experimental treatments (seven cocoa-milk mixture recipes) and three production replicates we have at least three alternative ways to define the cross-validation segmentation. We may, as before, cross-validate between the seven treatments ($M = 7$). Alternatively we may use leave-one-out cross-validation ($M = 21$), or cross-validate between the replicates ($M = 3$).

☐ DIFFERENT ESTIMATES OF RMSEP(Y) AT DIFFERENT VALIDITY LEVELS

Figure 10.6 compares different estimates of prediction error in **Y**, RMSEP(Y), for the same raw data, averaged over all the Y-variables. In all four cases the 2-PC solution is seen to be optimal.

*Mean data are more precise than the individual replicates:*   The two curves with triangle symbols in Figure 10.6 represent between-treatment cross-validation ($M = 7$). The lower, dashed curve was obtained from the previous analysis of the *means of the three replicates* (Figure 10.5, $N = 7$), while the upper curve was obtained with the individual replicate raw data ($N = 21$).

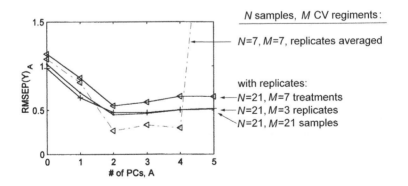

**Figure 10.6.**    Different validation levels. RMSEP(Y)$_A$ vs. $A = 0,1,...,5$ for the chemical/physical Y-variables predicted from the active sensory X-variables in Figure 10.5, but with different ways of using three sensory replicates for each of the seven treatments (cocoa-milk recipes).

This confirms that working on the mean data gives lower prediction error (and hence higher percentage explained cross-validation-variance) than working on the raw data with replicates, even at the same between-treatments validity level.

*With replicates present, leave-one-sample-out ≈ leave-one-replicate-out:* The middle two curves in Figure 10.6 show the RMSEP(Y) results of leave-one-sample-out ($M = 21$) and leave-one-replicate-out ($M = 3$) cross-validation. The results here seem to be virtually identical. This indicates that these two ways of cross-validating in this case represent rather similar types of validation. This will be more evident when looking at the full model and its stability, in the next figure.

❏  THE FULL MODEL: SAME SOLUTION AS BEFORE, BUT A LITTLE LESS PERFECT

Figure 10.7a shows the correlation loading solution of the full model, obtained with all $N = 21$ samples. Passive design indicator variables have been added also for the three replicates, named *r1, r2* and *r3*, along with the object indicators 1–7.

The same pattern is seen as in the previous model based on the sensory means (Figure 10.5b), except that PC2 has been flipped upside down. But detailed inspection shows that the present model has *slightly less perfectly described* data: Both the X-variables (×) and the Y-variables (triangles) are slightly further inside the ellipse of 100% explained variance. This is not unexpected, considering that the individual replicates probably have higher level of *random* measurement noise than their means. As long as we do not know anything about the type of between-replicate variations and measurement noise, the least bad assumption we can make is to regard their effect as random.

❏  THREE ALTERNATIVE STABILITY PLOTS FOR THE SAME FULL MODEL

The stability plot of the loadings for these three ways of segmenting the samples are shown in Figure 10.7b–d, respectively. In Figure 10.7b there are seven partial perturbation lines protruding for each variable, one for each treatment. In Figure 10.7c there are 21 partial perturbation lines, one for each individual row in the data matrix, and in Figure 10.7d there are three partial perturbation lines, one for each sensory replicate. Remember that the full model to be validated is the same: what differs is the way the cross-validation is performed and what the cross-validation is intended to show.

*To reproduce nature is more difficult than to reproduce ourselves:* The rough reliability range is shown for the same variable, *smooth-txtr,* in each case. It is seen to be larger for $M = 7$ than for $M = 21$ or $M = 3$. This illustrates that it is usually more difficult to attain good stability at the between-treatments level

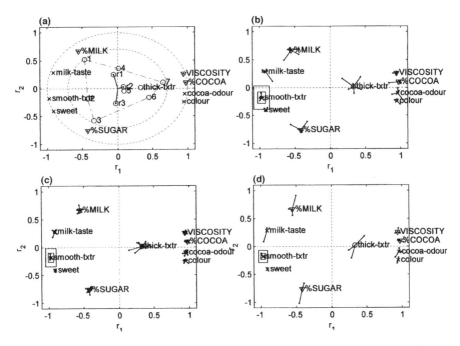

**Figure 10.7.** Different validation levels, continued. Predicting the $J = 4$ chemical/instrumental Y-variables and a passive variable from the $K = 5$ active sensory X-variables in the cocoa-milk samples (Figure 10.5). The jack-knifing results obtained from $N = 21$ samples, using $A_{Opt} = 2$ PCs, but with three different cross-validation segmentation schemes. (a) The full model to be assessed: correlation loading vectors $\mathbf{r}_1$ (abscissa) vs. $\mathbf{r}_2$ (ordinate), for the Y-variables ($\triangledown$), the X-variables ($\times$) and the passive variable ($\bigcirc$). Dotted circles, 50 and 100% explained variance (unvalidated). Numbers 1–7, sample indicators; r1–r3, replicate indicators. (b) Interpolation ability, $M = 7$ treatments. (c) Repeatability between samples, $M = 21$ samples. (d) Repeatability between replicates, $M = 3$ replicates.

than at the between-replicates levels: to depict the underlying systematic, possibly causal relationships in the data (with non-linearities, etc.) is more difficult than to just be able to repeat the same measurements.

*Many small perturbations = a few large perturbations:* When comparing Figure 10.7c and d, it may be noted that perturbing the model many times ($M = 21$), each time just a little, is similar to perturbing the model only a few times ($M = 3$), but each time a whole lot. For instance, the reliability range rectangles shown for *smooth-txtr* is about the same in both cases.

This similarity was also seen for RMSEP(Y) in Figure 10.6, and illustrates that when replicates are present as samples in a data set, the leave-one-out cross-validation is similar to the between-replicates cross-validation.

*Systematic differences between replicates:* For two of the Y-variables, in particular *%MILK* and *%SUGAR*, the bi-linear model seems to be sensitive to between-replicates variations ($M = 3$). This indicates that the products may differ for the three times they were produced. However, this is confounded with another explanation: over time, the sensory assessors may have changed their perception, or their description, of the milk/sugar variations. That is not unexpected, and that is why sensory experiments are usually designed so that each replicate includes the full set of treatments.

Much of this time-drift may be removed by replicate effect-correction, i.e. mean-centring each replicate separately, instead of mean-centring all the samples together.

Effect-correction (cf. Appendix A4.5) of e.g. replicates may be done by preliminary PLSR modelling, with the three replicate indicators *r1, r2* and *r3* (see eq. (8.3)) as **X** and the response variables to be corrected, as **Y**. The Y-residuals $F_A$, at maximum possible rank (here $A = 2$, since *r1 + r2 + r3 = 1*), represent the drift-corrected response variables. These $F_A$ residuals may replace the original values for the response variables, for use in **X** or **Y** in subsequent *normal* BLM.

Appendix A10.1 also outlines other, more advanced choices of cross-valida-tion segmentation to test linearity etc.

## 10.6. WHEN AND HOW

### 10.6.1. When to Validate

This chapter has outlined one approach to model validation, the graphically oriented cross-validation with jack-knifing of the model parameters.

The same approach is applicable for a wide variety of bi-linear modelling applications. These range from multivariate calibration of instruments, purely explorative analysis of empirical observations by PCA or PLSR, via classifica-tion/discrimination, to the analysis of designed experiments and dynamic processes.

### 10.6.2. How to Validate

The expert's interpretation in light of contextual background knowledge is the most important validation tool. It is most effectively implemented by following the rule *Plot A Lot*.

The second most important validation tool is to ensure sufficient predictive ability and model stability. In this book we advocate the use of one statistical tool for this: cross-validation with jack-knifing of the model parameters. These may be extended to formal statistical hypothesis testing (Chapters 11 and A16), but rough graphical reliability assessments are usually enough.

In the discovery process, focus should be on the *signal*, not on the *noise*. But it is important to avoid being fooled by noise. In the present context that primarily means to avoid overfitting, not accepting too many PCs in the final model.

The essence of model validation in this book may be summarised by the following maxime:

No interpretation without predictive ability
No prediction without interpretability

## 10.7. TEST QUESTIONS

1. What is the relationship between cross-validation and jack-knifing?
2. Why are the model perturbations in jack-knifing only *partial*?
3. Can the model fit error RMSEC(Y) be used for model validation?
4. How can cross-validation be used for assessing different levels of validation?

## 10.8. ANSWERS

1. Jack-knifing is part of the cross-validation, because it analyses the purposely deprived sub-models used by cross-validation.
2. The $M$ sub-models in jack-knifing share most of their calibration samples with each other, and hence their model parameter estimates are not independent of each other. Their perturbations are therefore only *partial* (except in the case of $M = 2$, the split half cross-validation). But by *summing* their perturbations, instead of averaging them, this lack of independence is compensated for.
3. No.
4. By choosing different ways to segment the available calibration samples.

# Chapter 11

# Experimental Planning Y? X!

*In this last method chapter, the importance of experimental planning is pointed out. Different planning approaches are compared. These include explorative designs based on natural (random) or stratified sampling, factorial designs for controlled experiments, and principal properties designs. Some basic factorial designs are explained. Finally, a general method for comparing alternative designs with respect to statistical power is explained and illustrated.*

## 11.1. INFORMATIVE DATA NEEDED, BUT HOW CAN WE GET IT?

### 11.1.1. Good Data is the Basis for Good Results

The importance of relevance and reliability in research data was discussed in Chapter 2.2.

Any data set may be analysed, as long as it can be put into a data table. But the quality of experimental data, and the odds for finding relevant and reliable information, may be greatly increased by experimental planning.

This is true for modelling of all kinds of empirical data, ranging from botanical observations, via medical records, psychological questionnaires or microbiological laboratory experiments to advanced electronic measurements or industrial process observation.

### 11.1.2. Experimental Design

Experimental design concerns choosing which *samples* to study. It is part of the more general experimental *planning*, which also includes other important topics: primarily the selection of which *response variables* to measure, and under which *conditions* to do the measurements.

Figure 3.1 showed the importance of basing the choice of samples and variables on the project purpose as well as on prior knowledge. Figure 3.2 outlined the important role of experimental planning in the research cycle (*Step II*).

The present chapter outlines some basic principles for experimental design, sufficient for the reader to start using, e.g. the experimental design part of commercial data analysis programs.

More introductory information about experimental design and the analysis of designed experiments may be found in Esbensen (2000). Technical details on planning of multivariate calibration experiments are given in Martens and Næs (1989). Further relevant design information may be found in Box et al. (1978) and Massart et al. (1997).

### 11.1.3. What is a Sufficiently Informative Data Set?

*On one hand, the experiment should not cost more than necessary:* The data generation process (experiment or data collection) must not be more expensive than necessary. This means that the number of samples, the cost of instrumentation and the amount of analytical labour must be kept down.

*On the other hand, the data must contain sufficient information:* Some ways of doing experiments give much higher chance of getting informative data than others. Ideally, each data set ought to have samples and variables that reflect all of the *relevant phenomena* (main types of variation) in the system, at least all of the *phenomena* that affect the chosen response-variables.

Each such *phenomenon* should be represented at *several levels* within a *variation range* that is sufficiently wide, yet relevant for the given purpose, and should be *spanned independently* of the other phenomena. Moreover, the observations should have *sufficient precision* compared to the sampling- and measurement-uncertainties. Finally the sample set should contain *sufficient redundancy* to allow reliability assessment by cross-validation.

There are several ways to design an experiment in order to balance the cost and benefit. Some are rather risky, others are very generally applicable, but expensive, and others yet require the ability to do controlled experiments, but may be very cost effective.

### 11.2. DIFFERENT APPROACHES TO EXPERIMENTAL PLANNING

❑ No Explicit Design: Analysis of Available Data

Sometimes, the practical data analyst may have to work with whatever data is available. This lack of design may be due to a preliminary nature of the project, to the fact that the design of the data is unknown, or just due to lack of design experience.

The multivariate data analysis may still be worthwhile, as a purely explorative endeavour. But without explicit design, the risk increases for not being able to find anything.

The example in Chapter 14 represents a case where the design of the calibration set was unknown initially, and found to be inadequate after the analysis. But

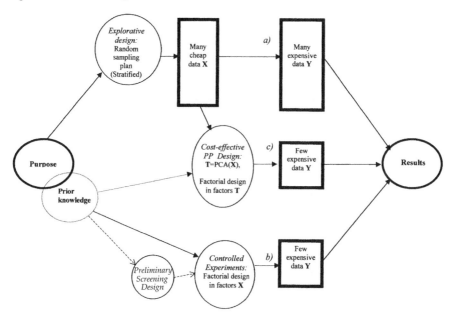

**Figure 11.1.**    Experimental design: three alternative approaches for planning how to get informative **X** and **Y** data.

by combining it with another sample set, the system could be spanned reasonably well.

□  EXPLICIT DESIGN APPROACHES

Figure 11.1 shows that there are several routes from project purpose via designed experiments to results. It outlines the three routes to be discussed and used in this book: (a) explorative designs, based on random or *natural* sampling, (b) controlled experiments with factorial designs, and (c) cost-effective principal properties (PP) designs, which is a combination of (a) and (b).

In the figure, the letter **X** symbolises the *given*: some easily measured properties, or known values of some controlled experimental factors. The letter **Y** symbolises the *observed*: usually expensive response-variables that presumably are relevant for the given purpose.

Once **X** and **Y** data have been obtained, the **X–Y** relationship may be studied by the BLM data analysis, with the validity of the results assessed by cross-validation and graphical interpretation. If so desired, this relationship may later be used in order to predict **Y** from **X** in new samples of the same general kind.

Which road to take, depends on several things, e.g. on the amount and type of prior knowledge, and the need for reducing the number of measurements, as well as the feasibility of performing controlled experiments and the real-world relevance of their results.

### 11.2.1 Explorative Design: Stratified Random Sampling

This approach (top of Figure 11.1) is taken if very little prior knowledge is available about the system at hand, or if there are variation phenomena that can neither be controlled nor easily observed.

In its simplest version, a representative and diverse sample set is chosen, more or less at random, in light of the general purpose of the project. The advantage of such random sampling is that naturally occurring samples may be used, and that it may be done without a lot of prior knowledge, at least for explorative purposes. But the disadvantage is that a lot of samples is usually required before the data become sufficiently informative.

> Sometimes it is difficult to define what is *representative*, because we often have trouble defining the target *population* to be represented. Moreover, due to unknown heterogeneities in this postulated *population*, we may run into serious sampling problems, as pointed out by Gy (1998). Therefore, if random sampling is to be used for developing an important quantitative model, e.g. for paying according to predicted quality, then a professional statistician should be consulted for the experimental design.

Easily obtained variables **X** are measured or otherwise obtained in the many samples, in order to support the data analysis. In addition, the important, but possibly expensive qualities **Y** have to be obtained in the same many samples.

*Stratified random sampling:*    If prior knowledge is available about the naturally occurring samples, this may be used for making the random sampling *stratified* and thereby more informative with fewer samples (Robson, 1993).

Various types of known variations are forced into the sampling plan. Thereby it is possible to ensure that, e.g. both men and women, or both high-and low-temperature measurements etc., are included in the sample set.

However, the knowledge about the known types of variation may be incomplete or erroneous, and there may be additional unknown variation types. Therefore there will almost always be additional sources of variation that create *sampling errors*. It is advisable to supplement the stratified sampling with a generous random sampling within each of the chosen stratification levels. This can also provide the necessary replication needed to cross-validate the conclusions.

Multivariate calibration of, e.g. near-infrared instruments is a typical application of stratified random sampling. In Chapter 12 it is used in order to ensure that wheat samples from different types of wheat plants, different growth conditions and different years were collected.

*Not enough natural variation?:* In some cases, stratified random sampling simply cannot provide sufficient variation to enable sensible data-driven modelling. This is for instance often the case in the process industry, when a process runs under tight control.

If it is important to learn how to optimise the process, then the normal process situations are not informative enough. It is necessary to generate more variability by conscious experimentation, unpopular as that may be with some process engineers. As usual, the chosen experimental conditions should not be so extreme that the results become irrelevant.

### 11.2.2 Design of Controlled Experiments

To do a *controlled experiment* means to control the level (value) of various design factors **X** that are suspected to affect the chosen response variables **Y**.

To elucidate causal relationships it is necessary to perform controlled experiments. The advantage of this controlled design approach is that if the prior knowledge is good, then the actual experiments can be made quite small while still being very informative. Hence, if the samples themselves, or the response measurements, are expensive, well-designed controlled experiments can be very *cost-effective*.

But the disadvantage is that controlled experiments may become rather unrealistic, if done in over-simplified model systems or at irrelevant levels of the experimental factors.

*Don't vary just one factor at a time!:* The traditional meaning of a *controlled scientific experiment* was to keep everything as constant as possible, and vary one experimental factor at a time. That makes it very easy to think about the design, and the results can be analysed on the back of a napkin.

But this traditional approach is not recommended, because it is very inefficient for discovering interactions between different experimental factors. It is certainly not more *scientific* than a well-designed factorial experiment!

*Factorial designs: systematically vary several factors:* Multivariate factorial designs are much more efficient: they span the information space efficiently, they can reveal interactions, and give a statistical *"multiplexing" advantage* and therefore reduce the total number of samples needed. With user-friendly BLM software, the analysis of factorial designs is almost as easy as the napkin-analysis above, and much more informative.

The bottom of Figure 11.1 shows that, based on the purpose of the experiment, prior knowledge is used for developing a design for a controlled experiment. This plan is then used for selecting or producing the samples to be analysed. The known, controlled design factors may be called **X**. They may be of continuous nature (e.g. age of people, or temperature of a process) or of discrete, categorical nature (e.g. gender of people, or compound *A*, *B* or *C*).

To ensure a cost-effective, informative experiment, various combinations of the design factors **X** are set up and defined as an experimental design plan. In accordance with this plan, the experiment is then performed, and the chosen response-variables (**Y**) are measured. The multivariate data analysis finally reveals how the response-variables **Y** relate to each other and to the design factors **X**. More detail on factorial design of controlled experiments is given below. It is illustrated in Chapters 8–10, in the present chapter (mixture design) and in Chapter 16 (factorial design).

*Designed preliminary screening experiments:* It is usually advantageous to perform some preliminary screening experiments before the main experiments, to remove seemingly uninteresting design factors and to ensure relevant variation ranges of the remaining ones. An example is illustrated in Chapter 16.

### 11.2.3 Principal Properties (PP) Design

The middle of Figure 11.1 shows that availability of cheap, multivariate but non-selective information **X** provides an interesting combination of the best of the controlled and the random design approaches.

In the analysis of technical or biological qualities, PP design is becoming increasingly important. It may also be used, e.g. in marketing research, for selecting a small but representative and informative set of people, for expensive, in-depth interviews.

A PP design process consists of

1. Making fast and cheap multichannel measurements **X**, or collecting literature data, in a large set of samples, selected by natural (random) sampling, with more or less stratification.
2. Summarising these X-data by PCA, extracting the main variation-types among the samples. If the X-variables are sufficiently informative, the first few PCs, $\mathbf{T} = [\mathbf{t}_1, \mathbf{t}_2, \ldots]$ will probably span all the important types of variation among these samples.
3. The first few PCs are then to be spanned, as if they represented design factors in a controlled experiment: a few of the samples are chosen to span the sub-space of the first few PCs, more or less in accordance with a chosen factorial design as in Figure 11.1b).
4. The small set of samples is submitted to the more expensive measurements **Y**.

A small example of PP design was illustrated in Chapter 7. The pioneering application of the PP design in chemistry was done in quantitative structure activity relationship (QSAR) research (Wold et al., 1986).

### 11.2.4. Good Experimental Practice

Irrespective the choice of experimental design, there are some general *good practice* rules which should be observed, with respect to how the actual experimental work is carried out. Two of the most important of these are *randomisation*, and systematic recording of *unexpected* events.

❏ RANDOMISE THE ORDER OF OBSERVATIONS

There will usually be sources of systematic variation that we do not know of *a priori* or cannot afford to investigate under controlled conditions. Unexpected drift in the analytical instruments during the experiment is one example of this. The effect of increasingly tired investigators explaining a questionnaire to respondents is another.

Unless handled correctly, such systematic variation-sources will create systematic *alias* errors, which could cause us to draw erroneous conclusions.

It is therefore good experimental practice to try to neutralise such unknown error sources as well as possible. This may be done by randomising the order of the actual measurements, so that the uncontrolled variations are equalled out, or at least somewhat decoupled from the design factors that we want to study.

❏ SYSTEMATICALLY RECORD THE UNEXPECTED

Uncontrolled and unexpected phenomena may become evident during an experiment. It is a good idea to record them systematically, based on the experimenter's own senses.

For hundreds of years, discovery in the natural sciences was based on simple human observation. The present fear of *subjective* human observation came relatively recently, after the advent of so-called *objective* measuring devices. While it is true that a camera may be an *objective* device, someone is usually holding it and choosing when to snap its button. Even when a camera operates automatically, it is smart if the owner make notes of unexpected changes in the light setting.

The researcher's own observations may be systematised, in analogy to how verbal descriptors and scales are used routinely in e.g. sensory science (cf. Appendix A2) and consumer studies. They may then be included as additional variables in the data analysis, and help explain otherwise strange results.

### 11.2.5. Design Approaches Used in this Book

Various methods to ensure that the sample set for a future experiment is informative, or at least has the *potential* for being informative were outlined above.

The use of the different design strategies in this book are shown below. The authors consider these strategies suitable for do-it-yourself data analysis, as long as the users try to understand what they are doing, apply their common sense and ask for help when in doubt.

---

Experimental designs used in the present book

*No explicit design: analysis of the available data*
Toxicity of polluting aromatic molecules (Chapter 14):
  Selection of molecule species: those with literature data available
Process quality monitoring (Chapter 15):
  Selection of samples: A given sugar plant for the first part of a production season in a given year

*Stratified random sampling (a, in Figure 11.1)*
Quality concept study (Chapter 2.3):
  Formal stratification between three countries
  Informal, random selection among scientists within each country
Protein quality of wheat by NIR (Chapter 12):
  Informal stratification between three genotypes and several growth locations
  Informal, random selection among available wheat samples within each
Working place quality (Chapter 13):
  Formal stratification between 34 company departments
  Informal, random (?) subset of employees responding

*Principal properties design (c)*
Simplification of LITMUS calibration (Chapter 7):
  Informal selection of samples for OD measurements
  Informal reduced design in the latent variables from OD, for chemical analysis
Consumer study of cocoa-milk drinks (Chapter 3.5)
  Formal controlled design for sensory analysis
  Informally reduced controlled design in the latent variables from sensory analysis, for pilot consumer study

*Factorial design (b)*
Microbiological food quality: mould growth in different environments (Chapter 16):
  Factorial $2^4$ screening design, extended to formal central composite design for response surface modelling
Sensory quality of cocoa-milk drinks (Chapters 1.4, 10–11):
  A reduced three-factor mixture design at two levels of texture stabiliser, produced and analysed in three full, independent replicates

---

However, we repeat that when planning dangerous or particularly costly experiments, or experiments where some parties have strong vested interests, it is advisable for the data analytical novice to leave the experimental design and data analysis to a professional statistician.

## 11.3. FACTORIAL DESIGNS FOR CONTROLLED EXPERIMENTS

### 11.3.1. Quality Criteria for Factorial Designs

A list of quality criteria for factorial designs of controlled experiments is given below. The concrete realisation of these criteria depends on the type of design involved. A plan which seriously fails to satisfy one of these cannot be considered a good enough design.

Quality criteria for an experimental design

1. The experimental design is *understandable*
2. It utilises the *limited resources* well
3. It spans *all experimental factors* that need to be investigated
4. Each factor is spanned *within its relevant level range*, not outside
5. Each factor is *spanned clearly*, to detect its main effect
6. Each factor is used at sufficiently *many levels*, to detect non-linearities
7. All important *pairs of factors* are used, to detect interaction effects
8. Each factor is spanned with sufficient *redundancy*, to allow cross-validation
9. The design has low enough statistical *risk*, i.e. good enough *power*

There is no such thing as *the best* design because the quality of an experimental design is based on several criteria: relevance criteria, cost criteria, statistical criteria, etc. Still, some designs are better than others.

One way to arrive at a *good enough design* is to set up a few design *alternatives*, compare them with respect to these criteria, and choose the one that appeals the most.

### 11.3.2. What Can the Factorial Designs Reveal?

Factorial experimental designs help us to find the effects that our design factors **X** have on our chosen response variables **Y**. More specifically, the designs can provide assessment of *main effects*, *curvatures* and *interactions*, and find interesting outliers.

Analysis of controlled experiments.

*Confirmative analysis* of factorial design experiments: in order to see how the Y-variables have responded to the design variables, the response data **Y** may be related to the design factors **X**: this use of PLSR is referred to here as *APLSR* (cf. Figure 3.4). *Explorative analysis* of factorial design experiments: if several response variables have been measured, then it may also be useful to swap the definition of **X** and **Y**. This allows us to check if the response variables (now used as **X**) show systematic unexpected correlation patterns beyond their ability to predict the design factors (now used as **Y**). This is referred to here as *DPLSR* (cf. Figures 3.4 and 9.5a).

A *main effect* is the linear relationships between a Y-variable and an X-variable. A *curvature effect* shows the relationships between a Y-variable and the squared version of an X-variable. An *interaction effect* shows the relationships between a Y-variable and the product of two or more X-variables.

*Outliers* in controlled experiments arise when the response variables **Y** cannot be adequately modelled by the design variables **X** in some samples. These deviations may be due to human mistakes or abnormal measurement errors. But they may also represent particularly interesting higher-order interactions or other unmodelled phenomena of great interest. Therefore, outliers should always be pursued, before they are corrected or deleted/declared as *missing values*.

Different factorial designs give different abilities of revealing the main effects, curvatures, interactions and outliers: that depends on the number of design factors and the number of levels in each design factor.

In screening designs the number of different levels per design factor is usually low. On the other hand, the number of different design factors may be relatively high, in particular for the more advanced types of screening designs.

### 11.3.3. Some Alternative Factorial Designs for Controlled Experiments

Figure 11.2 shows some factorial designs suitable for controlled experiments. They are illustrated in the 3D case, where there are three design factors, $X = [x_1, x_2, x_3]$. The sub-plots (a)–(c) represent *screening designs*, while sub-plot (d) represents a *response surface design*.

❑  WHAT ARE THESE DESIGN FACTORS $x_1$, $x_2$ AND $x_3$?

The *design factors* represent three variables that can be more or less accurately controlled at different levels. The purpose of the experiment is then to see if they have any impact on the response variables.

As example, consider a consumer study. The three design factors could be *age, gender* and *income group*. In experiments on the fuel efficiency of a car engine, the design factors could be the fuel *injection rate*, fuel *temperature* and *compression*. In food microbiology experiments, they could be the *pH, moisture* and *temperature* of the food products. Chapter 16 gives a worked-through example of a controlled food microbiology experiment.

Sometimes, some of the design factors simply represent uninteresting, but unavoidable error sources, to be assessed and corrected for during the data analysis. One example of this is the use of consecutive production replicates in sensory analysis (Chapter 10).

Taguchi (1988) has contributed greatly to the practical robustness of designed experiments, by insisting on including conscious perturbation factors. This allows industrial engineers to find *good, stable* production conditions that can be trusted, instead of *optimal*, but *unstable* conditions that may be more impressive, but cannot be trusted. However, robust experimental design is beyond the scope of this introductory book.

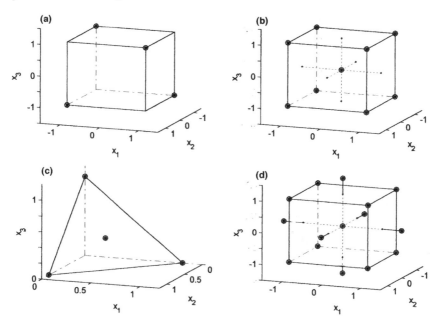

**Figure 11.2.** Some factorial designs for controlled experiments. Some ways to combine three design factors $x_1$, $x_2$ and $x_3$. (a) Fractional factorial $2^{3-1}$. (b) Full factorial $2^3$ with a centre point. (c) Simple mixture design, where $x_1 + x_2 + x_3 = 1$. (d) Response surface design: central composite.

Each different combination of design factor levels is called a *treatment* or an *experimental condition*. If a treatment is repeated, we expect to find the same responses, apart from sampling and measurement errors. Design factors **X** and any derivation thereof (e.g. square terms and interactions) are here collectively referred to as *design variables*.

☐ SCREENING DESIGNS

If the number of design factors $K$ is high, then an *all-combinations* factorial design creates a *combinatorial explosion*. This is so, even when there are only two levels of each design factor (i.e. a so-called $2^K$ design). It is even more so if there are several levels for each design factor; then the number of different treatments can really rise dramatically, making the cost of the experiment unacceptable.

*Fractional factorial design: $2^{K-L}$:* It is possible to bring down the size and cost of controlled experiments by so-called *reduced* experimental designs. The simplest reduced design is the so-called *fractional factorial design*.

This is in Figure 11.2a illustrated by the $2^{K-L}$ design for $K = 3$ two-level factors, where the number of samples is reduced by the factor 2, making it a $2^{3-1}$ design. The figure shows how the three design factors are spanned by only four samples, just like a dice can be held firmly by four of its corners.

Fractional factorial designs are very useful for screening experiments (Figure 11.1, bottom). But they create *confounding* patterns that may be a little confusing in the beginning. Therefore they are not used in this introductory book.

*Full factorial design:*    The upper right subplot of Figure 11.2b represents the $2^K$ full factorial design with *centre* point. $K = 3$ design factors $x_1$, $x_2$, $x_3$ are spanned by two levels each (high and low level), symbolised by the arbitrarily re-scaled values $+ 1$ and $- 1$. Together they create eight ($2^3$) samples.

It is called a *full* factorial because *all combinations* of factors are used. This full design allows the detection of the main effect of each design factor, as well as their interactions.

The extra centre sample in the middle may detect possible *curvature* in the how the measured Y-response(s) depend on the design factors.

> The centre-point is sometimes replicated a couple of times and the standard uncertainty between these replicates is used as an assessment of the repeatability of the system (cf. 5a and 5b in Figure 8.2). This is useful. But one should be aware that the standard uncertainty estimated from only these few centre-samples is in itself highly uncertain (cf. Chapter 16).

*Screening mixture design:*    Mixture designs constitute a particular type of factorial design. In mixture designs, two or more design factors are connected through some natural restriction (*closure*). The most common restriction is that the design factors *sum up to a fixed value*, e.g. 1.0 or 100%: When one factor goes up, the sum of some other design factors necessarily goes down. This closure has to be handled during the modelling and interpretation. More information on mixture designs may be found in Esbensen (2000): details are treated in Cornell (1981).

The mixture design shown in Figure 11.2c is so simple that it really represents a screening design. Like the full factorial design, it has a centre point that allows detection of curvature. But like the full factorial design, this one centre point cannot reveal *which* of the design factors are responsible for the non-linearity.

❐   RESPONSE SURFACE DESIGNS

Response surface designs are a little more ambitious than screening designs, but also more expensive. They span each design factor by three or more observations, in such a way that the design makes it possible to reveal details about curvature, and has enough redundancy to allow cross-validation.

*Central composite design:*    It is often interesting to extend a simple two-level factorial design like the $2^3$ (Figure 11.2b), to a slightly more extensive factorial design, in order to get more detailed information about how the responses depend on the design factors.

This allows us to develop a so-called *response surface model* by regressing the response variable(s) **Y** on the design variables **X**, including both the design factors and their square- and interaction terms.

Figure 11.2d illustrates one design suitable for response-surface modelling, the *central composite* design. In this case the $2^3$ design with nine treatments (Figure 11.2b) has been extended with six so-called *star* points. With a total of 15 treatments, this design allows more precise estimation of the main effects, square terms and interactions.

*Polynomial representation of curvatures and interactions*: The linear model summary for main-effects for the $K =$ three individual design factors may, as in eq. (6.1d), be written:

$$Y = b_0 + x_1 b_1 + x_2 b_2 + x_3 b_3 + \mathbf{F} \tag{11.1}$$

where regression coefficients $\mathbf{b}_k$, $(k = 1,2,...,K)$ are now called the *main* effects to be estimated. To obtain explicit modelling of possible response *curvature*, this may, for each design factor $k$, be done by defining, as a new X-variable, the centred square term $\mathbf{x}_{K+k} = (\mathbf{x}_k - \bar{x}_k)^2$ and by including it in eq. (11.1) in the term $\mathbf{x}_{K+k} \mathbf{b}_{K+k}$. Likewise, the *interactions* between two design factors $k$ and $m$ may be studied, by defining, as new X-variable number $n$, the centred product term $\mathbf{x}_n = (\mathbf{x}_k - \bar{x}_k)(\mathbf{x}_m - \bar{x}_m)^2$ and by including it in eq. (11.1), in the term $\mathbf{x}_n \mathbf{b}_n$ (cf. eq. (8.4) p. 389). The use of such second-degree polynomial modelling in BLM of a factorial design is illustrated in Chapter 16. It is the basis for successful modelling in Chapter 14.

*Non-metric representation*: An alternative to the polynomial representation is the use of indicator-variables (0/1 or $-1/1$) to represent the individual treatments. Thereby, no assumptions about linearity are made at all. For instance, in the COCOA example, the different mixtures 1–7 may be represented by seven indicator variables. This was in fact done by the object indicator variables in Figure 8.3, although there they were just passively fitted to the model instead of actively being used in **X** or **Y**.

*Response surface mixture design:*   Figure 11.3 shows the basic design of the COCOA data set used in Chapters 8–10. It represents a mixture design with sufficiently many different levels of each design factor that it qualifies as a response surface design. This design itself will shortly be assessed in more detail.

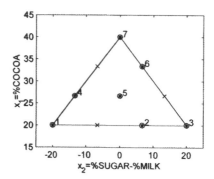

**Figure 11.3.** A full and a reduced mixture design. Mixture design for three design factors
(*%COCOA*, *%SUGAR* and *%MILK*), plotted in 2D for $x_1$ = *%COCOA*, $x_2$ = *%SUGAR* −
*%MILK*. The sum of the three design factors (*%COCOA*, *%SUGAR* and *%MILK*) equals
100%. Only samples 1–7 were actually produced and analysed, because three samples (×) had
to be skipped during experimentation.

## 11.4. WHICH DESIGN IS GOOD ENOUGH? POWER ESTIMATION

### 11.4.1. Cost-benefit Comparison

To make an experimental design for a controlled experiment requires the three
steps:

1. Set up various design *alternatives*, with different number of design factors,
   different levels of the design factors, or different number of replicates.
2. *Compare* the alternatives with respect to cost vs. benefit.
3. *Choose* the best of the alternatives as plan for the actual experiment.

The *cost* of an experimental design is generally a function of the number of
independent treatments and replicates. The *benefit* of a design reflects the sum of
the quality criteria summarised in Section 11.3.1. The last criterion, the statistical *power*, is the ability of the design to reveal interesting structures while at the
same time guarding against the effect of random input errors.

Which design would then be *good enough* for your purpose? That of course
depends on what you want to attain, and on the problem at hand. But it also
depends on how quickly you can finish one research cycle (Figure 3.2) and verify
the results in the next one, in a constructive research spiral. It even depends on
who is going to pay for the experiment!

Some ways in which *lack* of proper design often leads to bad experiments that
*fail* to give satisfactory information are:

1. *Lack of power:* An under-dimensioned design, having far too narrow span in the
   design factors, far too few treatments or far too little replication

2. *Analytical over-kill:* An over-dimensioned design, having far too many treatments or far too many replicates
3. *Irrelevant data:* A design with trivial, non-representative selection of samples, or with unrealistically wide span in the design factors

### 11.4.2. Planning to Deal with Uncertainty

Different experimental designs carry different costs and bring different benefits. Common to all designs is that they have to give adequate protection against uncertainty in the input data.

Uncertainty in the input data creates uncertainty in the estimated model parameters, which in turn creates uncertainty in our decisions and conclusions. In some experimental designs, with well-spanned experimental factors and many independently observed samples, the uncertainty in the input data may be quite high, and yet be averaged out to be rather low in the resulting model parameter estimates.

But when choosing low-cost designs, characterised by the use of few samples, the modelling results may become very uncertain. It is important to assess this model uncertainty. Otherwise we have no way of knowing what kind of risk we would be running with the alternative designs, and we would have no way of making a rational choice of which design to use in practice.

### 11.4.3. Example of Optimising a Design *Prior* to the Experiment

❑ THE COCOA MIXTURE DESIGN

Figure 11.3 shows the three-component mixture design, which was used for generating the data used in Chapters 8–10. There are three dry ingredients that constitute the powder mixtures, %COCOA, %SUGAR and %MILK, and their sum is always 100%.

A total of ten mixtures were originally intended to be produced and characterised, in three independent replicates, by the sensory, analyses. However, due to unexpected problems with the number of samples that could be analysed in the sensory laboratory, this number had to be reduced to only seven mixture treatments, named *1–7*. To retain the balance in the unexpectedly reduced design, one of the samples was dropped along each of the edges of the design triangle, as symbolised by ✕ in Figure 11.3.

When this problem became evident, the questions immediately arose: with only seven treatments instead of the intended ten, would it still suffice to produce and analyse only three replicates, or did the loss of treatments require an increase in the number of replicates? Is it possible to assess this prior to the experiment, before it is too late, or do we just have to hope for the best, and see what result the actual experiment brings?

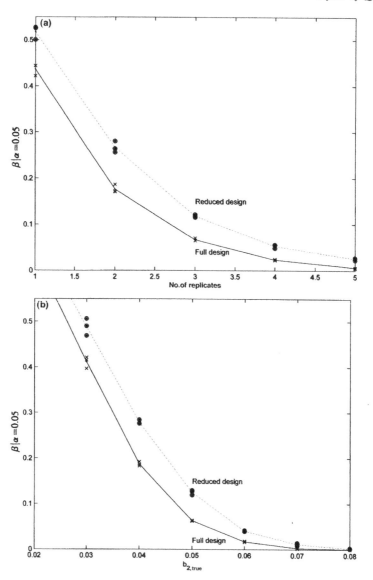

**Figure 11.4.** Power estimation for the mixture designs. The risk $\beta$ of overlooking a ''true'' effect $b_{2,\text{true}}$, the effect of $\mathbf{x}_2 = \%SUGAR - \%MILK$, assuming that the standard uncertainty of the input response, $s_{\text{err}}$, is 1 s.u. Solid curves, the original, full mixture design with ten treatments. Dotted curves, the final, reduced design with only seven treatments. (a) Risk vs. the number of replicates used, assuming $b_{2,\text{true}} = 0.05$. (b) Risk vs. the size of the assumed $b_{2,\text{true}}$, using three replicates. The risk $\beta$ of committing Type II error was estimated by Monte Carlo simulation at a risk of committing Type I error (i.e. the significance level) of $\alpha = 0.05$.

In brief, Figure 11.4 demonstrates that it is indeed possible to assess designs before they are used, by estimating their statistical *power*: when comparing the ten- and seven-treatment experimental designs, the conclusion is that even with only seven treatments, it is sufficient to use only three replicates. Moreover, it indicates that with three replicates and for the noise level expected in this type of data, we should be able to reveal *"true"* effects as small as two sensory units, when the most important design variable changed over its full range in the design. How this was done will now be outlined.

## ❑ POWER ASSESSMENT IN THIS BOOK

The power of an experimental design is a measure of its ability to distinguish real effects from noise. Traditionally, power estimation has been done with rather advanced statistical theory, or with *black box* look-up tables that are quite difficult to understand.

For this book, a simpler method has been developed, based on contemporary methodology for *Monte Carlo* (MC) computer simulation. This is a generic technique that may be used in the same way for all the BLM applications in this book, but it will only be demonstrated in design assessment for controlled experiments.

The MC method consists in *simulating* experiments in the computer according to various alternative designs, and comparing the results. Thereby, we can assess the designs before they have been used in practice.

The *simplicity* of the method used (Martens H. et al., 2000) comes from the fact that the simulated data are modelled just like the data from the real experiment will be modelled. There is very little theory involved. The simulation technicalities is just a software issue, representing just another *button* in the user's program menu. A similar method is employed in Lundahl (1992).

But certain *conceptual complexities* cannot be avoided: we have to make some *assumptions* about the future; otherwise the simulation cannot foresee how the designs will behave. We also have to face some realities concerning *risk*.

These assumptions and risks are being dealt with, one way or the other, every time an experimental plan is made. The following will explain how to do it explicitly instead of just intuitively.

## ❑ THE DESIGN ALTERNATIVES TO BE COMPARED

The reader is now invited to envision the situation just when the analytical capacity problems became evident, i.e. *before the actual experiment has taken place!*

The two mixture designs with ten and seven treatments (Figure 11.3) are now to be compared, for various numbers of production *replicates*.

❑  WHAT THE DESIGNS WILL BE COMPARED FOR

Various designs will be compared with respect to their abilities to reveal interesting effects, at the general uncertainty level expected in the input data. The comparison will be rather relaxed. However, some statistical terms are needed:

The ability of a design to reveal effects is estimated as the statistical *power* $1 - \beta$ when tested at a chosen *significance level* $\alpha$. More specifically, $\beta$ is the risk of giving *false negative* results (causing us to overlook a real and interesting effect, i.e. committing a *Type II* error) at a given risk $\alpha$ of giving *false positive* results (allowing ourselves to be fooled by noise: a *Type I* error).

The trade-off between the two types of risks $\alpha$ and $\beta$ is fundamental. We here intend to accept a risk of *false positive* results of roughly 5% ($\alpha \approx 0.05$). At this level of significance, we want the risk of *false negative* results to be reasonably low, say $< 15\%$ ($\beta < 0.15$). The probability of not committing Type II errors, $1 - \beta$, is the statistical power.

### 11.4.4. Assumptions to be Made *Prior* to Power Estimation

❑  HOW THE REAL DATA ARE GOING TO BE ANALYSED IN THE FUTURE

*The estimation of the effects of the design variables:*   Once the actual experiment, some time in the future, has been performed, the data will in this case primarily be modelled by APLSR: measured response variables will be defined as **Y** and regressed on the design variables $\mathbf{X} = [\mathbf{x}_1, \mathbf{x}_2]$ as in Figure 11.3.

As usual, we write the linear summary model (eq. (8.4)) for one Y-variable

$$\mathbf{y} = b_0 + \mathbf{Xb} + \mathbf{f}$$

Since **X** here has full rank, we shall use the bi-linear modelling with 2 PCs to obtain the estimated effects, $\hat{\mathbf{b}}$.

*Reliability assessment of the estimated effects: the decision criterion:*   The cross-validation/jack-knifing (Chapter 10) will provide an estimate of the standard uncertainty of each observed effect $\hat{b}_k$, $\hat{s}(\hat{b}_k)$. On this basis, the reliability ranges $\hat{\mathbf{b}} \pm 2\hat{s}(\hat{\mathbf{b}})$ will be drawn, as illustrated in e.g. Figure 10.4d. For simplicity we just focus on one element in **b**, $b_k$, corresponding to the X-variable $\mathbf{x}_k$ which is considered to be most important.

If the value $b_k = 0$ lies *within* the estimated reliability range $\hat{b}_k \pm 2\hat{s}(\hat{b}_k)$, then we intend to disregard $\hat{b}_k$ as a *negative result*, because it could just as well have been caused by incidental errors **f** in **y**.

There is always a risk of *false positive* results. With sufficiently many samples, and due to the so-called *central limit theorem*, the level of confidence (ISO, 1995a) of the chosen reliability range, $\pm 2$ standard uncertainties, corresponds to $\alpha \approx 0.05$, i.e. accepting approximately 5% risk of *false positives*, which is the level of confidence that we desire. For more details, see Martens H. et al., (2000).

The question then is what the risk of *false negatives* is: what is the probability that in analysing this data set, we shall disregard a real, interesting effect in the data, because its reliability range incidentally includes zero? That risk cannot be assessed from this one future data set alone. But it can be assessed here and now by simple computer simulation, long before the actual experiment has been performed:

*How the MC simulation results are going to be analysed:* Precisely the same regression modelling will be used in the present MC simulation as for the future analysis of the real data. But it will be *repeated thousands of times,* each time on a new *artificial set of data,* **y**. Thereby, we simulate the type of uncertainties which we expect to see in the data **y** from the real experiment, with chosen, *"true"* values of **b** and different random errors **f**.

This will be done for each of the designs to be assessed. The results will then be compared by a simple counting of the relative number of times we commit Type II errors, $\beta$, at the chosen risk of committing Type I errors, $\alpha \approx 0.05$.

### 11.4.5. Power Estimation in the COCOA Example

☐ REMOVAL OF CLOSURE IN **X** DUE TO THE MIXTURE DESIGN

The components in the present mixture design sum to 100%. As demonstrated in Chapter 8, this will be no problem in the data analysis, once the empirical data from the actual experiment arrive. But in order to simplify the explanation of the power assessment, let us first remove this closure.

The three-component design could have been drawn in 3D, as in Figure 11.2c. Instead, in Figure 11.3 it has been reduced to 2D. The first X-variable (ordinate) is the first design factor: $x_1 = \%COCOA$. But the second X-variable (abscissa) is now defined as the difference between the other two design factors: $x_2 = \%SUGAR - \%MILK$. The new X-matrix $\mathbf{X} = [\mathbf{x}_1, \mathbf{x}_2]$ has full column rank ($\mathbf{x}_1$ and $\mathbf{x}_2$ are more or less uncorrelated), in both the ten- and seven-mixture designs. This means that **X** in the present example requires a model with 2 PCs.

☐ THE DESIGN X-VARIABLE CONSIDERED AS MOST IMPORTANT

A topic of initial interest is here to see how the product changes when milk powder is replaced by sugar in the cocoa-milk mixtures. Therefore, we choose to focus on how well we can study the effect of $x_2 = \%SUGAR - \%MILK$.

☐ THE RESPONSE Y-VARIABLE CONSIDERED AS MOST IMPORTANT

The rather abstract power assessment becomes simpler if we consider only *one* particular response variable: the one that is most important for our purpose, instead of having to think about all the intended response variables.

Therefore, the sensory variable *colour* darkness is chosen as the variable of primary interest. The MC results will be equally valid for the other variables, as long as they fulfil the assumptions that we have to make.

*The noise level to be expected in the response variable y:* Prior experience with this type of measurement indicates that for our chosen response variable, **y** (sensory *colour* darkness), the standard uncertainty from one single replicate is $s_{err} \approx 1.0$ sensory units (s.u.), on the sensory 0–15 scale.

*The minimum true effect desired:* For the purpose of the project, it was decided that a design will not be considered useful, unless it is able to reveal a sensory change of minimum 2 s.u., when going from the one extreme treatment to the other.

This means that the *colour* darkness should increase more than 2 s.u. when replacing milk by sugar, going from 20%*SUGAR*, 40%*MILK* (sample *1*) to 40%*SUGAR*, 20%*MILK* (sample *3*), i.e. when $x_2$ changes from $-20$ to $+20$, (Figure 11.3). This corresponds to a minimum required effect of 2 s.u./40%, i.e. $b_{2,Min} = 0.05$ s.u./%.

Now we have made the assumptions needed in order to assess how alternative designs will behave when used in experiments, and on that basis choose the best design for our situation.

☐ DESIGN ALTERNATIVES: HOW MANY REPLICATES ARE REQUIRED?

Figure 11.4a shows the approximate risk $\beta$ of overlooking the *true* effect of $b_{2,true} = 0.05$, given the standard uncertainty of the input data **y** of $s_{err} = 1$ s.u. and being willing to accept about 5% risk of being fooled by random error, $\alpha \approx 0.05$.

The lower curve shows the risk $\beta$ for the original, full design (ten treatments), and the upper, dotted one for the new, reduced design (seven treatments), each as a function of the number of production replicates. Each MC simulation was performed three times, with 10 000 artificial input data sets **y** in each.

With the full design, the figure shows that the risk $\beta$ of overlooking the interesting effect $b_{2,Min} = 0.05$ falls from about 0.44 to $<0.005$ when the number of replicates increases from one to five. With three replicates it is about 0.07. This means that the power of this design is 0.93, i.e. it has 93% chance of revealing $b_{2,true} = 0.05$.

With the reduced design the corresponding risk $\beta$ falls from about 0.51 to 0.025; at three replicates it is about 0.12, i.e. a power of 0.88, so we have a 88% chance of revealing $b_{2,true} = 0.05$.

Figure 11.4a shows that both designs require more than two replicates. But while the full design would have been more desirable, even the reduced design has reasonable power with three replicates. On the other hand, with four replicates we get a $\beta$ of about 0.05, which is considered unnecessarily good for this purpose. Using five replicates would certainly represent an analytical over-kill.

Consequently, we choose the reduced design with three replicates to be used in practice.

❐ How Small Effects can we Hope to be Able to Reveal?

Figure 11.4b is similar to Figure 11.4a except that the number of replicates is now kept constantly at three, while the presumed value $b_{2,true}$ is varied along the abscissa. The figure shows, as before, that the risk of overlooking $b_{2,true} = 0.05$ is about 12%.

If the *true* effect is as large as 0.08 (i.e. changing as much as 3.2 s.u. from sample 1 to 3), the risk of overlooking it at this noise level is virtually nil: the power is almost 100%, even for the reduced design.

On the other hand, if the true effect is as low as 0.03 (i.e. changing only 1.2 s.u.), risk of overlooking it is almost 50%!

So we realise that the chosen experimental design cannot be expected to do wonders. But its resolving power seems adequate for our purpose.

### 11.4.6. Choosing the Best Design from the Power Estimation

By MC simulation we have been able to show that the loss of three treatments from the experimental mixture design did lower the power of the design, but it was not a big disaster. Three replicates still seem to be adequate for the ambition level of the project.

If the true effect of $x_2 = \%SUGAR - \%MILK$ on a response $y$ (e.g. sensory *colour* darkness) were 0.05, then the risk of overlooking it with the reduced design and three replicates would be about 12%. If the effect were as high as 0.08, the risk would be reduced to virtually nil, while if it were as low as 0.03, there would be 50/50 risk of not being able to distinguish it from noise effects.

### 11.4.7. Results from the Actual Experiment, Based on the Chosen Design

The reader is now invited forward again in time, to the actual experiment. The reduced mixture design, with three production replicates, has been used as plan for the actual experiment, leading to the results presented in Chapters 8–11.

Figure 11.5 shows the actual results for the relationship between

$$y = \text{sensory } colour \text{ darkness}, \mathbf{X} = [\mathbf{x}_1 = \%COCOA, \ \mathbf{x}_2 = \%SUGAR - \%MILK]$$

using leave-one-sample-out cross-validation ($M = N = 21$ segments, i.e. at the *repeatability* level of validity), for data from file COCOA-III.

Figure 11.5a shows the 21 partially perturbed effect estimates for the two X-variables. It indicates that for $\mathbf{x}_2$ the cross-validation segments show no outliers.

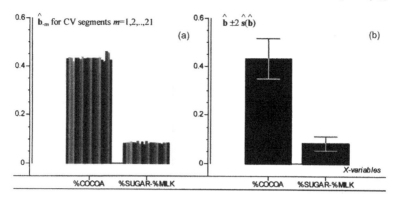

**Figure 11.5.** Results from the actual experiment. Estimated effects of $x_1 = \%COCOA$ and $x_2 = \%SUGAR - \%MILK$, for $y =$ sensory *colour* darkness, $\hat{\mathbf{b}}$, using three replicates of the reduced mixture design in Figure 11.3, i.e. $N = 21$ samples. (a) Partially perturbed effect estimates, $\hat{\mathbf{b}}_{-m}$ for the $M = 21$ cross-validation segments. (b) Full-model effect estimate and its expanded reliability range estimate, $\hat{\mathbf{b}} \pm 2\hat{s}(\hat{\mathbf{b}})$.

Figure 11.5b shows the estimated effects and their reliability ranges $\hat{b}_k \pm 2\hat{s}(\hat{b}_k)$. Design variable $x_2$ is seen to have an effect $\hat{b}_2 = 0.08 \pm 0.03$.

This means that it is positive, in both meanings of the word: the sign of $\hat{b}_2$ is positive, and the effect is clearly non-zero. The reliability range, in fact, is so narrow that the risk that this could be a *false positive* result is virtually nil.

For $x_1$ the segments $m = 19$ and 20 (i.e. the two first replicates for mixture 7, which has the highest cocoa content) we see larger deviations than for the others in estimated effect of $\%COCOA$. This may indicate a non-linearity in how the assessors perceived the upper end of the colour scale in the first two replicates.

☐ CRITIQUE OF THE ASSUMPTIONS USED DURING THE POWER ESTIMATION

The obtained effect estimate, $\hat{b}_2 = 0.08$, is somewhat higher than the minimum, *"true"* effect required during the preceeding experimental design phase, $b_{2,Min} = 0.05$, but the latter lies just inside the estimated reliability range of 0.05–0.11.

The standard uncertainty of prediction, RMSEP(Y) is 1.2, which is also slightly higher than the assumed standard uncertainty of the Y-variable, $s_{err} = 1.0$. But the between-samples variability (eq. (10.5i)) indicates that RMSEP(Y) could in fact lie anywhere between 0.9 and 1.5. Hence, the assumed $s_{err}$ was not too unrealistic, either.

RMSEP(Y) and $s_{err}$ do not measure exactly the same thing; RMSEP(Y) includes an extra contribution that reflects model errors.

In summary, the MC assumptions corresponded acceptably with the results from the actual experiment. If, in the *next* cycle of the research project, another experiment is to be made under the same general conditions, then more or less the same assumptions could be used in the MC part of the new round of experimental planning.

More details about the MC power-estimation are given in Appendix A11, cf. Figures 11.6 and 11.7. The method will be used in Chapter 16 for planning screening and response surface designs.

## 11.5. WHEN AND HOW

### 11.5.1. When to Plan

Conscious experimental planning is worth the time it takes, in all research projects, both for purely explorative ones and for more confirmative ones. Data analysis of sample sets with clear *deltas* usually have a high *Wow!*-factor of excitement, or at least high *Puh!*-factor of release.

❑ MAXIMISE THE CHANCE OF DISCOVERING WHAT IS IMPORTANT

In *explorative* projects, based purely on empirical observation, a mild form of experimental planning is used, for steering the observation process towards situations or objects that *differ* in all relevant qualities.

❑ MAXIMISE THE CHANCE OF CONFIRMING PRIOR HYPOTHESES

Before a *controlled experiment* is started, the hierarchy of purposes, the hypotheses and open questions, the measurement opportunities and the resource limitations are specified. Concise experimental designs can then make the experimental effort very effective.

### 11.5.2. How to Plan

The main principle in experimental planning is to ensure that every important variation-phenomenon has a good chance of being clearly detected by the chosen variables in the chosen samples. The selection of samples and variables should be *generous*, in order to ensure sufficient redundancy to counteract sampling and measurement uncertainty, and to allow cross-validation.

❑ GO BEYOND THE NATURAL VARIATION

In purely explorative projects it is useful to apply background knowledge in order to *stratify* the random sample selection, to ensure that all known variation sources are well spanned, and to avoid spending too much energy on non-informative *typical* samples.

❑ MAXIMISE THE CHANCE OF CLEAR OBSERVATION

For controlled experiments, *small, space-spanning screening-designs* are used for initial optimisation and simplification of the experimental conditions, while *larger, space-filling response surface designs* are used for detailed studies.

If lots of randomly chosen samples can be characterised by cheap, multivariate data, a *principal properties design* helps select a small subset of samples for more expensive analysis.

❑ OPTIMISE THE RISK VS. COST

Compare several alternative designs with respect to statistical power and experimental cost, and choose a cost-effective one.

## 11.6. TEST QUESTIONS

1. What are the main risks in data analysis, and how can they be minimised?
2. What is the main difference between the different design approaches: random sampling design, screening design, response surface design and principal properties design?
3. What is the similarity between these design approaches?
4. What is Monte Carlo simulation?

## 11.7. ANSWERS

1. The four main risks in data analysis are: (a) The risk $\alpha$, which reflects the probability of being fooled into believing in observed effects that might have been caused just by random noise (*false positives*, Type I error). (b) The risk called $\beta$, which reflects the probability that, for fear of being fooled by noise, we overlook a true and interesting effect, (*false negatives*, Type II error). (c) The risk that the data analyst does not understand the data. (d) The risk that the data analyst does not understand the data analysis.

> Type I errors may be minimised by using a combination of statistical model validation (cross-validation/jack-knifing with test level $\alpha$), plus interpretational scepticism. Type II errors may be minimised by conscious and generous experimental planning, and by ensuring that the chosen experimental plan has sufficient statistical power $(1 - \beta)$. Risk (c) may be reduced if the *owner* of the data performs the data analysis. Risk (d) may be reduced if this *owner* chooses only one of the many approaches to data analysis, and learns it well.

2. The differences between these design strategies mainly concern the amount of prior knowledge available, and the amount of new knowledge desired from the design (Figures 11.1 and 11.2).

3. The basic underlying concepts are the same in all conscious designs: trying to get a sample set with sufficient variability to allow every important phenomenon in the system to be observed independently of each other.

4. Monte Carlo simulation is a wonderful tool from contemporary computer science. It is a powerful alternative to traditional studies of theoretical properties of statistical methods. We just specify the situation that we want to study, and then simulate this situation a number of times (typically, 10 000) and study the outcomes. In this book, the MC technique is used for assessing the statistical power of experimental designs, by an approach that is independent of the design used and of the type of noise distribution expected.

# PART THREE:
# APPLICATIONS
# (Chapters 12–16)

This part gives five real-world examples of how *soft modelling* is used in practice. The applications come from different research fields. Together, they are intended to cover the main topics of importance in applied multivariate data analysis. All the data sets are authentic, taken from recent research projects in universities or industry. An overview of all the examples was given in Table 3.1. Table 3.2 related these to various quality goals.

The reader is invited to browse through all the examples, even those foreign to the user's own field; the general data analytical problems in different research fields are not as different as one might think. Each example follows the framework in Layman's guide in Chapter 3, Figure 3.2a,b. All the examples use the same data analytical method: cross-validated BLM analysis, but for quite different types of input data, and for quite different purposes.

Chapter 12 shows how to get fast, quantitative determination of quality from information-rich, precise, but as such meaningless measurements.

Chapter 13 demonstrates the analysis of *soft* qualitative questionnaire data in a total quality management (TQM) framework.

Chapter 14 relates purely computer-generated descriptors (molecular quantum chemistry) to biological measurements of toxic effects.

Chapter 15 concerns multivariate statistical process control (MSPC) for monitoring the quality of an industrial production process.

Finally, Chapter 16 shows how to plan, perform and interpret controlled experiments, based on a select sequence of factorial designs.

The corresponding technical appendices in Part Four cover the following:

Appendix A12 explains why diffuse multi-wavelength near infrared (*NIR*) spectroscopy is so information rich as a quality monitoring tool, and yet so different from traditional methods in academic analytical chemistry.

Appendix A13 briefly outlines the beneficial *consequences* of the actual project concerning the quality of the working environment described in Chapter 13.

Appendix A14 briefly summarises the bi-linear models of the toxicity/quantum chemistry relationships in three different *classes* of aromatic molecules.

Appendix A15 describes how the BLM method may be used for *forecasting* in the *time domain*.

Finally, Appendix A16 shows that, if so desired, statistical *significance testing* of effects in designed experiments by the BLM method (cross-validated PLS regression) give results very similar to those from classical analysis of variance (ANOVA based on the full-rank OLS regression (MLR)).

# Chapter 12

# Multivariate Calibration: Quality Determination of Wheat From High-speed NIR Spectra

*This first full example demonstrates how the BLM method is used for extracting reliable and relevant information about quality from inexpensive, information-rich measurements obtained with multichannel fingerprinting analysers. A number of high-precision, but low-selectivity variables are measured and combined statistically into precise predictions of quality. The example concerns near-infrared (NIR) light reflectance (X) determination of the economically important quality criterion for wheat, % protein (y).*

*This example shows how unexpected patterns in the data may be discovered by BLM plots. Automatic outlier detection is also demonstrated.*

## 12.1. INTRODUCTION

### 12.1.1. Purpose of the Project

The goal of the actual application is to use high-speed, inexpensive quality measurement of wheat samples for quality screening purposes. The most economically important quality of wheat is the *protein* content, which is analysed for here.

> The data (file WHEAT-NIR) were generously supplied by L. Munck and J. Pram Nielsen from the Royal Veterinary and Agricultural University, Copenhagen. The data have not been modified, but the explanation of the experimental procedure has been slightly modified and simplified, for didactic purposes.

❐ FINDING NEEDLES IN THE HAYSTACK

Modern instrumentation, e.g. light measuring devices, can provide torrents of information-rich but non-selective and as such meaningless measurements of our every-day world. This first application example concerns how to extract meaningful information from such high-speed measurements.

In classical chemistry, analysis is performed on the basis of specific molecular species and reactions. Modern analytical techniques such as light reflectance spectroscopy enables a complementary strategy, by collecting a *global finger-print* of the physical and chemical state of the samples.

Multivariate calibration has recently made is possible to separate out relevant and reliable co-variation patterns (bi-linear PCs) from the non-systematic, useless background noise in these *global fingerprints*. This can be done, even for incompletely understood biological systems such as food. The bi-linear regression modelling thus allows us to find needles in the haystack: Small, but highly informative co-variation types can be extracted and converted to quantitative information about quality.

❐ THE QUALITY OF WHEAT

Protein content is the most important chemical quality variable in wheat flour for bread baking; it determines the volume of the bread. In addition, it is nutritionally valuable. But the protein content varies a lot with wheat plant genes, with climate and growth conditions, etc.

Therefore it is important to be able to grade wheat samples with respect to quality: in this case protein percentage. In order to be able to pay the farmers according to the quality they deliver, the protein content of the produce must be measured. Traditionally, protein content has been measured by the so-called *Kjeldahl-N* method. However, while being well established, this method is rather imprecise, uses noxious chemical reagents and is too slow for routine analysis of individual truckloads of agricultural produce arriving at the local grain elevator.

The sample set represents the *classic* application of NIR instrumentation, which is still very important, commercially. But the present data set is peculiar in that it represents a considerable human cultural history as well.

The samples include three widely different *genotypes* of wheat.

1. 42-chromosome wheat: a number of varieties of today's highly productive wheats.
2. 28-chromosome wheat: a historically important intermediate wheat type, now surviving, e.g. as durum wheat (used for pasta production).
3. 14-chromosome wheat: the primitive ancestor variety, believed to have formed the economic basis of the first human cities in the Middle East in pre-history.

In addition, the samples have considerable *phenotypical* variation, since they had been sampled from several distinct locations over several years.

The original way of obtaining NIR-spectra is illustrated: shining near-infrared light at wheat flour samples, and measuring the intensity of the diffusely reflected light at the different wavelengths.

## 12.1.2. Purpose of the Data Analysis

The purpose of the present data analysis is three-fold: First, the analysis of very high precision, but non-selective data is demonstrated. Secondly, the principles behind calibration and prediction are explained through real world data, and thirdly this example shows the *surprise* factor characteristic of explorative multivariate data analysis.

◻ HIGH PRECISION DATA

The present example involves extremely precise physical measurements, where, e.g. the difference between readings 1.0000 and 1.0005 carries important information. Readers more interested in *soft* human data may find it intriguing that the present example comes from a scientific field aggressively ignored by many traditional *hard* natural scientists in chemistry and physics, in spite of its well established scientific basis and its huge practical success. The thought model behind the method is quite different than the one-to-one causality thinking in classical *hard* science, and probably closer to traditions in psychology, social science and economics.

◻ CALIBRATION AND PREDICTION

As demonstrated in Chapter 7 a multivariate calibration project consists of two phases, CALIBRATION and PREDICTION. First, data from an informative training set of samples are used for estimating a bi-linear calibration model that converts NIR measurements $X$ into relevant information $Y$ about the sample quality. Later, this calibration model is applied to NIR measurements from new samples, $X_{New}$, in order to predict their quality, $\hat{Y}_{New}$. In each phase, the data analytical *Steps I–VI*, (cf. Figure 3.2) are followed.

## 12.2. CALIBRATION

*I. Purpose of the calibration phase.* The goal of the present data analysis is to establish a calibration model for converting the NIR spectra ($X$) into estimates of *%Protein*($y$) in wheat samples, and to look for patterns among the samples.

## II. Experimental planning

### 12.2.1. Selection of Calibration Samples

Wheat varieties are collected from a gene bank with a wide variety of genetic properties, from cultivars adapted to different climatic conditions. Since the experiment had to take place under various organisational and practical limitations, it did not follow a strict statistical experimental design, like the one followed in Chapters 8–11 or 16. Some combinations of genotype and phenotype were thus not available, and the samples were not produced at the same place in the same year. Instead, the experimental design took place in the mind of the researchers, who used their background knowledge as plant breeders to select an interesting calibration sample set, representative of a wide class of wheat samples, spanning many different types of genotype and phenotype variability. Based on this stratified random sampling, a total of $N = 151$ of these samples was available for the present calibration modelling.

### 12.2.2. Selection of Variables

*NIR spectroscopy*: The light measurement was planned to be determined on ground samples, in a commercial instrument (NIR Systems 6500) in the upper NIR range (1000–2500 nm at 2 nm intervals, in total 750 wavelengths channels). The repeatability error of this instrument is very low compared to the biological sampling error.

*Reference method*: The protein content was to be determined by measuring total nitrogen percentage (Kjeldahl N% × 5.7) and expressed in weight percent of the samples (*%Protein*). The repeatability error $s_{err}$ of this reference method is about 0.1% protein.

***III. Experimental work***. The 151 samples of wheat were ground to flour with particles <0.5 mm diameter. The amount of reflected NIR light (*reflectance*) was measured with samples analysed in a random order. The raw spectra were stored in a database.

The *%Protein* was likewise measured in a random order; measurements from two replicates were averaged, and merged with the NIR data base of spectra.

## IV. Pre-processing and QC of data

### 12.2.3. Making data Tables

The measurements were collected from the database into a large table of raw data. Informative names were given to the variables and samples. Additional information about the samples (number of chromosomes, location of growth, year, etc.) was added as *category variables* (text variables), to be used as help in the graphical interpretation of the resulting models.

### 12.2.4. Data Pre-treatment

The measured reflectance spectra were transformed into optical density (OD) by the conventional transform OD = log(1/reflectance) (like in Chapter 4.2). To reduce the amount of raw data, five adjacent wavelengths were averaged, resulting in 150 X-variables. (The number of variables, 150, is similar to the number of calibration samples, 151, but that is purely incidental. Contrary to traditional multiple linear regression, bi-linear regression modelling is rather insensitive to the relative number of rows and columns in **X**).

### 12.2.5. Graphical Inspection of Calibration Data

❑   NIR DATA (**X**)

The OD spectra of the samples are shown (Figure 12.1a). No obvious outliers are evident. The peaks where the major constituents of the samples are known to absorb light, are indicated by the first letter; water (*w*), carbohydrates (*c*, mainly starch) and protein (*p*). For instance, the characteristic water absorptions at 1450 and 1940 nm and carbohydrate absorption near 2100 nm are clearly evident. The protein peaks are less clear; the characteristic *fingerprint contributions* from, e.g. the N–H bonds in protein are rather weak.

   In data analysis it is the *variation* between the samples that is of interest. The main visible difference between the samples is seen to be a large general vertical shift variation. This is due to changes in light scattering (whiteness), which is known to be caused by changes in particle size of the wheat flour.

❑   CHEMICAL DATA (**y**)

Figure 12.1b shows the histogram of the distribution of *%Protein* (**y**). The samples are nicely distributed around 15% and look OK. Hence, the calibration data appear suitable for multivariate modelling.

### *V. Data analysis*

### 12.2.6. Weighting of the Variables

Since the 150 variables within **X** are given in the same unit (OD) with presumably comparable noise levels, they will be used *unweighted*. The variable *%Protein* (**y**) will also be kept unweighted, since the relative noise levels between **X** and **y** are more or less irrelevant.

### 12.2.7. Regression Modelling

The Y-variable *%Protein* was regressed on the 150 X-variables in the $N = 151$ samples by bi-linear PLSR, using $M = 16$ sorted cross-validation segments.

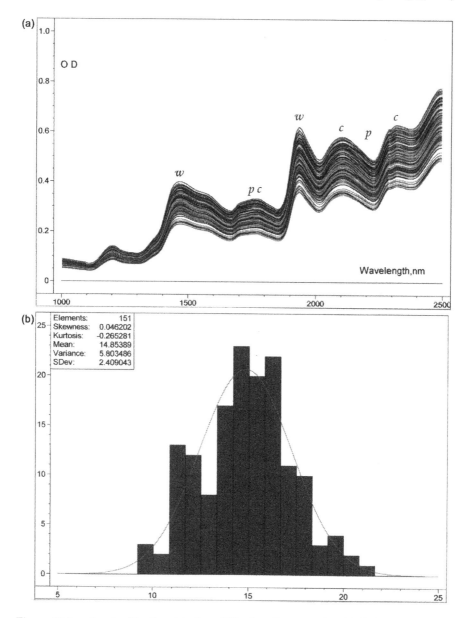

**Figure 12.1.** Input calibration data from 151 wheat flour samples. (a) X-data: NIR spectra of the ground wheat samples representing a wide range of genotypes and growth conditions. The spectra were measured by diffuse reflectance ($R$) and linearised into OD units by the transform $\log(1/R)$. w,p,c: see text. (b) Y-data: distribution of the quality criterion *%Protein* (**y**) to be calibrated for. This was measured chemically by a slow and noxious reference method (Kjeldahl-N) for the 151 calibration samples.

Even before *%Protein* had been measured by the slow and noxious traditional method, the NIR data were checked by PCA (Chapter 5). No serious outliers were detected. The PCA model is not shown here, since it is similar to the present PLSR model, using the NIR data as **X**. However, the PCA of the NIR data could in principle have been used for reducing the number of samples to be submitted to the traditional measurement of *%Protein* (principal properties design, cf. Chapters 7 and 11).

The number of segments, 16, was chosen instead of full cross-validation, in order to save computer time. Prior to the analysis, the samples had been sorted according to increasing values of **y** = *%Protein*. The samples were split systematically into 16 cross-validation segments according to the scheme 1,2,3,...,15,16; 1,2,3,...,15,16,...., (cf. Appendix A10.1 for technical details) to ensure that the segments could be expected to have reasonably similar distributions to the analyte **y**.

## 12.2.8. Prediction Error in y

Figure 12.2a shows how the prediction error in **y** in each of the $M$ cross-validation segments $m = 1,2,...,16$, develops with increasing number of PCs, $A = 0,1,...,15$. There are about $N_m = 9$ samples in each of the segments. Each curve summarises the prediction errors in the $N_m$ samples $RMSEP(Y)_{m,A}$ (eq. (10.5c), p. 404). The figure shows that after about 3 PCs the segments have rather similar prediction errors.

Figure 12.2b shows how the model's over-all $RMSEP(Y)$ develops. This summarises the 16 curves in Figure 12.2a. The dark curve in the middle represents the $RMSEP(Y)$ obtained from the cross-validation, the two grey lines mark its reliability range, eq. (10.5i). The figure shows that the first PC has little or no predictive ability for **y**=*%Protein*. But already with 3 PCs the X-variables are able to predict *%Protein* with relatively low prediction error. Thereafter the prediction ability improves little by little.

The minimum of the estimated prediction error is attained with 14 PCs. But for practical purposes an optimal model complexity of $A_{Opt} = 8$ is chosen; the figure shows that the exact number of PCs is not very important in the present case. This 8-PC model has an estimated prediction error $RMSEP(Y) = 0.40\%$ protein, with an reliability range of (0.36–0.45%).

Reliability ranges are most easily computed in terms of squared deviations (variances, MSE), although the results in the end are more easily understandable in the unit of the input data, (standard deviations, RMSE). Let mean square error of prediction of **y** in segment number $m$ be called $MSEP(Y)_m = RMSEP(Y)_m^2$. After 8 PCs the $M = 16$ segments have a mean $MSEP(Y)$ of 0.163 and a standard deviation $s$ of 0.073.

The uncertainty of the mean from the $M$ segments is of course smaller than the uncertainty in the $M$ individual segments; the standard error of the mean MSEP, $s_{Mean}$ is $s/M$, i.e. 0.073/16 = 0.018. Hence, the approximate reliability range of the mean $MSEP(Y)$ (eq. 10.5i), estimated as $\pm 2$ standard errors of the mean $MSEP(Y)$, is $(0.163 - 0.018, 0.163 + 0.018) = (0.145–0.181)$.

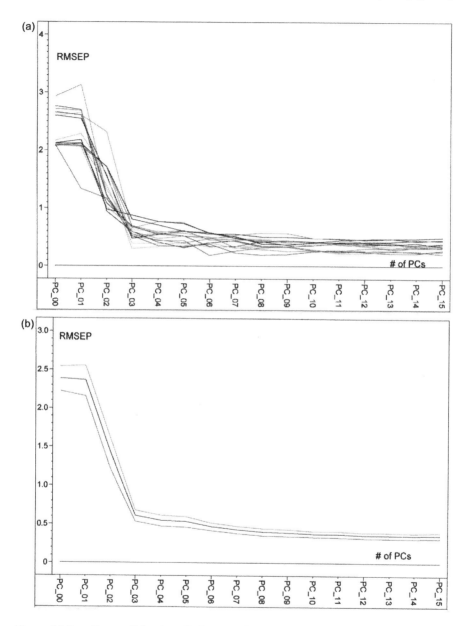

**Figure 12.2.** Cross-validated prediction error in **y**. (a) Prediction error RMSEP(Y)$_{m,A}$ for the $M = 16$ individual cross-validation segments, plotted against the number of PCs, $A = 0,1,2,...,15$. (b) Prediction error RMSEP(Y)$_A$, averaged over all the cross-validation segments, shown with its reliability range corresponding to $\pm 2$ standard uncertainties (grey), vs. number of PCs, $A = 0,1,2,...,15$.

The corresponding range of RMSEP(Y) is the square root of this, 0.403, with a reliability range (0.355–0.446). Finally, the last digit is truncated, since it is not significant. The error is given in measurement units of **y**, % of sample weight, hence $RMSEP(Y)_{A=7} = 0.40\%$ protein, with an uncertainty range of (0.36–0.45%) of sample weight.

## 12.2.9. Graphical Interpretation of the Bi-linear Model

Figure 12.3a shows the scores for the first two PCs, $t_1$ vs. $t_2$. Each of the 151 samples is represented by the category variable naming its number of chromosomes: *14, 28, 42* or *m* (for missing). A relatively clear clustering of the samples into chromosome groups is evident, although some overlap is seen.

Figure 12.3b shows the same score plot, but this time the 151 samples are named by another category variable, indicating the name of their growth locations. Also in this respect a relatively obvious grouping of the samples is evident. But this phenotypic clustering is primarily vertically along PC2, as opposed to the genotypic clustering in Figure 12.3a, which was primarily horizontally along PC1. This might call for more detailed analysis of the groups, (DPLSR or SIMCA modelling, Chapters 3 and 8), but for simplicity that is ignored here.

Figure 12.4 shows how the model develops for the 150 X-variables (NIR wavelength channels). Figure 12.4a shows the mean spectrum $\bar{x}$, which is subtracted, in analogy to Figure 5.2. Figure 12.4b–e shows the loadings of the X-variables for some PCs ($p_1$, $p_2$, $p_3$ and, finally $p_8$).

The peaks in PCs 1–3, which dominate the modelling, indicate that there is solid causal basis for why NIR calibration works in this case. After 3 PCs, more than 99.9% of the variance in the NIR input data has been accounted for.

The figure shows that the first PC, ($p_1$) is similar to the mean spectrum $\bar{x}$. It accounts for most of the light scattering variation, which here acts as a clearly evident interference type that needs to be accounted for before *%Protein* can be modelled properly.

The second PC ($p_2$) shows some very distinct negative peaks as well as several positive ones. The positive peaks are probably due to light absorption by proteins. But it seems that the negative peaks represent the beginning of accounting for one of the two major *unknown* interferences in wheat, the water variations of the samples: Pure water is known to absorb light around 1450 and 1940 nm. Since the sum of the constituents is 100%, *protein* will tend to be negatively correlated with the water content, which can vary considerably between wheat samples.

The third PC ($p_3$) shows a distinct negative peak at 2100 nm, where the second major interference in wheat, starch, is known to absorb light. Also, it shows a complex positive peak between 1900 and 2000 nm, which may reflect free vs. bound water.

However, 3 PCs were not enough to give adequate modelling of the protein-relevant variations in the NIR data. Since the NIR data are so precise, it is still possible for the PLSR model to extract many more PCs with valid information.

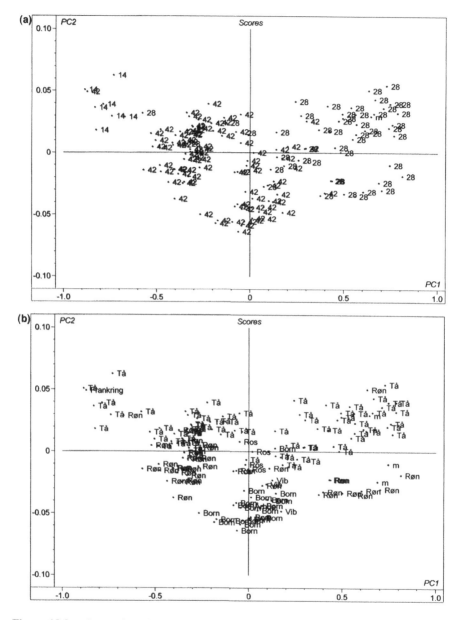

**Figure 12.3.** Inspection of the samples' scores for the calibration model's first two PCs. Latent variables $t_1$ (abscissa) and $t_2$ (ordinate) for the 151 calibration samples. (a) Naming of the samples: *genotypic* information (14, 28 or 42 chromosomes). (b) Naming of the samples: *phenotypic* information (growth location, abbreviated names).

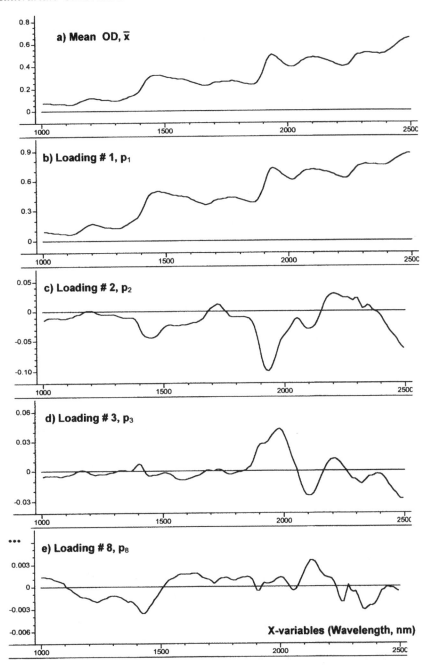

**Figure 12.4.** Inspection of the loadings of the X-variables (150 wavelength channels). The mean spectrum x̄, and the loadings for PC numbers 1, 2, 3 and 8.

Later PCs are very small, and they acquire increasingly complex structure, as exemplified by the 8th PC (Figure 12.4e, $\mathbf{p}_8$). Presumably these later PCs correct for a number of minor effects, like quality variations within the proteins, temperature variations between the samples, non-linear instrument responses, remaining light scattering effects, etc.

Figure 12.5 shows the resulting prediction model and its performance for *%Protein*. Figure 12.5a shows the estimated regression coefficient spectrum, expressed for the model deemed to be about optimal, using 8 PCs ($\hat{\mathbf{b}}_{A=8}$). The shape of the regression coefficient spectrum in Figure 12.5a is rather complicated. But it demonstrates how the multivariate calibration has automatically found a set of positive and negative weights to ascribe to the different wavelength channels in order to remove various interferences in the NIR data.

These interferences may be unknown to us, but show sufficient variations in the calibration set to allow the model to detect them and account for them by the latent variables. The result allows us to predict $\mathbf{y}$ (*%Protein*) from NIR ($\mathbf{X}$): $\hat{\mathbf{y}} = 0.36 + \mathbf{X}*\hat{\mathbf{b}}_{A=8}$, (where the scalar $\hat{b}_0 = 0.36$ is just an offset found to optimise the predictions).

Figure 12.5b shows the apparent predictive ability of this model: The *%Protein* predicted with 8 PCs during the cross-validation $\hat{\mathbf{y}}_{m,-m}$ (ordinate) is plotted against the measured *%Protein*, $\mathbf{y}$ (abscissa). Given the wide range of sample qualities calibrated for in the present case, with no explicit attempt at improved handling of light-scattering, instrument linearity, etc., this is considered a satisfactory predictive performance.

Figure 12.5c shows the spectral X-residuals after the calibration modelling. $\mathbf{E}_{A=8}$ shows the OD after having subtracted the 8-PC model. It indicates that, while there seems to be some non-random, but minor NIR variations left, the residual levels are very low (most of them within below 0.0005 OD units). Future samples of the same general kind should also fit the calibration model about as well. Otherwise they have to be treated as abnormal outliers for which the calibration model does not apply.

### 12.2.10. Bi-linear Modelling Used as a Hypothesis-generating Tool

The light scattering variation picked up by PC 1 (Figure 12.4b) is usually due to variations in particle size of the samples after milling. These, in turn, is known to reflect variation in another commercially important quality of the wheat- the so-called *hardness*. Looking back at the scores of this factor, $\mathbf{t}_1$ (abscissa, Figure 12.3a), it may seem that the hardness is somehow related to the number of *chromosomes*. However, the strange order (14–42–28) along PC 1 indicates that the relationship is not straightforward. How could this be?

Historically, today's wheat was developed by crossing the original, the 14-chromosome *einkorn* wheat, whose genes may be termed *AA*, with a *weed* with genes *BB*, to yield the 28-chromosome wheat with genes *AABB*. This, in turn, was crossed with another *weed* with genes *DD*, to yield the modern 42-chromo-

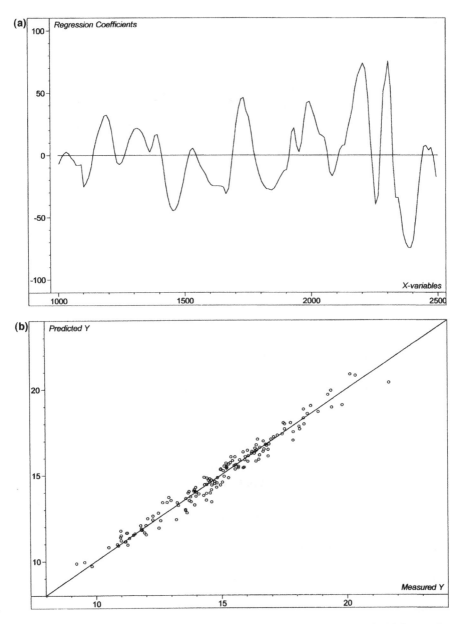

**Figure 12.5.**    Prediction from the final calibration model for *%Protein* (**y**). (a) Regression coefficients $\hat{\mathbf{b}}_{A=8}$, estimated with $A = 8$ PCs. (b) Measured (**y**, abscissa) vs. NIR predicted *%Protein* during cross-validation with $A = 8$ PCs, ($\hat{\mathbf{y}}_{m,-m}$, ordinate). The cross-validated prediction error is RMSEP(Y)$_{A=8} = 0.41$ *%Protein*. (c) Residual NIR OD after 8 PCs, $\mathbf{E}_8$.

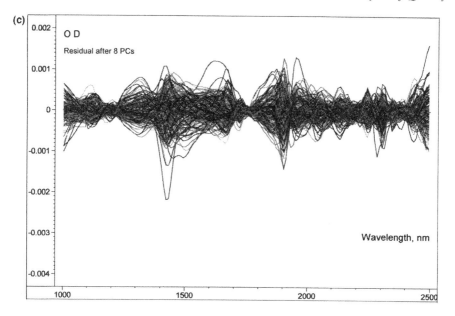

**Figure 12.5.** (*continued*)

some wheat with genes *AABBDD*. Thus, the 14, 28 and 42 chromosome wheat varieties, with genes *AA*, *AABB* and *AABBDD*, have a percentage of *BB* genes of 0, 50 and 30%. Could it be that the *hardness* of the wheat samples depends on the percentage of the *BB* genes?

When the hardness of the present samples were actually measured (not shown here), it indeed showed a clear correlation with the amount of *BB* genes ($r = 0.7$; Pram and Munck, pers. commun.).

This is an example of how visual inspection of bi-linear models may give rise to a new hypothesis. However, it should also be noted that the genotypic and the phenotypic variations are somewhat confounded in the present data set. Hence, this genetic correlation may possibly be an indirect one. So this is also an example of the need for subsequent hypothesis *verification* by new experiments, to try to disentangle these effects.

Moreover, a couple of samples in Figure 12.3a seem to be mis-classified with respect to the number of chromosomes. One 28-chromosome sample and one 42-chromosome sample appear to fall inside the 14-chromosome class. The reasons for this are not yet clear.

### 12.2.11. Attempts at Improving the Calibration Modelling

*More extensive cross-validation:* The present protein calibration model may in principle be optimised in various ways. First of all, full cross-validation might be used, with $M = 151$ segments instead of just $M = 16$, to make sure that the choice of segmentation do not affect our conclusions. In practice, the RMSE-P(Y)$_{A=8}$ then becomes 0.40% protein, with its reliability range (0.36–0.45%). This is virtually identical to the results with $M = 16$ segments, so the original cross-validation was good enough.

*Removal of useless X-variables:* Attempts at removing non-relevant wavelengths from **X** on the basis of the jack-knifing (Chapter 10) did not improve the predictive ability of the present model (as opposed to, e.g. Chapters 15 and 16). The reason is probably that almost all the wavelengths contributed to the prediction of *%Protein*. However, if it were desirable to reduce the number wavelength channels, e.g. in order to make a new and simpler NIR instrument with only a few fixed wavelength filters, this can be done. But the ability to detect abnormal outliers is then reduced.

*Less drastic data reduction in the pre-processing stage:* The original 750 wavelength channels, sampled every 2 nm, were used in **X** instead of the reduced set, averaged over 10 nm intervals. This did not improve the prediction. The reason is probably that the spectral shape of the actual chemical and physical phenomena in the NIR spectra are 10 nm or wider, so no essential information was lost by the data reduction.

*Calibrating for other properties too:* It would have been possible to predict a number of other quality-relevant properties of the wheat samples from the same NIR spectra. For instance, it is common to predict other main chemical constituents like water and starch, together with protein. It is also common to predict the mechanical milling properties like *hardness*.

*Removing the effect of light scattering:* The protein content is not expected to be correlated with the particle size. Hence, the large, non-additive light scattering effects in the spectra may be removed by multiplicative signal correction (MSC) or by taking spectral first or second derivatives (Martens and Næs, 1989). In practice, MSC did remove the first PC from the data, but this did not improve the present calibration model enough to be reported here.

*Explicit non-linearity modelling:* Non-linearities between the Y- and X-variables may be modelled more explicitly. A simple approach within the framework of the methods in this book is to include the squares and cross products of the OD values at some of the most salient wavelengths as new X-variables. Another one is to perform a PCA of the NIR data first, and include the squares and cross products of the first few PCA score vectors as additional X-variables for the final PLSR calibration.

**VI. Conclusion of the calibration phase.** The obtained calibration model is now ready to be applied to NIR spectra from new, unknown samples of the same general type.

## 12.3. PREDICTION

*I. Purpose of the prediction phase.* As new samples arrive, we intend to predict their economic and nutritional quality in terms of *%Protein* (**y**), from the high-speed NIR reflectance measurements, via the obtained calibration model.

**II. Experimental planning**. A new set of 30 new samples arrived, presumably representing wheat samples of the same general kind calibrated for. We would now like to grade these new samples with respect to their quality level, i.e. their NIR-predicted protein content.

**III. Experimental work**. The samples were ground and their NIR reflectance spectra were measured under the same conditions as during calibration, and put into the database.

**IV. Pre-processing and QC of data**. From the database, the NIR reflectance spectra were put into a data table, linearised into OD and averaged down to 150 X-variables, just like the calibration spectra were treated above.

**V. Data analysis**. Figure 12.6a shows the NIR OD spectra of the 30 new samples. They resemble the calibration samples in Figure 12.1; no obvious abnormalities can be seen.

Figure 12.6b shows the predicted level of *%Protein*, $\hat{y}$ (ordinate), obtained for each of the 30 samples (abscissa) from its NIR spectrum via the calibration model in Figures 12.4 and 12.5a. The bar around each predicted value gives an indication of the outlier-sensitive uncertainty (eq. (10.6)) of this value. Samples with much higher error bars are likely to be outliers that need special attention. The figure shows that the first 26 samples have generally increasing predicted *%Protein*, with reasonably low uncertainty. But samples 27–30 are tagged automatically as outliers. Why?

Figure 12.6c shows the NIR OD residuals for the four outlier samples (dark), after 8 PCs, $E_8$. For comparison, the light grey curves represent the basis for the outlier test, in terms of the calibration residual level, $\pm 1$ RMSEP(X)$_k$, and the corresponding reliability range $\pm 2$ RMSEP(X)$_k$ at the different wavelengths (cf. Figure 12.5c). The four samples show residuals which, at many wavelengths, are more than five times higher than expected; at the first wavelength channel four of them have residuals more than 20 times higher than expected (dotted, outside the figure scale).

Apparently, the NIR data for these four samples do not fit the eight-dimensional calibration model. Therefore we do not know if their *%Protein* predictions can be trusted or not. This illustrates the advantage of the automatic outlier detection: The NIR instrument can help the important scientific induction process.

When inspected in more detail, Figure 12.6a indicates that there has been an instrument or file transfer error of some kind at the lowest wavelength channel near 1000 nm. Closer inspection of the raw data confirms this. In addition, clear OD residuals are seen in many regions of the spectra, particularly a positive-and-negative structure around 1940 nm where it is known that a water peak shifts position with changing sample temperature or degree of water binding. Hence, it seems that we have at least two types of problems in these last samples: a slightly damaged instrument as well as an unexpected sample temperature or sample type. The former phenomenon is an analytical error that should be corrected, or at least avoided (e.g. by excluding the first few wavelength channels). The latter phenomenon may also be an error. But it may alternatively be something that makes these outliers the most interesting of all the samples!

Upon checking more closely into the origin of these four samples, it became evident that they indeed were outliers. First of all, they had been measured under slightly different experimental conditions. Moreover, sample 27 was confirmed to be a wheat sample. But samples 28–30 were not wheat samples at all: they were in fact *barley* samples.

### 12.3.1. Checking the Model

Given the uncertainty raised by the outliers, it was decided to verify that the calibration model is still valid.

*More experimental work*: The 30 new samples were submitted to the slow, noxious reference analysis method (Kjeldahl N).

*New data analytical modelling*: Figure 12.7 shows the predicted vs. the measured %*Protein* for these 30 samples, with the line representing the ideal target line. Except for the erroneous wheat prediction sample (*Wheat?*), the predictions are fine: even for the three barley samples (named *b.1–b.3*). Ignoring the *Wheat?* outlier, the RMSEP(Y)$_{Testset}$ for the other 29 samples, using 8-PC model, is only 0.31% protein. This is even slightly lower than the reliability range estimated in Figure 12.2b) ($<0.36, 0.45\%>$)!

Closer inspection revealed that the new set had a more narrow total range of variation in %*Protein* (total initial standard deviation around the calibration mean of 2.1, instead of 2.4% in the calibration set). So the most probable reason for the unexpectedly good prediction is that the new set was *too easy* for the model! This illustrates that if an *independent test set* is set aside in order to be used for assessing the success of a model, it has to be statistically representative and sufficiently large, otherwise the informative value of the resulting RMSEP(Y)$_{Testset}$ is questionable. In this book, we advocate that during the actual development of a calibration model, cross-validation is used instead.

However, at *later points in time* it is important to check the long-term stability of a calibration model with respect to instrument drift, etc. This is particularly important if the sample quality is expected to change systematically over time. In, e.g. agricultural crops like wheat, this may be expected due to yearly climatic variations. When NIR is used in industrial process control instead, time drift must be expected due to changes in raw materials and processing equipment, and due to ageing of the analytical instruments. In either case, it is possible to update the calibration models. This will be demonstrated in Chapter 15.

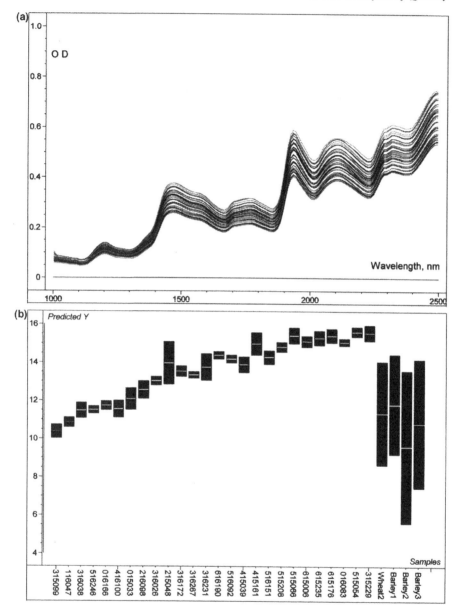

**Figure 12.6.** New samples for prediction of *%Protein* from NIR. (a) Input NIR spectra **X** of 30 new samples, in OD units. (b) *%Protein* $\hat{\mathbf{y}}$, predicted from the **X** data in the 30 samples, using the 8-PC calibration model. The bars represent the approximate uncertainty (c) Residual NIR spectra $\mathbf{E}_8$ (dark) for the last four samples, Barley 1, Barley 2, Barley 3 and Wheat? (=w), after fitting to the 8 PC model, compared with the residual error levels expected from the cross-validation during calibration, $\pm 1$ and $\pm 2$ RMSEP(X)$_{k,A=8}$ (light grey).

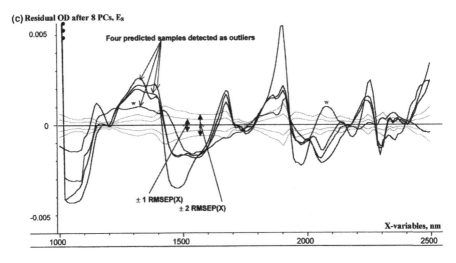

**Figure 12.6.** (*continued*)

***VI. Conclusion of the prediction phase***. The calibration model obtained in Section 12.2 was successfully applied to predict protein content from the NIR data in a new set of samples.

## 12.4. CONCLUSIONS

In summary, a selection of 151 relevant calibration samples was used as an information source, from which high-speed NIR measurements as well as slow, noxious *%Protein* measurements were taken. These were converted to data tables **X** and **y**, and used for building a multivariate calibration model. The optimal model complexity, as well as the predictive ability and its uncertainty range, were estimated by cross-validation.

The calibration model with 3 PCs displays spectral patterns in good correspondence with what we expect for NIR data of these constituents. Even the subsequent PCs gave small improvements to the apparent predictive ability, and they also gave reasonably smooth loadings; this indicates that they do contribute small, but valid pieces of information. A pragmatic choice of 8 PCs was taken as *optimal*. Thus, the calibration led to a complex, but scientifically plausible model, with good predictive ability within the wide range of samples calibrated for.

In the prediction phase, the calibration model was successfully applied to predict protein content from the NIR data from a new set of samples. Four of the samples were automatically tagged as outliers, since their X-data did not fit the calibration model. Closer inspection revealed these problems to reflect both instrument errors and abnormal sample types. When compared to traditional

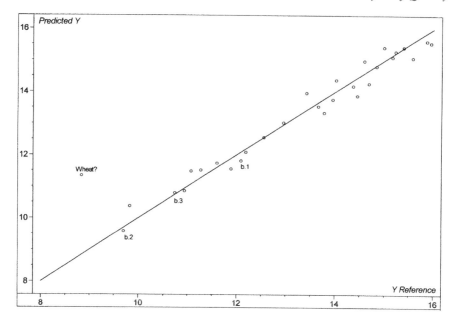

**Figure 12.7.** Checking the calibration model's predictive ability against long-term drift. %*Protein* measured (**y**, abscissa) vs. NIR predicted during cross-validation with $A = 8$ PCs, ($\hat{y}_{A=8}$, ordinate) for the 30 new samples.

protein measurement, the NIR-predictions for normal new samples were confirmed to be satisfactory.

## 12.5. TEST QUESTIONS

### 12.5.1. Questions About the Calibration Phase

1. How can we see that the NIR data in Figure 12.1a will be suitable for quantitative determination of %*Protein*?
2. Are there other chemical wheat constituents, for which these NIR data seem to allow quantitative determination?
3. What is the effect of the more or less unidentified light scattering effects?

### 12.5.2. Answers

1. We cannot see with our eyes that the NIR raw data contain protein information because the protein peaks are too weak.

2. Since water and carbohydrate peaks are visible, one should expect the NIR data to allow determination of these.
3. Unless the light scattering effects somehow are dealt with explicitly in some pre-processing step (Martens and Næs, 1989, p. 345), they will be modelled implicitly in the PLSR, probably dominating its first PCs. However, as long as their spectral profiles and sample-to-sample variations are different from those of protein, they will be mathematically separated from the protein information during the modelling, just like the other interferences from, e.g. water and carbohydrate variations. Thus, the light scattering represents a nuisance, but also a useful source of information.

### 12.5.3. Questions About the Prediction Phase

1. What is required of the new samples in order for the calibration model to apply?
2. How would the new NIR measurements have to be done in order for the calibration model to apply?
3. What would happen if (1) and (2) were not satisfied?

### 12.5.4. Answers

1. The new samples should ideally be of the same general kind as the calibration samples. They should not have chemical or physical variations that the calibration-model has not been trained to detect and correct for.
2. The NIR measurements should be performed with the same experimental procedure, the same instrument and the same pre-processing: The final X-data must not contain experimental effects in the NIR range that the calibration-model has not been trained to detect and correct for. (But calibration models may be transferred mathematically between instruments.)
3. Un-modelled effects in the NIR measurements may generate large errors in the predicted *%Protein*. However, such abnormalities in the samples or in the NIR measurements often give rise to X-residuals and thus to automatic outlier-warnings.

### 12.5.5. General Question

1. Can you list other possible uses of the same data analytical approach for converting precise and indirect measurements into meaningful information?

### 12.5.6. Answers

1. Some examples:

a. Chapter 15 illustrates the use of a different type of *colour-sensitive electronic eye*, ultra-violet *auto-fluorescence* spectroscopy, for monitoring an industrial process in the time domain. Colour cameras represent an extension, in which also the spatial domain is brought into play.

b. Computer-generated *images* such as multichannel or time-resolved medical images from fMRI extends the use of indirect measurements even further, predicting, e.g. $Y$ = brain activity or cancer from $X$ = image data.

c. But the input data may be of a totally different kind as well. For instance, sound spectra are an excellent source of complex, but useful information. Think only of how the sound from a tea kettle changes when its water begins to boil, or from a car when the speed changes! Mechanical vibration patterns from industrial processes have proven very useful as indirect measurements of process-states.

d. More speculatively, one could even envision money as indirect measurements of something else. For instance, a fraud-detection system in banking might be constructed, if the available pieces of information on every electronic money transfer were fed into a multivariate fraud-prediction model already calibrated based on known fraud cases and expert assessments.

# Chapter 13

# Analysis of Questionnaire Data: What Determines Quality of the Working Environment?

*This example demonstrates the BLM analysis of soft psychosocial data in a total quality management (TQM) framework. It shows how variations in people's perceived job satisfaction (**y**) in a large company can be analysed in terms of other types of information (**X**). The study reveals that variations in management style and in type of work seem to be responsible for the perceived quality variations.*

*First, the quantitative response-averages from 34 different company departments are analysed. Then the more qualitative, binary raw data input from 649 individual employees are analysed, with similar results.*

## 13.1. INTRODUCTION

### 13.1.1. Purpose of the Project

The working environment is in focus in many companies today. Are employees happy with the quality of their working environment? If not, why not? What can be done to improve it? A topic relevant within total quality management is, among others, to develop criteria for good leadership and to study factors influencing motivation of employees to do a good job.

The overall goal of the study was to improve the working environment in a Norwegian company that handles most of the countries' transactions between bank accounts, by

- investigating the working environment quality with the main focus on the psychosocial aspects;
- using these results to plan for actions;

- using the study as a fundament for leadership development in the company.

The present data sets (file JOB-QUALITY-I, -II) were kindly submitted by Westad (1997).

## 13.1.2. Purpose of the Data Analysis

This example aims at illustrating two ways of analysing the same data set, as a few, quantitative data or as many, qualitative data. Chapter 3.2 stressed that it is important to simplify the data analysis, at least initially, by reducing the size of the input data table, but not so much that important information is likely to be lost. In the present case, that will be attained by averaging the responses for each department in the company (*quantitative data*). This reduces the number of samples to 34 departments (file: JOB-QUALITY I). Then, the conclusions are checked in the original input data for the original 649 individuals  irrespective of which department they work in (*qualitative data*) (file: JOB-QUALITY-II).

◻ LOW PRECISION DATA

Chapter 12 dealt with high precision data. This example will show how *soft*, low precision questionnaire data may be summarized efficiently by the same BLM method. *Low precision* here means lower repeatability and reproducibility in the measurements. It also includes the *qualitative* character of data often used in psychometrics literature (e.g. Hoffman and Young, 1983), in the present example referring to binary 1/0 data representing yes/no responses.

One intriguing aspect of the present questionnaire example is that it demonstrates how multivariate modelling of *quantitative* and of *qualitative* data can give almost identical conclusions. This opens up a data analytical continuum between *the two cultures*, i.e. between *hard* quantitative measurements and *soft* narratives from people.

## 13.2. WORKING ENVIRONMENT STUDIED ON AVERAGED DATA

*I. Purpose of quantitative data.* The goal of the first data analytical step is to compare the 34 different departments within the company with respect to how their employees describe their working place, and to find explanations for possible differences. In particular, the goal is to model the reported job satisfaction from the 26 other descriptors.

## II. Experimental planning

### 13.2.1. Selection of Samples

All employees in the company were invited to participate.

### 13.2.2. Selection of Variables

The questionnaire with 27 questions is shown in Table 13.1. Questions 1–26 concern details with respect to the working environment, with yes/no answers requested, i.e. *qualitative* binary responses. Question 27, asking for degree to which people generally liked their working place, here named *JOBSATISFAC-TION*, was recorded on a *quantitative* interval scale from 1 (dislike very much) to 10 (like very much).

***III. Experimental work.*** The questionnaire was properly filled out and returned by 649 employees, which corresponded to about 80% of the total number of employees. The recordings were transferred from the paper sheets to a computer data base. Some of the respondents had left a few of the questions unanswered; these were recorded and treated as missing values (cf. Appendix A5.6); the variables had between 1 and 5% missing values.

## IV. Pre-processing and QC of data

### 13.2.3. Making Data Tables

Since the working environment at different departments was of main interest, the responses were first grouped into departments and summarised. Departments with less than 15 employees were ignored in order to ensure a minimum of precision in the averaged data. The final number of departments to be modelled was $N = 34$. For each of the 26 detailed questions, the *percentage* 'yes'-responses was computed within each of the departments, $i = 1,2,...,N$. The 27th question in the questionnaire was averaged within each of the 34 departments.

### 13.2.4. Inspection of Input Data

Figure 13.1a gives a summary of the distribution of the 34 departments with respect to the 26 detailed questions, in terms of the 0, 25, 50, 75 and 100 percentiles. It shows, for instance, that most of the departments mainly offered routine work (question 4: *WorkRoutine*), although some departments (probably the development departments) had very little routine work. Most departments had low frequency of complaints like *BossDemanding*, *BossConfusing* and *Harassment*, and reported high frequencies of helpful colleagues (*Coll.Helpful*)

**Table 13.1.** Questionnaire. Question 1–26: Yes/No answers.

| Number | Full text | Abbreviation |
|---|---|---|
| 1 | Can you decide the speed of your work yourself? | *PaceSelf* |
| 2 | Does your boss give clear requirements? | *DemandsClear* |
| 3 | Do you get enough information to do good work? | *InfoEnough* |
| 4 | Is your work characterized as uniform and routine? | *WorkRoutine* |
| 5 | Does your boss put too strong demands on you? | *BossDemanding* |
| 6 | Is your boss fair? | *BossFair* |
| 7 | Is your work interesting? | *WorkInteresting* |
| 8 | Do you have enough time to do your job in a good way? | *TimeEnough* |
| 9 | Do you take decisions in your work? | *DecideSelf* |
| 10 | Is your work demanding creativity? | *CreativeNeeded* |
| 11 | Are your colleagues helpful? | *Coll.Helpful* |
| 12 | Do you feel that you have a safe working place? | *WorkSafe* |
| 13 | Does your boss communicate contradictory demands? | *BossConfusing* |
| 14 | Do you feel that you have success in your work? | *SuccessFelt* |
| 15 | Does your work offer you possibilities to learn something new? | *LearnChance* |
| 16 | Have you got sufficient training to do a good job? | *TrainEnough* |
| 17 | Does your boss inspire you to do a good job? | *BossInspiring* |
| 18 | Is the slogan 'The employees are the most important resource' real? | *SloganReal* |
| 19 | Does bullying exist in your department? | *Harassment* |
| 20 | Are you given any special responsibility in your work? | *Responsibility* |
| 21 | Do you receive recognition when you do a good job? | *Recognition* |
| 22 | Do your working tasks give you job satisfaction? | *TasksOK* |
| 23 | Do you have possibilities to develop? | *DevelopChance* |
| 24 | Does your boss give you precise and clear feedback? | *FeedbackClear* |
| 25 | Is the collaborative working climate in your department good? | *Collaboration* |
| 26 | Are you given the authority you need to do a good job? | *Empowered* |
| 27 | Give a score expressing how much you like/enjoy your work (from 1 (dislike very much) to 10 (like very much)) | *JOBSATISFACTION* |

and *Collaboration*, but some variation is evident, indicating management problems.

Figure 13.1b shows the distribution of the variable *JOBSATISFACTION* in more detail, in terms of a normal probability plot. It shows department number $i = 16, 26, 21, 23$ and 34 to reflect the lowest mean *JOBSATISFACTION* ($<6.5$), and $i = 14, 30, 33$ and 9 the highest ones ($>8.5$). The departments fall along a reasonably straight line in the normal probability plot, showing that *JOBSATIS-FACTION* is nicely distributed, without apparent outliers.

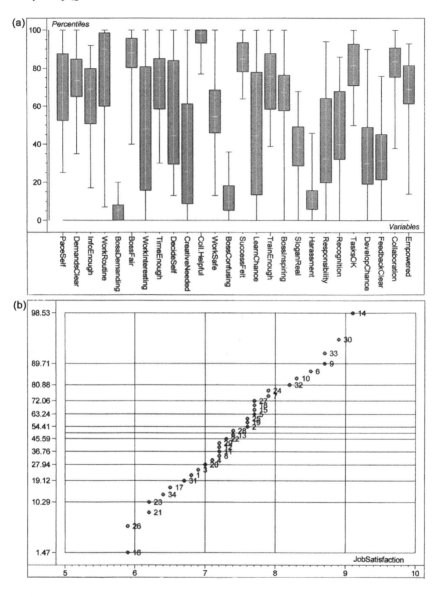

**Figure 13.1.** Quality of the working environment in different departments of a bank transaction company: summary of input data. (a) Distribution of samples (bank departments) for the 26 questions. The percentage of employees that answered yes to each question in each of the 34 departments. The percentile plot shows the distribution of the 34 departments with respect to minimum, lower quartile, median, upper quartile and maximum percentage. (b) Distribution of samples (bank departments) with respect to *JOBSATISFACTION*. Normal probability of the mean value of *JOBSATISFACTION* (abscissa), showing the distribution for the 34 departments (ordinate).

*V. Data analysis*

### 13.2.5. Modelling JOBSATISFACTION (y) from 26 Detailed Questions (X)

The data table of yes-frequencies for the 26 detailed questions is used as **X** ($34 \times 26$), and the mean *JOBSATISFACTION* is used as **y** ($34 \times 1$). Each variable was for simplicity standardised before the analysis, in order to ensure that each X-variable had an equal chance of contributing to the modelling of **y**.

The X- and y-data were submitted to PLSR. Full leave-one-department-out cross-validation (i.e. $M = N = 34$ segments) was used in order to assess the model. $A_{\text{Max}} = 7$ PLSR components were computed: this was considered *more than enough*.

The first component described 60% of the CV-variance in **y** (66% of the non-validated variance). Thus, there is obviously a clear enough X–Y relationship to warrant modelling. This first component is clearly valid, in the predictive sense.

The second component contributed an additional 2% explained CV-variance (6% non-validated variance). This was considered enough to be included in the modelling, although it is rather small.

All subsequent components just decreased the predictive ability again, so the optimal number of components, $A_{\text{Opt}}$, is 2.

### 13.2.6. The Bi-linear Model of the 34 Departments

Figure 13.2 displays the main information from the PLSR model; the first two components. Figure 13.2a shows the configuration of 34 departments in the score plot, and Figure 13.2b shows the corresponding configuration of 26 X-variables and the Y-variable, in the correlation loading plot.

It is good data analytical practice to look for meaningful interpretations, not only along the horizontal and vertical PC axes, but also in other, off-axis directions in the PC plots (Chapter 8.2.6). Inspection of the loading plot in Figure 13.2b reveals that all the variables in the north-west/south-east direction concern *the type of the work itself*, while all the variables in the south-west/north-east direction concern *psychosocial aspects* of the working environment.

The north-west/south-east direction spans the contrast between low-skilled work operations (e.g. *WorkRoutine*) and high-skilled work (e.g. *DesideSelf*) and a cluster of related variables (*CreativeNeeded, Responsibility, LearnChance, WorkInteresting* and *DevelopChance*). Departments towards the south-east in the score plot (e.g. 33, 30 and 27) are expected to represent the most highly skilled work, and objects in the opposite diagonal direction (e.g. 25, 16 and 20) are expected to represent the least highly skilled work.

People in the high-skill departments report that they perceive their employment as being more secure (*WorkSafe*) than people in the low-skill departments. This is not unexpected, since the low-skill manual jobs are in constant danger of being replaced by automatic optical character reading (OCR) machines.

The south-west/north-east direction spans the contrast between negative

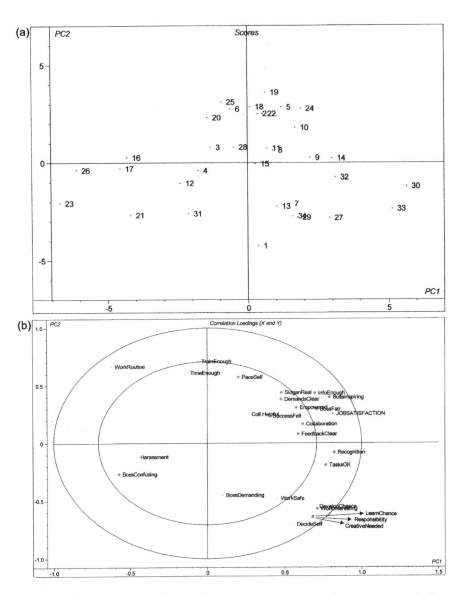

**Figure 13.2.** Summary of differences between the departments with respect to work place quality. Characterisation of the two significant PCs 1 and 2, found when regressing *JOBSA-TISFACTION* (**y**) on 26 detailed questions (**X**) over the 34 departments by PLSR. (a) Score plot for the 34 departments. (b) Correlation loading plot for the 26 X-variables (questions) and the Y-variable *JOBSATISFACTION*. The ellipses represent 50 and 100% explained variance.

The X-variables found to give significant contributions to **y** are symbolised by small circles.

psychosocial variables (*BossConfusing, Harassment*) and a number of positive psychosocial variables (e.g. *InfoEnough, BossInspiring, BossFair, SloganReal, DemandsClear, Empowered, Collaboration, FeedbackClear,* etc.). All the important descriptors along this direction may be interpreted as reflecting the quality of the management in the departments, irrespective of the type of work in the department. Departments to the north-east (e.g. 19, 24, 10 and 14) are expected to have successful management, and those to the south-west (e.g. 21, 23, 26, 17 and 31) are expected to have the least successful management.

### 13.2.7. Prediction of Job Satisfaction

How do the 26 detailed questions relate to the overall quality criterion *JOBSA-TISFACTION*? In the loading plot of Figure 13.2b we find our Y-variable in the east-north-east direction that we already have associated with good management.

Figure 13.3 summarises the 2-PC prediction model and shows its predictive ability. Figure 13.3a shows the standardised regression coefficients $\hat{b}$, i.e. the estimated contribution of each of the 26 detailed questions to the prediction of *JOBSATISFACTION*.

Care must be taken when interpreting such regression coefficients, estimated with data from non-designed experiments. What the regression coefficients show is the *predictive contributions* from each X-variable to **y**, not the X–y correlations, and not any necessarily *causal* effects. Still, the standardised regression coefficients $\hat{b}$ can be quite informative, especially when interpreted together with the bi-linear correlation loadings.

The figure confirms that *BossConfusing*, *Harassment* and *BossDemanding* are the most important negative contributions to the predicted *JOBSATISFACTION*. *BossFair, BossInspiring* and *DemandsClear* are the most important positive contributions, although several other positive contributions are also seen. Some questions, like, e.g. *WorkRoutine* and *WorkSafe*, do not seem to contribute to the predictive ability for *JOBSATISFACTION* in this case.

The reliability ranges (2 × the standard uncertainties of $\hat{b}$, estimated in the cross-validation) show that many of the positive contributions are highly reliable in the sense that their relationship to *JOBSATISFACTION* is similar in the departments. Among the negative contributions, only *BossConfusing* seems to be consistent in two or more departments.

Figure 13.3b shows the estimated performance of the prediction model by cross-validated predictions of *JOBSATISFACTION* for the 34 individual departments within the company. The abscissa represents the actually recorded mean *JOBSATISFACTION*, $y_i$, and the ordinate represents the *JOBSATISFACTION*, $\hat{y}_{i,-i}$, predicted from the 26 detailed questions $x_i$ for each of the individual department $i = 1,2,...,34$ when it had been kept *secret* during the cross-valida-tion (cf. Chapter 10). The straight target line represents where predictions of *JOBSATISFACTION* would have fallen if they had been perfect.

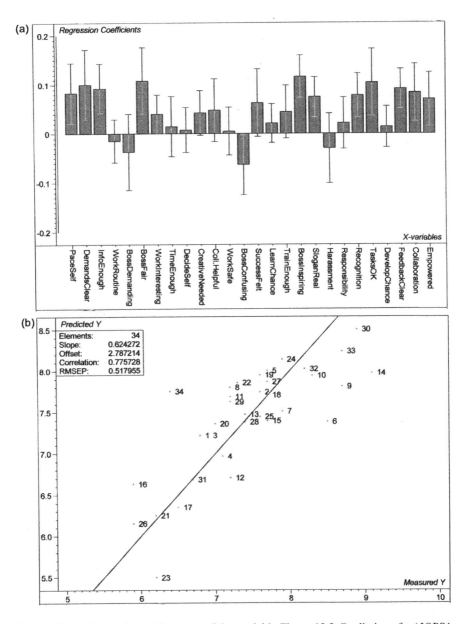

**Figure 13.3.** Predictive performance of the model in Figure 13.2. Prediction of **y** (*JOBSA-TISFACTION*) from **X** (26 detailed questions) via the 2-PC model from Figure 13.2. (a) Standardized regression coefficient vector $\hat{\mathbf{b}}$, showing the contribution of each X-question to the prediction of Y-variable *JOBSATISFACTION*; error bars represent $\pm 2$ estimated standard uncertainty from the cross-validation. (b) Predicted (cross-validated) vs. recorded mean JOBSATISFACTION **y** for the 34 departments (the computer plot has too many digits).

The model predicts the *JOBSATISFACTION* data for some departments better than others. In particular, department number 34 is seen (vertically) to be more off the target line than the others. If we remove this department as an outlier, the predictive ability of the model for the remaining 33 departments of course improves a little. But we have no external knowledge that could explain why this department does not fit as well as the others, and we could find no easy explanation from the data themselves. Therefore, we leave department number 34 in, and accept that the predictive model is good enough for our purpose, but not perfect.

### 13.2.8. Possible Model Optimisation Steps

❐  REMOVING USELESS VARIABLES?

Apparently useless X-variables (X-variables found to be un-correlated to **y** or yielding relations to **y** found to be unreliable in the jack-knifing) may be deactivated in the final PLSR model. In the present case only a couple of X-variables were found to be useless, and deactivating these had only a small improvement effect for **y**. Therefore this is not pursued in the present example.

❐  POSSIBLE CURVATURE?

It could be argued that a systematic curvature may be observed in Figure 13.3b, and that improved modelling could be attained by explicitly handling this curvature. This curvature handling could be done either by extending **X** to include the square of some of the original X-variables (Appendix A8.1, eq. (8.4)) or by linearising **y**. The minimum value of **y** is about 5.5. One possible linearising function is to use $\mathbf{y}_{new} = \sqrt{(\mathbf{y} - 5.0)}$. This improved the prediction ability by 3% in the first PLSR component. However, the loadings were not appreciably affected. Since the purpose of our modelling is simply to develop general quality criteria rather than to develop a predictor with maximal precision, we do not proceed with linearising enhancements in this example.

❐  POSSIBLE INTERACTIONS?

It could be that certain combinations of phenomena contribute multiplicatively rather than additively to the *JOBSATISFACTION*. This could be checked by extending **X** to include cross products of some of the original X-variables (eq. (8.4)). For simplicity, that is not pursued here. But if we had chosen to include many squares and cross products in **X**, then a final jack-knife based deactivation (Chapter 10) of the least reliable of these would be recommended.

### ☐ DIFFERENT WEIGHTING?

In the present case, each of the variables was, for simplicity, standardised *a priori*. Could this have over-amplified some X-variables ($k$) with small total initial standard deviation $s_k$, like *BossDemanding* (Figure 13.1a)? Using unweighted instead of standardised X-variables only caused the modelling to deteriorate. But re-estimating the model with the X-weights changed (eq. (8.6), p. 391) from just $1/s_k$ to $1/(s_k + 5)$ indeed improved the predictive ability for **y**. But this gain was very little (from 62.4 to 63.0% CV-variance explained), and the interpretation did not change. So this is not pursued here.

### 13.2.9. Tracing the Conclusions Back to the Raw Data

It is good data analytical practice to verify that the main conclusions from the modelling are also evident in the raw data. With 26 X-variables, there are $26 \times 25/2 = 325$ different pairs of X-variables to look at. But now that we know where to look, it suffices to plot just a few of them. In the present case, we would expect, e.g. the most extreme X-variables along the south-west/north-east direction to be strongly intercorrelated: *BossInspiring* and *InfoEnough* ought to be positively correlated, and *BossInspiring* and *BossConfusing* should be strongly anticorrelated. Department numbers 24 and 14 should have high levels of *BossInspiring* and low levels of *BossConfusing*, and department numbers 23, 26 and 21 should have the opposite pattern. Inspection of the raw data (not shown here) indeed confirms this.

Having found the main patterns of variation, it is time to go into more detail about the individual departments. Considering the *worst* departments in the score plot of Figure 13.2a, all of them are of the low-skill routine type. The main reported problem in department number 26 is harassment, lack of help from colleagues and lack of collaboration, plus a management perceived to be unfair. In department number 21 the main reported problem is uninspiring management who issued confusing orders instead of giving enough information. In department number 23 the main reported problem is uninspiring, unfair management who did not consider people important and who did not give people control over their own work.

Looking at the *best* departments, numbers 30 and 33 were particularly high-skilled, while, e.g. numbers 9 and 10 were low-skilled. So it is possible to have good *JOBSATISFACTION* even in low-skill working places, but then the management must really be good.

**VI. Conclusion quantitative data.** In summary, the modelling of the 34 departments showed relatively good predictive ability for **y** = *JOBSATISFAC-TION* from the *yes*-frequencies of the 26 other questions (**X**). Two different patterns of these variables emerged: one apparently associated with management style and one associated with the type of work in the department.

## 13.3. WORKING ENVIRONMENT STUDIED ON INDIVIDUALS' DATA

*I. Purpose of qualitative data*. The goal of the present data analysis is to look for variation patterns among the 649 individuals at the raw data yes/no level, and compare these to the differences found previously between the averaged data from the 34 departments.

*II, III. Experimental planning and work*. Same as in Section 13.2.

*IV. Pre-processing and QC of data*

### 13.3.1. Making the Data Table

If we unwittingly had made some mistakes in the preparation of the previous data for the departments, the modelling of these data might, in the worst case, pick up these mistakes as a large and apparently *valid* component. Therefore, another aspect of good data analytical practice is to try to ensure that the structure found in the data table is not just the result of mistakes in the pre-processing of the raw measurements.

We now let the $N = 649$ individuals constitute one row each. The columns for the X- and Y-variables are the same as above, but the resolution and precision of the individual data is of course expected to be much lower for the raw data than for the department averages. The X-matrix now consists of only qualitative (binary) 1/0 data, i.e. yes = 1 and no = 0. Although the data table includes a considerable number of missing values, we ought to be able to observe the same latent phenomena as above.

*V. Data analysis*

### 13.3.2. Data Analysis of all the Individuals

The same PLSR modelling was repeated, but this time on the table of raw responses from all 649 respondents, irrespective of department. In order to save computation time, a *leave-some-random-people-out* cross-validation (Appendix A10.1) was used, based on 100 randomly selected subsets of people.

In the cross-validation the first two components explained only 30.9% of the CV-variance in **y**. This is considerably lower than in the department mean model (62%). Repeating the cross-validation with $M = 100$, 20 and five other, randomly selected cross-validation segments gave 31.4, 30.8 and 30.7%

explained CV-variance in **y**. It appears that the decreased predictive ability is not due to incidental problems in the cross-validation segmentation.

Instead, it appears that the lack of resolution and precision in the individual raw data has reduced the apparent predictive ability and amount of interpretable structure. Still, the model clearly reveals structure in the data. The corresponding explained CV-variance in **X** is about 25% in these cross-validations.

Figure 13.4a shows the scores of the 649 individual respondents along the first two PCs. The grey-tone of the circles indicates their reported level of *JOBSA-TISFACTION* (**y**), ranging from very high (10, light grey) to very low (1, black). The highest *JOBSATISFACTION* is generally found to the right.

Figure 13.4b shows the corresponding correlation loading plot. This structure in the correlation loading plot from the raw data is very similar to that obtained from the department averages (Figure 13.2b). The general management and the skill level directions are still the same, and connected to the Y-variable as before.

Thus, the individual qualitative binary raw data yielded the same patterns as the more quantitative department averages. Since the two models are so similar, the pre-processing (e.g. averaging without the missing values) seems to have been correct.

The people who report low *JOBSATISFACTION* (black circles), are mainly found in the lower left quadrant of the score plot in Figure 13.4a. But not all people in this quadrant are unhappy with their working environment, and there are some unhappy individuals also on the other quadrants. Therefore, a logical next step could be to try to find groupings among the individuals, in terms of the 26 detailed questions.

### 13.3.3. Segmentation of People into Groups

Such a segmentation of the people into groups based on their responses to questionnaires or other responses is of interest in medicine, marketing, etc. This may be done by separate PLSR or PCA modelling of, e.g. various groups of people, or by discriminant PLSR, with indicator variables for the various groups (cf. Figure 3.4, eq. (8.3)).

An example of this is the following. When repeating the modelling separately for the people who reported doing *routine work* and those who did not, some interesting differences appear: Figure 13.5 compares the regression coefficients $\hat{\mathbf{b}}$ for predicting **y** = *JOBSATISFACTION* from **X** = 26 detailed questions, for the 471 people reporting *WorkRoutine* = 1 and the 151 people reporting *WorkRoutine* = 0. The two groups of people show the same general patterns in the regression coefficients, e.g. with respect to *BossInspiring* and *BossConfusing*. But there are interesting differences. The routine workers ascribe significant importance to the questions *BossFair, WorkInteresting, Harassment,*and *TasksOK* for predicting *JOBSATISFACTION*, while the others do not. Conversely, the non-routine work respondents ascribe predictive importance to the question *TrainEnough*. People with missing values for *WorkRoutine* or for *JOBSATIS-*

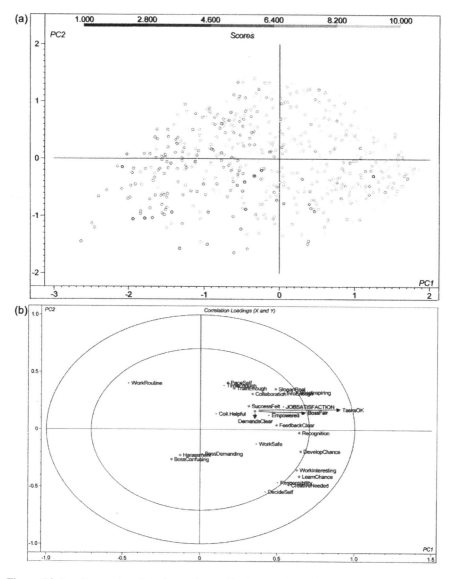

**Figure 13.4.** Regression directly on the qualitative questionnaire raw data. (a) Score plot for the 649 respondents. The symbol intensity reflects the *JOBSATISFACTION* (**y**), ranging from very high (10, light grey) to very low (1, black). (b) Correlation loadings of the variables. The questions *DemandsClear*, *BossFair* and *TasksOK* fall almost on top of each other near *JOBSATISFACTION*.

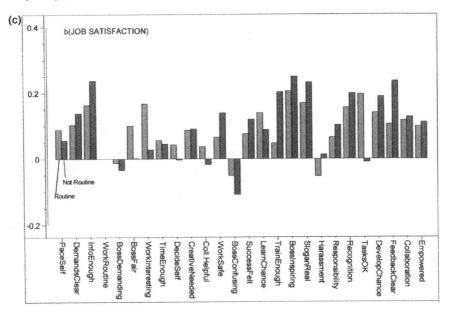

**Figure 13.5.** Comparison of routine-workers and non-routine workers with respect to prediction of *JOBSATISFACTION*. Regression coefficients $\hat{b}$ estimated for the two groups separately.

*FACTION* were ignored. Cross-validation in both cases showed 1 PC to be optimal.

Similarly, separating the people into three groups according to their level of *JOBSATISFACTION* (low = 1–4, middle = 5–7, high = 8–10) and repeating the modelling of *JOBSATISFACTION* for each group (not shown here) revealed some clear differences: For instance, within the low-satisfaction group the positive contribution of *BossFair* was the strongest; in the middle-satisfaction group the positive contribution of *LearnChance* was the strongest, and in the high-satisfaction group the positive predictive effect of *SloganReal* was the strongest.

***VI. Conclusion qualitative data.*** In summary, the modelling of the qualitative yes/no (1/0) responses from the 649 individual respondents gave the same general conclusions as the more quantitative responses in the 34 department averages.

The individuals were grouped according to whether they reported to do routine work or not, and each group was modelled separately. Some nuances could then be seen in what appeared to be important for predicting their *JOBSATISFACTION*. Likewise, grouping the individuals according to their reported *JOBSATISFACTION* indicated interesting detailed differences.

## 13.4. CONCLUSIONS

The following general conclusions were found.

- The valid latent variables revealed two main ways in which the different banking departments differed: one concerning psychosocial/management aspects and one concerning the type of work done and the skill level required.
- The psychosocial variables, primarily those associated with management style, were the main predictors for the general *JOBSATISFACTION*.
- High-skill and low-skill departments differed in how people perceived their working place, but both types of departments could have very high *JOBSA-TISFACTION*. However, the departments with lowest *JOBSATISFACTION* were all of the low-skill routine-type.
- People in low-skill departments had the highest fear of losing their jobs.
- A few departments stood out as having employees with particularly low degree of *JOBSATISFACTION*.

Are these trivial conclusions? Well, to say that good management is important and that people in low-skill jobs fear to be replaced by computers, hardly surprises anyone. But at least it gives a nice feeling that the empirical data confirm the generally expected patterns. Also, it is easy to exclaim, *after* the data analysis, that 'I could have said this before we started'. It is much more difficult actually to say that, *before* the data analysis!

Interesting new details were indeed revealed. By the multivariate analysis, we quickly homed in on the main structure in the data, displayed it graphically for efficient interpretation, and validated the models by checking their predictive ability. Now, we have reasons to claim that we seem to *know* something about these departments' working environment.

For instance, *before* we had analysed the data, we might have speculated that *JOBSATISFACTION* could be particularly low for people who have demanding bosses, and if they were given insufficient time and training to do a good job. Now, *after* the data analysis, we have good reasons to say that these were not the main problems reported here.

Statistically, the data analysis has only revealed reliable correlations, not causal relationships. However, the correlation patterns found seem to allow causal *interpretation*, upon which it is possible to act. Appendix A13 tells more about that.

## 13.5. TEST QUESTIONS

1. Is it possible to combine quantitative and qualitative data?
2. Which methodological considerations are relevant for questionnaire data?
3. How are the results validated?
4. Can you list possible uses of the same data analytical approach for studies on people's attitude or behaviour?

## 13.6. ANSWERS

1. Yes, there is no problem in combining different types of variables to explore their relationships or predictive ability. The BLM data analysis may be used for interpreting any *combination* of, e.g. (1) quantitative measurements of a set of comparable *items (work places, cars, food products, music performances or paintings, etc.)*, (2) intersubjective, quantitative sensory expert assessments of the same items, (3) semi-quantitative consumer studies on the same items, (4) purely qualitative narrative verbal descriptions about how the same items are perceived, coded as zeroes and ones for words used.

2. Some methodological weaknesses:
   - Not all of the employees returned questionnaire sheets. Were these employees systematically different from the ones who did? Not knowing these details, we can only hope that the systematic sampling errors are small.
   - The return percentage might have been included as an extra X-variable for the different departments. Additional variables such as economic pressure, sick leave frequencies, educational background, sex, age, etc. might have been included, to complete the picture.
   - Finally, it should be mentioned that allowing only yes or no as possible answers may have caused some problems for individuals who want to express themselves on a more detailed scale. Also there may be a difference between what people say in a questionnaire, and what they actually feel or do in reality.

3. The validity of the modelling results depends on several factors, e.g.:
   - The validity of the input data. Survey techniques like the one used here are often valuable, but they are known to have some methodological weaknesses (see point 2 above).
   - The internal data analytical validation technique. Here cross-validation is used, ensuring statistically reliable results.
   - General internal validity, meaning that the results we get conform to the purpose or goal of the measurements, which is fulfilled in the present case.
   - External validity, meaning that the results may be generalised to a broader population or setting, and/or given a causal interpretation. The present experiment should be repeated in other companies or at another time, to fulfil this level of validity. However, an external measure of decreasing sick-leave days in the banking company as a result of actions based on this project (cf. Appendix A13), may add to the total validity of the results.

4. There are several other possible uses of the same data analytical approach to study data from people, e.g.:
   - Chapter 2.3 showed another example of analysis of questionnaire data

from people, in that case concerning the meaning of *quality* for different items. Like the present example those data came from untrained people.

- The BLM method is used extensively in consumer and market research, as illustrated by the mini-example in Chapter 3.5 and in the literature (Helgesen et al., 1997; Vrignaud, 1999).

- Likewise, it may be used for studying individual differences between various *people,* from quantitative measurements of, e.g. brain activity to various qualitative reports of emotion and cognition, be it for sick or for healthy people.

# Chapter 14

## Analysis of a Heterogeneous Sample Set: Predicting Toxicity From Quantum Chemistry

*This example shows how hard, computer-generated descriptors from quantum chemistry can be related to soft, but relevant quality measurements of biological effects. In this quantitative structure-activity relationship (QSAR) study, expensive cancer-related toxicity (y) is to be replaced by cheap quantum chemical computations (X) in a wide class of pollutants (aromatic hydrocarbons) that are encountered in our environment. The BLM analysis sheds light on possible molecular mechanisms.*

*This example demonstrates how predictive modelling may be improved by the inclusion of square terms in X, and by jack-knifed elimination of useless X-variables. It illustrates some problems that arise when the samples are of quite different kinds, and how this may be reduced by splitting the samples into more homogeneous classes.*


## 14.1. INTRODUCTION

### 14.1.1. Purpose of the Project

The purpose of the project is to characterise toxic components in our environment by learning to predict biological effects from computed data.

The data (file: TOX-QUANTCHEM) were kindly submitted by Endre Anderssen and Kay J. Steen from the Norwegian University of Science and Technology, Trondheim, Norway.

## ❏ THE ATTRACTION OF COMPUTER-GENERATED MEASUREMENTS

Computer-based *"experiments"* are faster and cleaner than old-fashioned *dirty* experiments. There is a contemporary trend in science, towards replacing real-world experimental observations by theory-based, computer-generated descriptions of the physical world.

This is a valuable trend, for several reasons. But it may be overly tempting, because the researcher's feeling of being in control may be deceptive: If the theories are incomplete or the computational algorithms have problems, then the results may be misleading. The need to relate the computer-generated results to real-world observation is fundamental.

The present application demonstrates how the output from contemporary *computational chemistry* may by checked, and furthermore enhanced in relevance, by being related to real-world biological experimental results.

## ❏ THE CHEMICAL QUALITY OF OUR ENVIRONMENT

The data set concerns the quality of our environment. It comes from the research field of QSAR. A certain family of molecules (in chemistry called *aromatic* compounds), which are more or less toxic, is studied here. The *toxicity* in this case concerns the mutagenicity, i.e. the tendency for causing cancer-related mutations in living organisms.

To measure *toxicity* with sufficient precision and in a way that is relevant to humans, is difficult and expensive. The most common method to measure mutagenicity today is to check the rate of genetic alterations in certain micro-organisms. Could this be replaced by computer calculations? Moreover, it is difficult to understand precisely how different molecules cause mutation. Maybe that can be revealed by the molecular quantum chemical descriptors?

## ❏ BUILDING ON PREVIOUS STUDIES

Benigni and Richard (1996) published a paper on the problems of relating *toxicity* to molecular descriptors over a wide class of related aromatic compounds. Their conclusions were not very optimistic, but the reason for the problem was not clear.

Their reported data on the molecules' *toxicity* will here be used as target variable **y**, after a rescaling. But their molecular descriptors will be replaced with descriptors computed anew, and their segmentation of the molecules into calibration and test set will be done differently.

### 14.1.2. Purpose of the Data Analysis

The goal of the present data analysis is to establish calibration models for converting the *molecular descriptors* in **X** into estimates of the molecules' *toxicity*, **y**. Graphical interpretation of the modelling results may then indicate a causal molecular basis for their *toxicity*. Moreover, the model may later be used to predict the unknown *toxicity* from the computed *molecular descriptors* in the hundreds of other aromatic compounds that have not yet been tested biologically, in analogy to Chapter 12.

The challenge of the data analysis is to be able to handle such a heterogeneous sample set as the present one, involving three distinct classes of aromatic molecules, and spanning a wide range of *toxicity*. Modelling the data set in one joint model vs. in separate classes of samples will be compared.

## 14.2. CALIBRATION AND PREDICTION OF THE WHOLE DATA SET

*I. Purpose of a joint model.* A calibration model for later prediction of *toxicity* from *molecular descriptors* is to be explored and optimised. Before knowing the behaviour/effect of the different molecules, the whole data set is first studied.

*II. Experimental planning*

### 14.2.1. Selection of Calibration Samples

A total of $N = 51$ molecules were selected by Benigni and Richard (1996) from the genetic activity profiles (GAP) in the EPA/IARC GeneTox data bases, to represent chemicals listed as hazardous air pollutants under the Clean Air Act in USA. The set comprises three different, but related classes of molecules: 23 aromatic amines (AA), 22 nitro-aromatics (NO2A) and six polycyclic aromatic hydrocarbons (PAH). Benigni and Richard grouped the 51 molecules into two presumably representative sub-sets: 25 calibration samples (molecules here named *K1–K25*) and 26 test samples (here named *t26–t51*). Table 14.1 gives some details on the molecules.

The original paper did not provide further detail on the experimental design. Therefore, the data set represents an example of having to deal with a set of samples designed by someone else (cf. Chapter 11.2).

### 14.2.2. Selection of Variables

The biological *toxicity* data, in terms of the lowest effective dose (LED), are to be taken from Benigni and Richard (1996), based on the GAP database.

Seventeen *molecular descriptors* were selected to be computed in HyperChem™ (www.hyper.com, 2000) using the AM1 semi-empirical method (Dewar et al., 1985). The descriptors concern various molecular properties that might have an impact on the *biological effect*. They include size- and shape-factors, a variety of energetic properties plus other descriptors. More details on the X-variables are given in Table 14.2.

**Table 14.1.** Summary statistics of samples[a].

| CAS. no. | | Class | Log(LED) | Mass | y | 2 CV | K/t | Number |
|---|---|---|---|---|---|---|---|---|
| | | 1$'$ | 2$'$ | 3$'$ | 4$'$ | 5$'$ | 6$'$ | 7$'$ |
| 42397-67-8 | 1 | 1: NO2A | −1.480 | 292.3 | 3.946 | 1 | K | 1 |
| 75321-20-9 | 2 | 1: NO2A | −1.460 | 292.3 | 3.926 | 2 | K | 2 |
| 42397-65-9 | 3 | 1: NO2A | −1.410 | 292.3 | 3.876 | 1 | K | 3 |
| 05522-43-0 | 4 | 1: NO2A | 0.190 | 247.3 | 2.203 | 1 | K | 4 |
| 63041-90-7 | 5 | 1: NO2A | 0.030 | 297.3 | 2.443 | 2 | K | 5 |
| 00053-70-3 | 6 | 2: PAH | 0.720 | 278.4 | 1.725 | 1 | K | 6 |
| 00050-32-8 | 7 | 2: PAH | 0.760 | 252.3 | 1.642 | 2 | K | 7 |
| 00206-44-0 | 8 | 2: PAH | 0.960 | 202.3 | 1.346 | 2 | K | 8 |
| 00218-01-9 | 9 | 2: PAH | 1.140 | 228.3 | 1.218 | 2 | K | 9 |
| 00056-55-3 | 10 | 2: PAH | 1.140 | 228.3 | 1.218 | 1 | K | 10 |
| 00053-96-3 | 11 | 3: AA | 1.150 | 223.3 | 1.199 | 2 | K | 11 |
| 00092-67-1 | 12 | 3: AA | 1.040 | 169.2 | 1.188 | 1 | K | 12 |
| 00091-94-1 | 13 | 3: AA | 1.120 | 253.1 | 1.283 | 1 | K | 13 |
| 01582-09-8 | 14 | 1: NO2A | 1.260 | 335.3 | 1.265 | 2 | K | 14 |
| 00092-93-3 | 15 | 1: NO2A | 1.450 | 199.2 | 0.849 | 1 | K | 15 |
| 00086-57-7 | 16 | 1: NO2A | 1.400 | 173.2 | 0.838 | 1 | K | 16 |
| 00581-89-5 | 17 | 1: NO2A | 1.550 | 173.2 | 0.688 | 2 | K | 17 |
| 00063-25-2 | 18 | 3: AA | 1.460 | 201.2 | 0.844 | 2 | K | 18 |
| 00119-90-4 | 19 | 3: AA | 1.410 | 244.3 | 0.978 | 2 | K | 19 |
| 00092-87-5 | 20 | 3: AA | 1.480 | 184.2 | 0.785 | 1 | K | 20 |
| 00101-77-9 | 21 | 3: AA | 1.510 | 198.3 | 0.787 | 2 | K | 21 |
| 00101-14-4 | 22 | 3: AA | 1.460 | 267.2 | 0.967 | 1 | K | 22 |
| 00062-53-3 | 23 | 3: AA | 1.870 | 93.1 | 0.099 | 2 | K | 23 |
| 00095-53-4 | 24 | 3: AA | 2.330 | 107.2 | −0.300 | 1 | K | 24 |
| 00121-14-2 | 25 | 1: NO2A | 2.380 | 182.1 | −0.120 | 2 | K | 25 |
| 07496-02-8 | 26 | 1: NO2A | −0.180 | 273.3 | 2.617 | 1 | t | 26 |
| 20589-63-3 | 27 | 1: NO2A | −0.220 | 297.3 | 2.693 | 2 | t | 27 |
| 00059-87-0 | 28 | 1: NO2A | 1.090 | 198.1 | 1.207 | 1 | t | 28 |
| 00607-57-8 | 29 | 1: NO2A | 1.230 | 211.2 | 1.095 | 1 | t | 29 |
| 00067-20-9 | 30 | 1: NO2A | 1.280 | 238.2 | 1.097 | 2 | t | 30 |
| 00466-86-6 | 31 | 1: NO2A | 1.860 | 277.3 | 0.583 | 2 | t | 31 |
| 00099-55-8 | 32 | 1: NO2A | 1.910 | 152.1 | 0.272 | 2 | t | 32 |
| 00443-48-1 | 33 | 1: NO2A | 1.940 | 171.2 | 0.293 | 1 | t | 33 |
| 00056-75-7 | 34 | 1: NO2A | 2.350 | 323.1 | 0.159 | 1 | t | 34 |
| 00100-02-7 | 35 | 1: NO2A | 2.400 | 139.1 | −0.257 | 2 | t | 35 |
| 00606-20-2 | 36 | 1: NO2A | 2.500 | 182.1 | −0.240 | 1 | t | 36 |
| 00082-68-8 | 37 | 1: NO2A | 2.530 | 295.3 | −0.060 | 1 | t | 37 |
| 00086-88-4 | 38 | 3: AA | −0.820 | 202.3 | 3.126 | 2 | t | 38 |
| 23214-92-8 | 39 | 3: AA | −0.080 | 497.5 | 2.777 | | t | 39 |
| 77094-11-2 | 40 | 3: AA | 0.080 | 212.3 | 2.247 | 2 | t | 40 |
| 76180-96-6 | 41 | 3: AA | 0.230 | 198.2 | 2.067 | 2 | t | 41 |
| 00320-67-2 | 42 | 3: AA | 0.480 | 244.2 | 1.908 | 1 | t | 42 |
| 77500-04-0 | 43 | 3: AA | 0.560 | 213.2 | 1.769 | 2 | t | 43 |

Table 14.1 (*continued*)

| CAS. no. | | Class | Log(LED) | Mass | y | 2 CV | K/t | Number |
|---|---|---|---|---|---|---|---|---|
| 00494-03-1 | 44 | 3: AA | 0.800 | 268.2 | 1.628 | 1 | *t* | *44* |
| 00119-93-7 | 45 | 3: AA | 1.370 | 212.3 | 0.957 | 2 | *t* | *45* |
| 00134-32-7 | 46 | 3: AA | 1.490 | 143.2 | 0.666 | 1 | *t* | *46* |
| 00091-59-8 | 47 | 3: AA | 1.500 | 143.2 | 0.656 | 2 | *t* | *47* |
| 00069-53-4 | 48 | 3: AA | 1.950 | 349.4 | 0.593 | 1 | *t* | *48* |
| 04342-03-4 | 49 | 3: AA | 1.950 | 182.2 | 0.311 | 2 | *t* | *49* |
| 00569-61-9 | 50 | 3: AA | 2.090 | 323.8 | 0.420 | 1 | *t* | *50* |
| 00193-39-5 | 51 | 2: PAH | 0.180 | 276.3 | 2.261 | 1 | *t* | *51* |

[a] Row sample name: molecule identifier (CAS no.). Column 1: molecular class. Column 2: published biological effect: *log(Lowest Effective Dose,* μg/ml). Column 3: the molecular weight (*mass*). Column 4: toxicity on molar basis, $y = log(1/D,$ μgmol/ml), where $D = LED$. Column 5: sets for split-half cross-validation. Column 6: sets originally published for calibration (K) or test (t). Column 7: molecule (sample) number.

## III. Experimental work

### 14.2.3. Molecular Descriptors

For each of the 51 molecular species, the molecular structure was defined, based on their internationally acknowledged structure code (CAS no., Table 14.1). From this structure, 17 different *molecular descriptors* were computed (Table 14.2). All the computations were done within 12 h on a personal computer.

Being empirically calculated, the *molecular descriptors* have some uncertainty, due to possible model oversimplifications or algorithm oversimplifications. But the size of their uncertainty is unknown.

### 14.2.4. Toxicity

The biological mutagenicity was measured by the so-called *Ames test. Salmonella* bacteria were exposed to different concentration levels of each of the 51 compounds, and the frequency of observable mutations in the genetic material in the bacteria was recorded. From these measurements, the LED was calculated.

Since the different molecular species had been analyzed by different scientific groups, using different species of *Salmonella* bacteria, Benigni and Richard (1996) chose to report the mean logarithmic LED over whatever Ames test conditions were available for each molecular species. The originally published mean logarithmic LED values are shown in Table 14.1, column 1.

The precision of the reference input data is not known. However, since the experimental conditions are not the same for the different molecules, these *toxicity* data must be expected to have considerable uncertainty.

**Table 14.2.** Summary statistics of input variables[a].

|  |  | Min. | Max. | Mean | Std. Dev. |
|---|---|---|---|---|---|
|  |  | 1$'$ | 2$'$ | 3$'$ | 4$'$ |
| *TotE.* | 1 | −145322.50 | −22612.99 | −60643.30 | 20681.96 |
| *Dipole* | 2 | 0.00 | 8.29 | 3.46 | 2.39 |
| *HeatOfForm.* | 3 | −304.42 | 105.62 | 18.14 | 77.00 |
| *BindingE.* | 4 | −6803.34 | −1406.41 | −3058.68 | 967.52 |
| *AtomE.* | 5 | −138519.20 | −21135.52 | −57584.62 | 19976.17 |
| *Core–CoreE.* | 6 | 74704.16 | 1201223.00 | 328881.28 | 176656.95 |
| *HOMO* | 7 | −11.14 | −7.43 | −9.00 | 0.88 |
| *LUMO* | 8 | −2.48 | 0.62 | −1.04 | 0.77 |
| *HOMO–LUMO* | 9 | −9.27 | −6.02 | −7.96 | 0.78 |
| *Volume* | 10 | 367.17 | 1255.12 | 661.04 | 153.90 |
| *Hyd* | 11 | −21.49 | −0.66 | −8.73 | 4.65 |
| *LogP* | 12 | −0.64 | 6.06 | 2.86 | 1.76 |
| *Ref.* | 13 | 30.76 | 126.21 | 66.09 | 18.77 |
| *Pol.* | 14 | 11.79 | 48.89 | 26.10 | 8.05 |
| *Mass* | 15 | 93.13 | 497.50 | 232.15 | 70.11 |
| *Oval.* | 16 | 0.23 | 0.27 | 0.24 | 0.01 |
| *Area* | 17 | 258.78 | 706.36 | 408.93 | 77.39 |
| *Log(1/D,molar)*, **y** | 18 | −0.30 | 3.95 | 1.29 | 1.07 |

[a] Minimum, maximum, mean and standard deviation over the 51 input samples, for the 17 molecular descriptors and for the *toxicity* **y**. X-variables: *Tot.E.*, total molar energy; *Dipole*, dipole moment; *HeatOfForm*, heat of formation; *Binding E.*, binding energy; *Atom E.*, sum of atoms' energies; *Core-CoreE.*, sum of atoms' core energies; *HOMO*, highest occupied molecular orbital; *LUMO*, lowest unoccupied molecular orbital; *Volume*, molecular volume; *Hyd*, hydratisation energy; *LogP*, lipophilicity (partition coefficient between octanol and water); *Ref.*, molar refractive index; *Pol.*, polarizability; *Mass*, molecular weight; *Oval.*, ovality (1 = sphere, 0 = plane); *Area*, molecular surface area. Y-variable: *Log(1/D,molar)*, logarithmic$_{10}$ inverse of the lowest effective dose (μmol/ml).

## IV. Pre-processing and QC of data

### 14.2.5. Transforming the Input Variables into Comparable Units

It is an advantage to have the input data in comparable units. The original *toxicity* responses were published in a logarithmic$_{10}$ unit related to the *weight* concentrations (LED, μg/ml sample) of each compound, (cf. Table 14.1, column 2). But the *molecular descriptor* values were calculated for individual molecules. Since the compounds have different molecular weight, the units of the input *toxicity* and the *molecular descriptors* are not directly comparable.

The toxicity data were transformed in three steps. (1) The LED data were converted from a microgram basis to a *molar* μmol basis. (2) A *logarithmic* transformation was applied to the molar values, in order to compare values that range over several orders of magnitude. (3) To ensure that high toxicity is represented by high values instead of low values, the *inverse* toxicity (i.e. the negative logarithm) was used.

The transformed *toxicity* response is thus defined as the logarithmic inverse of the lowest effective dose, termed log(1/D, molar), and is given in the unit $\log((\mu\text{mol/ml sample})^{-1})$, which may also be written $-\log(\mu\text{mol/ml sample})$. The transformed input response data are shown as Table 14.1, column 4.

### 14.2.6. Making Data Tables

The 17 *molecular descriptors* from the quantum chemistry computation software were collected into a $51 \times 17$ data table **X**. Table 14.2 gives summary statistics for the 17 descriptors (columns in **X**, transposed), as well as the *toxicity* **y** (log(1/D, molar).

It should be noted that the 17 X-variables are meant to represent the quantum-chemical *information cloud* that act in the biological cells, irrespective of molecular class. Therefore, there are no explicit indicators of molecular *class* among the X-variables, only general molecular descriptors.

### 14.2.7. Graphical Inspection of Input Data

An overview of the table of the input data is shown in Figure 14.1a. The figure displays the transposed $51 \times 18$ data table, consisting of 17 X-variables and one Y-variable after standardization over the $N = 51$ molecules. The 51 molecules have been sorted according to **y**. It may be noted that sample *t39* stands out with particularly large deviations from the mean. Hence, we may expect this sample to represent a potential outlier in the subsequent modelling.

Figure 14.1b shows more detail on the target variable **y** = *toxicity*. This distribution is assessed as being suitable for multivariate modelling, with no apparent outliers. The range of **y** is seen to be from $-0.3$ to 3.9, which means that the LED, on a molar basis, ranges from $10^{0.3} = 2$ μmol/ml (not very toxic) to $10^{-3.9} = 0.00013$ μmol/ml (extremely toxic).

### V. Data analysis

### 14.2.8. Weighting of the Variables

Table 14.2 shows that the X-variables have very different ranges of variation. Hence, they were standardised so that they all had a mean of 0 and a standard deviation of 1 in the $N$ samples being modelled. The variable *toxicity* (**y**) will for simplicity be kept un-weighted, since the relative noise levels between **X** and **y** is irrelevant.

A small series of PLSR models will now be developed, in order to find an adequate way of handling outliers, non-linearities and sample class differences.

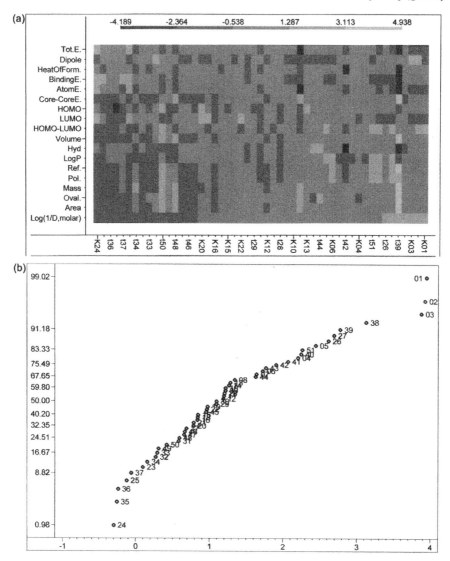

**Figure 14.1.** Input calibration data for the samples. (a) X- and Y-data (transposed). Overview of the 51 molecules (every 2nd is named) for the 17 molecular descriptors **X** and the target variable **y** *(toxicity)* = *log*(1/D, *molar*). (b) Y-data. Distribution of the quality criterion variable **y** in the 51 molecules, shown in normal probability plot.

Each model is assessed by cross-validation, and the development in RMSEP(Y) as a function of the number of PCs is used as a quality criterion.

First, the three classes of molecules will be modelled together. In Chapter 14.3 each of the molecule classes will be modelled separately.

### 14.2.9. Modelling With Only Linear Descriptors in X

The Y-variable *toxicity* is first regressed on the 17 X-variables by PLSR, with full cross-validation, i.e. $M = 51$ segments. The prediction error in **y** from this preliminary model is given by curve 1 in Figure 14.2. It shows a clear reduction in the cross-validated estimate of the prediction error RMSEP(Y), from the initial value of about 1.1 to about 0.86 with 2 PCs. Thereafter, the prediction error slowly increases again.

The *optimal* 2-PC model explains 37% of the CV-variance in **y** from **X**. This is not considered a satisfactory modelling. The input data have to be inspected in more detail in order to find the reason for this.

*Trying to find the reason for the unsatisfactory modelling*: In the modelling process, sample *t39* was automatically identified as an extreme outlier in **X**. Re-estimating the model after having removed outlier *t39* (not shown here) reduces the optimal rank from 2 to 1 PC, but reduces RMSEP(Y) only marginally, so outlier *t39* did not have much impact on this first model. What could then be the problem?

Figure 14.3 shows the input data for **y** vs. one of the important X-variables, the computed molecular *Volume* (shown without sample *t39*). The plot indicates a possible *curvature* in the relationship between these two variables, even though the pattern is far from clear. Perhaps improved modelling of **y** could be attained by including some non-linear terms in **X**, to account for this curvature?

### 14.2.10. Modelling With Linear and Squared Descriptors in X

In order to check the possibility of modelling X–Y response curvature, the initial X-matrix was now expanded with 17 new X-variables, the square terms $(\mathbf{x}_k - \bar{x}_k)^2$ for the molecular descriptors $k = 1,2,...,17$ (eq. (8.4), cf. Chapter 11.3).

The PLSR modelling was then repeated for all $N = 51$ samples, but now with 34 X-variables (all of which are standardised). The cross-validated ($M = 51$) prediction error RMSEP(Y) is given by curve 2 in Figure 14.2. The optimal prediction error is now attained already after 1 PC, and is reduced to RMSEP(Y) $= 0.85$, corresponding to 38% explained CV-variance in **y**. Hence, the inclusion of explicit curvature information in **X** has improved the prediction of **y**, although not much. After 1 PC the prediction error increases sharply again, so the 1-PC model is considered optimal for now. But this result is still not considered satisfactory.

Figure 14.4 shows the so-called *influence plot*, where the abscissa represents the leverage, i.e. the uniqueness of the samples *inside* the optimal model (eq. (9.3)), and the ordinate represents the cross-validated residual level RMSEP(X) (eq. (10.3a)), i.e. the variation remaining *outside* the optimal model (cf. Chapter 9). Sample *t39* was again tagged as an extreme outlier in **X**. Could removal of this outlier now improve the modelling?

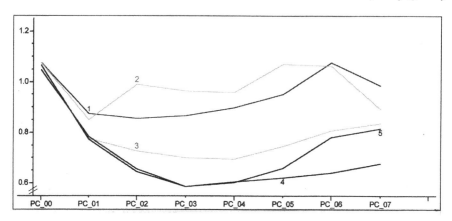

**Figure 14.2.** Root mean square error of prediction in **y**, RMSEP(Y)$_A$ for five different models, as functions of the number of PCs, $A$, estimated by full leave-one-molecule-out cross-validation. Model 1: $N = 51$ samples, 17 linear terms in **X**. Model 2: $N = 51$ samples, 17 linear and 17 square terms in **X**. Model 3: $N = 50$ samples (minus t39), 17 linear and 17 square terms in **X**. Model 4: 2nd iteration, 12 linear and seven square terms in **X**. Model 5: 3rd iteration, ten linear and five square terms in **X**.

### 14.2.11. Removal of a Gross Outlier Sample

Closer inspection shows that molecule *t39* (adriamycin) is quite special, primarily because it is a much larger molecule than the others. For instance, its molecular weight (*mass*) is as much as 498, while the others lie in the 90–350 Da range. Moreover, this sample is known as antibiotic, which may destroy the cell wall of the bacteria used in the Ames test to measure *toxicity*.

This is considered a good enough reason to remove this outlier *t39* and then re-estimate the model again. Four PCs could now be extracted. The optimal prediction error is now reduced to RMSEP(Y)$_{A=4}$ = 0.69 (curve 3 in Figure 14.2). This corresponds to 58% of the CV-variance in **y** predicted from **X**.

### 14.2.12. Removal of Useless or Detrimental X-variables

Only 47% of the CV-variance in **X** is explained by the 4-PC model. Could it be that **X** contains some variables that are *useless*, because they are irrelevant for **y**, and therefore could be ignored? Could it even be that some of these y-irrelevant X-variables are *detrimental*, because they are somehow correlated to other, more y-relevant X-variables, and are thereby drawn into the PLSR model and degrade it?

Re-estimating this model with only those X-variables that show a relatively significant regression coefficient ($P < 0.1$, estimated from the cross-validation) further reduces the prediction error, to RMSEP(Y)$_{A=3}$ = 0.58 (curve 4). This model has 12 linear and seven squared descriptors in **X**. Seventy percent of the

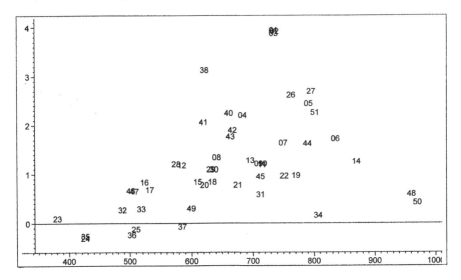

**Figure 14.3.** Checking the input data. The toxicity **y** (*log(1/D,molar)*; ordinate, plotted against one molecular descriptor, X-variable *Volume*, abscissa, for the 51 molecule samples. The numbers represent molecule number.

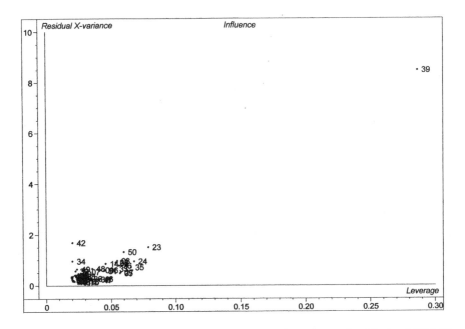

**Figure 14.4.** Detection of outlier (*t39*) in **X**. Mean square error of prediction MSEP(X) (ordinate) and the leverage $h_i$ (abscissa) for the 51 molecule samples after 1 PC in the model (model 2, Figure 14.2). The numbers represent molecule number.

CV-variance in **y** is explained by this 3-PC solution. It accounts for 51% of the variance in the remaining variables in **X**. To avoid over-fitting, this is taken as the resulting model, when all three molecular classes are modelled jointly.

Repeating this cross-validated elimination of useless X-variables once more (curve 5 in Figure 14.2) gave only a very small improvement of the predictive ability (RMSEP(Y)$_{A=3}$ = 0.57, 71% explained CV-variance in **y**). Repeating the cross-validated elimination yet another time did not further improve the predictive ability.

### 14.2.13. The Final Joint Calibration Model

*Graphical interpretation of the joint model*: Figure 14.5 presents the resulting model. Figure 14.5a shows the main pattern in score vectors, $t_1$ vs. $t_2$ for the 50 molecules (named by their number, cf. Table 14.1, column 7). The first PC, $t_1$, is seen to span primarily samples 23, 24 and 35 against samples 05 and 27, while PC 2 primarily spans samples 48 and 50 vs. the others.

The corresponding main relationship between the remaining variables is shown in Figure 14.5b in terms of the correlation coefficients between each X- and y-variable and $t_1$ (abscissa) and $t_2$ (ordinate). The X-variable *HOMO–LUMO* is seen to correlate most directly with the y-variable *Log(1/D,molar)*. Moreover, a general correlation pattern that relates *Area, Volume, CoreCoreE, Ref* and *Pol* against *TotalE, AtomE* and *BindingE* is apparent in the north-west–south-east direction, and a set of squared terms in the south-west direction.

*A misunderstanding corrected*: Given these patterns in Figure 14.5, one might at first glance expect for instance samples 50 and 48 to have higher-than-average *Area*, etc. and lower-than-average *Binding Energy, Total Energy*, etc. Inspection of the input data confirms the expectation for *Area*.

However, the inspection alerts us to the fact that *lower-than-average* does not always mean *weaker-than-average*. All the *Energy* variables, except the *Core–Core Energy*, have a negative sign in the input data, because they represent physical *heats of formation*. This means that the large molecules 50 and 48 do not have *weaker-than-average* energy. What they have is stronger-than-average negative energy. This misunderstanding illustrates the importance of checking conclusions from the modelling back into the input data.

One should likewise expect, e.g. molecules 5 and 27 at the upper right corner in Figure 14.5a to have both higher-than-average *toxicity* levels, *Log(1/D,molar)* and higher-than-average levels of the negative variable *HOMO–LUMO* (i.e. weaker *HOMO–LUMO*). This is indeed the case.

### 14.2.14. Prediction of y From X With Different Cross-validation Schemes

*Cross-validation with M = 50 segments*: Figure 14.5c shows the regression coefficient estimates (model 4, Figure 14.2), with the estimated reliability ranges: $\hat{\mathbf{b}} \pm 2\hat{s}(\hat{\mathbf{b}})$. The regression coefficients summarise the 3-PC PLSR model. The standard uncertainty $\hat{s}(\hat{b})$ were estimated from the cross-validation with $M = 50$ segments, as explained in Chapter 10.

The figure shows that the clearest and strongest predictors for toxicity are the *HOMO–LUMO*, *LUMO* and *HOMO*, plus the *HOMO**2* and *LUMO**2* curvature effects. Detailed molecular interpretation will not be pursued here, except for the most obvious effects. The most important contributions to the prediction of toxicity are provided by various versions of the two electron-energy variables *HOMO* and *LUMO*. The input variable *LUMO* (lowest unoccupied molecular orbital) has *negative* sign (except for some molecules where it is weakly positive). Its *negative* contribution ($\hat{b}_{\text{LUMO}} = -0.18 \pm 0.14$) therefore indicates that the more strongly negative *LUMO* is, (the stronger negative energy it has), the more toxic it tends to be. But the positive contribution of *LUMO**2* indicates an upward-faced curvature of this tendency.

Likewise, since *HOMO–LUMO* has *negative* sign, its high *positive* contribution ($\hat{b}_{\text{HOMO–LUMO}} = 0.24 \pm 0.13$) means that the closer to zero this property is, i.e. the more easily excitable electrons it has, the more toxic the molecule tends to be. When the input data for **y** were plotted vs. *LUMO* and vs. *HOMO–LUMO*, these model interpretations were confirmed.

*Representative split-half cross-validation, with $M = 2$ independent test sets:* Figure 14.6a gives a different estimate of the reliability of the same solution. After being sorted in descending values of the input variable $\mathbf{y} = Log(1/D, molar)$, the samples were here split into two independent, but presumably representative sample sets, numbered 1 and 2 with 25 molecules in each (Table 14.1, column 5), by separating the odd samples 1,3,5,... from the even samples 2,4,6,..., cf. Appendix A10.1. Again, outlier *t39* was kept out from the modelling. The figure shows the regression coefficient estimates obtained from this split-half cross-validation: $\hat{\mathbf{b}}_{-m}$, $m = 1, 2$. The two independent coefficient estimates are quite similar.

When calibrating on the first set and then using this model $\hat{\mathbf{b}}_{-2}$ to predict **y** from **X** in the second set, the optimal prediction error in the even set is $\text{RMSEP(Y)}_{A=3, \text{set2}, -2} = 0.63$. Conversely, when calibrating on the second sample set, the optimal prediction error in the first set is $\text{RMSEP(Y)}_{A=3, \text{set1}, -1} = 0.54$. The root-mean square average of the two sets is $\text{RMSEP(Y)}_{A=3} = 0.59$. This is about the same as the prediction error estimated with $M = 50$ segments ($\text{RMSEP(Y)}_{A=3} = 0.58$, model 4) in Figure 14.2.

Figure 14.6b summarises the model in terms of the predicted toxicity $\hat{\mathbf{y}}$ (ordinate), obtained from the two-segment cross-validation, plotted against the measured toxicity **y** (abscissa). The figure shows a clear predictive ability. The correlation coefficient between measured and predicted **y** is 0.83. However, for a few molecules the agreement between measured and predicted toxicity is bad. This could indicate that the *toxicity* is influenced by some additional *lurking* phenomena not represented among the 17 available quantum chemical descriptors.

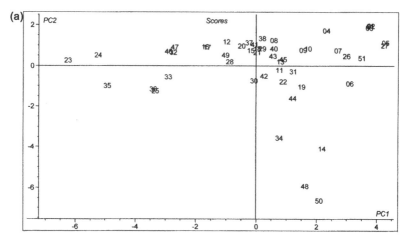

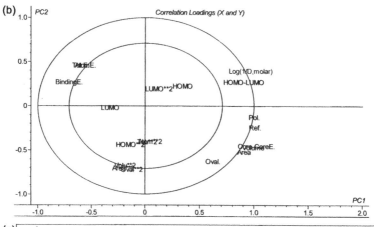

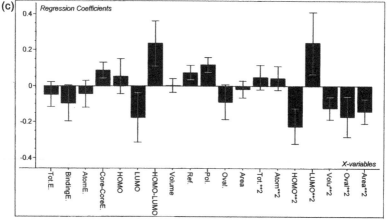

### 14.2.15. Comparing the Originally Published Calibration and Test Sets

Now we repeat this modelling (Figure 14.6a,b), but with the two-segment cross-validation based on the two sample set $K$ and $t$ prescribed by the original authors (Table 14.1), (instead of on the two systematically chosen segments 1,3,5,... and 2,4,6,...). The minimal prediction error now increases from the previous 0.58 to $RMSEP(Y)_{A=3} = 0.80$.

The reason is probably that these original sets differ considerably from each other in how well $\mathbf{y}$ is distributed; set $K$ has a wider, but more uneven distribution than set $t$. This will later be studied in more detail for each class separately.

### 14.2.16. Too Much Heterogeneity for Joint Modelling?

In the analysis presented above, the three molecular classes (PAH, AA and NO2A) were modelled together in one model. In their original paper, Benigni and Richard (1996) worried that the three classes were too different to allow meaningful prediction of *toxicity* from a joint calibration model; their mechanism of toxicity may be different.

When again repeating the modelling with $N = 50$ calibration samples (Figure 14.6a,b), this time with the three molecular classes as $M = 3$ cross-validation segments, the optimal prediction error further increased to $RMSEP(Y)_{A=1} = 0.94$ (which means only 23% of the explained CV-variance in $\mathbf{y}$). The predicted results are shown in Figure 14.7a. The increased prediction error indicates that the original authors were right: the three classes appear to differ somewhat in how the *molecular descriptors* $\mathbf{X}$ relate to the *toxicity* $\mathbf{y}$.

*VI. Conclusion for the joint model.* Three different classes of molecules were revealed. Each of the three classes will in the following be modelled *separately* (cf. Chapter 9).

### 14.3. THREE CLASSES OF SAMPLES MODELLED SEPARATELY

*I. Purpose of separate models.* The present sample set appeared to be very heterogeneous. The three classes of molecules will be analysed separately

---

**Figure 14.5.** Overview of model 4 ($M = 50$ CV segments). (a) Score plot of the samples for the first 2 PCs. Latent variables $\mathbf{t}_1$ (abscissa) and $\mathbf{t}_2$ (ordinate), for the 50 molecules (numbered). The numbers represent molecule number (cf. Table 14.1). (b) Loading plot of the variables for first 2 PCs. Correlation loadings, i.e. correlations between the input X- and y-variables and the latent variables $\mathbf{t}_1$ (abscissa) and $\mathbf{t}_2$ (ordinate). The concentric ellipses represent 50 and 100% described variance, respectively. (c) Regression coefficients with reliability ranges $\hat{\mathbf{b}}_{A=2} \pm 2\hat{s}(\hat{\mathbf{b}}_{A=2})$.

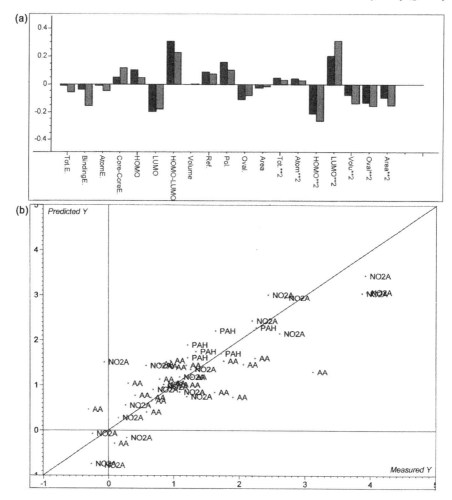

**Figure 14.6.** Summary of model 5 ($M = 2$ representative split-half CV segments). (a) Regression coefficients: two independent estimates: $\hat{\mathbf{b}}_{A=2}$ for each of the two segments (samples 1,3,5,.., and samples 2,4,6,.., after sorting for increasing $\mathbf{y}$). (b) Predictive ability for *toxicity*: $\hat{\mathbf{y}}_{A=2}$ (ordinate) vs. measured $\mathbf{y}$ (abscissa). The samples are named according to their molecular class.

in order to improve the prediction of *toxicity* $\mathbf{y}$ from the 34 *molecular descriptors* $\mathbf{X}$.

***II–IV. Experimental planning, work and pre-processing and QC of data***. These steps were the same as for the whole data set analysed above. The data set was split into three molecular classes, i.e. NO2A, PAH, AA.

***V. Data analysis of each class separately***. In each class a full leave-one-mole-cule-out cross-validation was used, and the modelling was repeated once after having removed X-variables whose jack-knifed regression coefficients were not significantly different from 0. Closer inspection of the three class models (not shown here) revealed **X–y** correlation patterns quite similar to those in Figure 14.5b. However, the classes differed in some nuances (cf. Appendix A14).

Figure 14.7b–d shows the predictive performance for each of the three classes modelled separately. The ordinate shows the cross-validated *toxicity* $\hat{\mathbf{y}}$, and the abscissa, the measured *toxicity* **y**.

### 14.3.1. Class 1: Nitro-aromatic Compounds

In class 1 (NO2A) the leave-one-molecule-out cross-validation between the 22 molecules showed $A_{Opt} = 3$ PCs. In Figure 14.7b the predictive error RMSE-P(Y)$_{AOpt=3}$ is now as low as 0.49 (88% explained CV-variance). Graphical inspection (not shown here) indicated the model to be molecularly plausible.

But the figure shows that the distribution of samples within the originally published calibration set (*K*) and test set (*t*) is different. The three most toxic molecules (*K01*, *K02* and *K03*) all belong to the calibration set, while the two least toxic molecules (*t35* and *t36*) belong to the test set. Hence the original splitting of samples in this class was rather odd; the K and t sets cannot be expected to represent each other very well.

### 14.3.2. Class 2: Poly-cyclic Aromatic Hydrocarbon Compounds

In class 2 (PAH) there is, unfortunately, only six molecules. The leave-one-molecule-out cross-validated prediction error RMSEP(Y)$_{AOpt=2}$ (Figure 14.7c) is here as low as 0.06 (98% explained CV-variance). This could indicate that the toxicity could be very easily predicted from molecular descriptors in PAH molecules. It is surprising that the cross-validation modelling is able to predict the most extreme sample *t51* so well from the other five samples.

*Problems*: With so few samples the risk of incidental effects is so high that one cannot really rely on statistical validation. Perhaps the six molecules (sampled for *toxicity* assay out of the large class of PAH molecules) happened to be particularly easy to model? After all, the range of *toxicity* among the chosen PAH molecules (Figure 14.7c) is rather narrow, compared to the other two classes, so non-linear relationships would hardly be observable.

Moreover, the regression coefficient $\hat{\mathbf{b}}$ gives an interpretation problem. The *dipole moment* of the PAH molecules is found to be the most dominating X-variable for the prediction of **y**. However, the dipole moment in PAH molecules is known to be very small, and covers a very small range in the PAH data set. To make things worse, these small dipole moment values are known to be particularly difficult to calculate in the computer program that generated the X-data, so the small variations observed may be particularly erroneous, reflecting something else other than the real dipole moment.

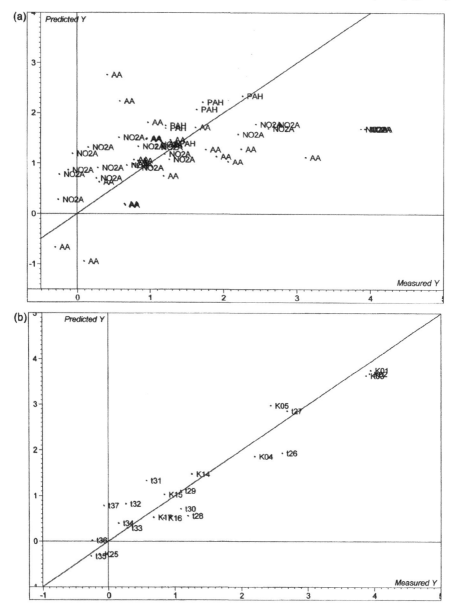

**Figure 14.7.** Predictive ability between and within the three classes of samples. $X =$ *molecular descriptors*; $y = $ *toxicity, log(1/D,molar)*. Abscissa: *Toxicity* $y$ measured. Ordinate: *Toxicity* $\hat{y}$ predicted in cross-validation at optimal rank. (a) All three classes; each class is predicted from the two other classes: $r_{prediction} = 0.47$. (b) Class 1 (NO2A): $r_{prediction} = 0.93$. (c) Class 2 (PAH): $r_{prediction} = 0.99$. (d) Class 3 (AA): $r_{prediction} = 0.94$. The predictions for four outliers *t39, t48, t49, t50* are also given. cf. Appendix A.14.

Through the standardisation of the X-variables, the small dipole variation was blown up to the size of the other input variations, so it is given a chance of affecting the model. The almost perfect predictive ability could then be due to an incidental combination of sampling errors and errors in the dipole data.

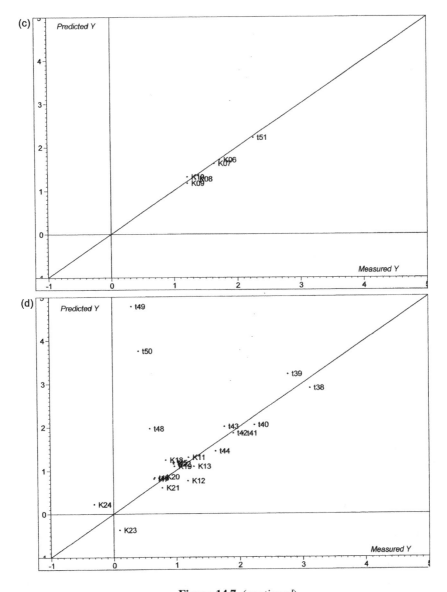

**Figure 14.7.** (*continued*)

Thus, the extremely good results from this small PAH class may be correct, but they may also be incorrect. The conclusions need to be supported by more empirical data.

Even in this PAH-class, the original selection of calibration (K) and test (t) molecules is seen to be odd: there is only one test molecule (*t51*), and this is extremely *toxic* compared to the five calibration samples.

### 14.3.3. Class 3: Aromatic Amines

When class 3 (AA) was modelled alone, the structure in the data could be scrutinised more closely. In addition to the gross outlier already identified, *t39*, three other samples were identified as outliers in this class. Abnormal X-data were found in *t48*, *t50*, moderately abnormal X-and y-data were found in *t49*. These were therefore also removed from the calibration modelling after some trial-and-error. A leave-one-out cross-validated RMSEP(Y)$_{AOpt=2}$ of 0.26 (88% explained CV-variance) was then found for the remaining 19 AA molecules.

The *toxicity* of the outliers was predicted by submitting their X-data to the obtained AA model afterwards. Figure 14.7d shows them together with the cross-validated predictions for the remaining 19 molecules. It confirms that outliers *t48*, *t49* and *t59* do not fit the obtained class model; they may possibly constitute a class of their own.

However, it comes as a surprise that sample *t39*, originally tagged as a *gross* outlier, was actually quite well predicted in Figure 14.7d. This may be due to the fact that the X-variables that made *t39* abnormal (those related to molecular size) had been more or less eliminated as unreliable in the calibration process.

As for the previous two classes, the original selection of calibration and test set samples is seen to be rather odd. In fact, the two sets *K* and *t* hardly overlap in Figure 14.7d. This is not a problem for our present analyses. But it indicates why the original authors found their calibration model to have low predictive success in the test set.

*VI. Conclusion for the separate classes*. Data analysis of the separate classes improved the predictive ability and contributed to a more causal understanding of the relationships. However, the separate data sets were too small to draw strong conclusions.

### 14.4. CONCLUSIONS

The data analysis is considered reasonably successful, particularly after splitting the sample set into three separate molecular classes. Calibration models were attained with good predictive ability and plausible interpretation, in spite of uncertain input data for *toxicity* y, and in spite of a rather heterogeneous and oddly distributed set of molecules. The correlation coefficient between measured *toxicity* **y** and the CV-predicted *toxicity* **ŷ** increased from 0.83 in the original, joint model to 0.95, 0.99 and 0.94, respectively within the three individual classes NO2A, PAH and AA, respectively.

In general, it seems that the molecular descriptors *LUMO* and *HOMO–LUMO* are particularly important for predicting the toxicity. In addition, the dipole moment was particularly important in the AA and PAH classes, but not in the NO2A class. The importance of some squared X-terms indicate somewhat curved **X–Y** relationships in the NO2A and AA classes; the PAH class was too small for this to be studied.

More data is needed in order to elucidate why the AA model did not fit outliers *t48, t49 and t50*. Moreover, the PAH class was very small ($N = 6$) and showed a strong effect of dipole moment that was difficult to understand. Hence, the good PAH results are regarded with considerable scepticism.

The original split between the calibration (K) and test (t) set yielded two sample sets not representative of each other. This is an experimental design problem, and may in part be responsible for the unsatisfactory test set results originally published.

For the NO2A and AA classes, the obtained models are probably good enough to be used for predicting unknown toxicity **y** from computed molecular descriptors **X** in new molecules of the same kinds, at least for screening purposes. But the predicted toxicity values should be verified experimentally before being relied upon in critical applications.

> It can be problematic to use the same statistical testing method repeatedly both for *optimising* and for *assessing* a model. The estimate of the prediction ability may then look better than it should be. In the present case we have used *jack-knifing* for identifying X-variables to be eliminated, and the associated technique *cross-validation* for estimating the final prediction error RMSEP(Y). In principle, our results may therefore be over-optimistic.
>
> In the present case, the total number of X-variables is low (<40), so this is not considered a problem. The risk of spurious correlation effects in the cross-validated BLM is then relatively low. Other error sources such as sampling problems, errors in the X- and Y-data and non-linearities in the **X–Y** relationships are probably far more important.
>
> However, this *optimise & assess* topic is important if the data set has hundreds or thousands of very noisy, irrelevant X-variables, and few samples. The final, jack-knife-optimised PLSR model may then be contaminated by several irrelevant X-variables that by chance happen to correlate ''significantly'' to **Y** in the cross-validation/jack-knifing sense. With such input data it is important *a priori* to reduce the number of independent X-variables in the input data, e.g. by averaging or smoothing (Figure 3.2), or simply by scaling down or passifying/removing all the least promising X-variables.

## 14.5. TEST QUESTIONS

1. What does this example tell about the saying 'math is cheaper than physics'?

2. How was the present data set designed, and how could it have been improved?
3. How was the initial model improved?
4. What are the limitations of the final results?
5. Which uses of the same data analytical approach can you think of?


## 14.6. ANSWERS

1. Since the molecular descriptors were computed over-night on a PC, based on the molecular structure formula, while the toxicity data were obtained by laborious biological assays, the X-data are much cheaper than the Y-data.
2. The design of the present data set is unclear, since the molecule identification and the Y-data were taken from a published toxicology paper, where very little is said about the experimental design. An alternative would be to use principal properties design (Chapters 7 and 11). (a) Compute the molecular descriptors **X** for a much larger set of molecules from the NO2A, PAH and AA classes. (b) Perform a PCA on these X-data, and select a subset of molecules that span and cover the score space for the first few PCs. (c) Measure toxicity of the selected molecules by biological assay (preferably with one single experimental procedure, not by different procedures for different molecules).
3. The initial model was improved in several stages: (a) removal of a gross outlier, (b) inclusion of square terms in **X**, (c) elimination of useless X-variables and (d) splitting the sample set into the *a priori* known classes.
4. The limitations of the final results are primarily (a) that class AA is too heterogeneous to be modelled as one class, and (b) that the number of molecules in class PAH was very low.
5. This way of analysing data has already proven very useful for interpreting data and making predictions in applied quantum chemistry. However, it may also be used for other computer-generated data. Example: computer simulation models are increasingly used for studying the performance of various complex systems, ranging from computer chips and central telephone switches to power plants and weapons systems. However, just because something is *computer-generated* of course does not mean that it is *right*! The present multivariate soft modelling may offer a simple way to understand and verify complex computer simulation systems.

# Chapter 15

# Multivariate Statistical Process Control: Quality Monitoring of a Sugar Production Process

*This example shows how the BLM method may be used for analysis of time-continuous systems. It concerns multivariate statistical process control (MSPC) for monitoring the quality of an industrial production process, and for detecting unexpected drift problems.*

*A sugar refinery is followed throughout the difficult start-up period for a new production season. A high-speed multichannel instrument (auto-fluorescence spectrophotometer, X) is calibrated and then used for predicting the main quality criterion of the sugar: its ash content (y). After some time, the bi-linear model warns for serious non-stationarity in the process. These unexpected problems are studied, and the model is updated.*

## 15.1. INTRODUCTION

### 15.1.1. Purpose of the Project

This application example concerns the quality of the product from an industrial process: in this case the purity of sugar from a sugar refinery.

The data (file: MSPC-SUGAR) have kindly been lent us by Bro (1998) and the DANISCO company (Bro, 1999). The project background has been adapted didactically).

❐ THE SUGAR REFINING PROCESS

Sugar for ordinary food use is a chemical compound (sucrose) produced in very high quantities with very high purity at a low price. In Europe, sugar is produced from sugar beet. The production takes place mainly during the autumn, when the beets are harvested.

The refining process is quite complex. The sugar-containing beet juice passes back and forth through a maze of pipelines and a series of filtering and centrifuging steps, whereby the non-sugar components of the juice are gradually removed. Finally, the almost pure sugar solution is dried and becomes the powder we buy in the shop.

Because of this complexity, at the start-up of the production each new autumn, it takes some time before the process stabilises to yield a sufficiently pure sugar. It is therefore important to have an easy way of assessing when the process has stabilised enough to yield sugar with sufficient purity. Even later on in the process this is important, because the raw material may unexpectedly change, e.g. due to freezing, and this calls for process changes.

The quality of the produced sugar is often assessed by measuring the amount of inorganic matter in the product. This is traditionally done by measuring the *Ash* content, expressed, e.g. in % of the dry matter. The level of this quality criterion *Ash* should be very low, in the range $<0.01\%$.

The main problem with this traditional *Ash* measurement is that it requires a chemical laboratory and takes too long, compared with the rate of change in the production process. Therefore, a faster, on-line method for assessing the *Ash* percentage would be nice.

❐ SENSITIVE FINGERPRINT: AUTO-FLUORESCENCE SPECTRA

It has long been known that impurities in sugar cause the product to emit light in the invisible ultra-violet (UV) and visible range, when subjected to (*excited by*) light at lower wavelengths in the UV range (Munck et al., 1998).

Readers who have visited exciting discotheques have seen auto-fluorescence with their own eyes. Commercial washing powder leaves a compound with auto-fluorescence, which makes white shirts look *extra white* in ordinary light. In the more intense UV light excitation of the discotheque, this compound makes the white shirts glow with an appalling blue fluorescence emission. This effect is called *auto*-fluorescence, because such compounds emit light automatically on their own, without any further additives.

The intensity of the emitted light depends on the concentration of the fluorescing compound. This makes auto-fluorescence a very attractive measurement principle: Contrary to optical density (OD) (Chapters 4–7 and 12) the emitted fluorescence light increases from *zero* when the concentration of fluorescing compounds increases from zero. Hence, fluorescence measurements can be much more sensitive than, e.g. NIR measurements.

In analogy to NIR spectroscopy, fluorescence spectroscopy gives information-rich *fingerprints*. Many different constituents may contribute to the measured spectrum. By multivariate reverse engineering by BLM, it is possible to *unscramble* the spectra into very useful information, even in *dirty*, ill understood systems. This is how the slow and cumbersome *Ash* determination will now be speeded up, and this is how unexpected process problems will be detected later.

## 15.1.2. Purpose of the Data Analysis

This example illustrates how quality may be monitored over time, how unexpected developments may be discovered and how models can dynamically be updated. It utilises the methodology illustrated in the stylised process example on LITMUS in Chapter 7.

The goal of the present data analysis is to establish a calibration model for predicting the *Ash* (**y**) content from the *auto-fluorescence* spectra (**X**) of the finished product from a certain sugar refinery. More specifically, a predictor of *Ash* is to be developed as quickly as possible during the start-up period of the refinery, to be used for rapid monitoring of how the process stabilises. Later, this predictor model will be updated, if necessary.

□ TIME SERIES DATA AND NON-STATIONARY PROCESSES

In this book, models are developed on the basis of empirical data from a range of samples of a certain kind. In the present example, the samples come from a certain sugar processing factory, measured by a certain methodology, obtained as *time series data*. Once developed, the models may be applied to data from later samples.

The empirical **X–Y** model is only valid as long as the new samples are of the same general kind as those in the calibration set. More specifically, variability types in **X** not seen clearly in the calibration set but occurring in new samples, will make the Y-predictions in the new samples unreliable. If the process or an analytical instrument changes systematically over time, we have a *non-stationary process* at hand, and the statistical concept of *the population* becomes even more difficult than it already is for stationary processes (cf. 11.2).

Luckily, the bi-linear modelling will often detect such problems automatically and give outlier warnings. Then it is important (a) to investigate what is changing (process or instrument) and (b) to update the **X–Y** model.

However, sometimes the non-stationarity of a process may occur in the nature of **Y** and its relationship to **X**, not in **X** itself. That cannot be detected from X-data alone. Therefore it is important also to keep on measuring **Y** once in a while, to check the predictive performance of the model, and update it if necessary.

## 15.2. CALIBRATION MODEL FOR THE FIRST TIME PERIOD

*I. Purpose of start-up calibration.* The goal for the first data analysis is to establish a calibration model for predicting the *Ash* (**y**) content from the *auto-fluorescence* (**X**) of the product for the start-up period.

*II. Experimental planning*

### 15.2.1. Selection of Calibration Samples

The sugar produced in a Danish sugar refinery was sampled at relatively even intervals for the first 44 hours (h) of the production campaign, totalling 60 calibration samples. They are termed sampling *Time 1–60.*

Due to lack of prior data, it was hoped that the unwanted, but unavoidable process variations in the start-up period were sufficient to span the variation necessary for giving a valid calibration model.

The process was likewise sampled for the subsequent 24 h, yielding a set of 46 new samples. The calibration model will be applied to these new samples. If necessary, the new samples will be used for updating the model.

### 15.2.2. Selection of Variables

*Process analysis:* The *auto-fluorescence* measurements were to be made in a Perkin Elmer PE LS50B spectrofluorometer.

The refined product is dissolved in non-buffered water at a concentration of 2.25 g/15 ml and the solutions measured in a 10 × 10 mm cuvette at a 90° angle between excitation and emission light.

*Product quality measurement:* The percentage of *Ash* was to be measured by the technique traditionally used in the sugar industry (conductivity).

Conductivity determination of ash: 28 g sugar is dissolved in 100 g deionised water. The conductivity $C$ of this solution as well as of pure deionised water are measured, and the *ash* content computed from: $\%Ash = 0.000576 \times (C_{\text{sugarsolution}} - 5 \times C_{\text{water}})$.

*III. Experimental work*

### 15.2.3. Process Analysis

The obtained sugar samples from the factory were analysed in the spectrofluorometer. The samples were *excited* by shining UV light on them at 230, 240, 290, and 340 nm. At each wavelength the light spectrum emitted in the 275–560 nm range was measured, at 0.5 nm steps, totalling 571 channels. In total 2284 excitation/emission channels were thus obtained and stored.

### 15.2.4. Product Quality Analysis

The obtained sugar samples were also sent for analysis of *Ash* percentage, which was entered into the database.

## IV. Pre-processing and QC of data

### 15.2.5. Making Data Tables

The four excitation × 571 emission channels were defined as 2284 columns. One table was made with the first 60 calibration samples as rows. The raw spectral data were inspected visually to check for gross errors. Later, another table was similarly made for the subsequent 46 *new* samples.

### 15.2.6. Initial Data Reduction

The shape of the raw spectra, supported by prior theory about emission spectra of this type, indicated that no information would be lost by reducing the number of wavelength channels (Figure 3.2). Therefore, for each sample the average was taken of ten adjacent emission wavelength channels, yielding a total of 228 excitation/emission channels.

Furthermore, emission wavelengths near or below the excitation wavelengths are useless due to scattered excitation light. Therefore, at 290 nm the wavelength range 275–312 nm was eliminated, and at 340 nm the range 275–372 nm was eliminated.

These reduced auto-fluorescence spectra, with 200 wavelength channels, were defined as the $K = 200$ input variables $X$ for the $N = 60$ calibration samples. The corresponding *Ash* data were defined as target variable $y$.

Later, the same was done for the subsequent $N_{new} = 46$ samples.

### 15.2.7. Graphical Inspection of Calibration Input Data

$y$: Figure 15.1a shows the quality criterion *Ash* percentage in the sugar, for the first 60 samples at the beginning of the production campaign. The mean of these data is $1.3 \times 10^{-2}\%$, and the total initial standard deviation is $0.58 \times 10^{-2}\%$.

The figure shows a general decrease in the impurities until after about 25 sampling times. Thereafter it appears to stabilise, although not completely. At *Time 7* the *Ash* content is particularly high; this sample may be expected to behave as an outlier in the subsequent analysis.

$X$: Figure 15.1b shows how four typical wavelength channels develop over the first 60 points in time. The questions are now:

1. To what extent is it possible to predict $y = Ash$ from $X = auto\text{-}fluorescence$?

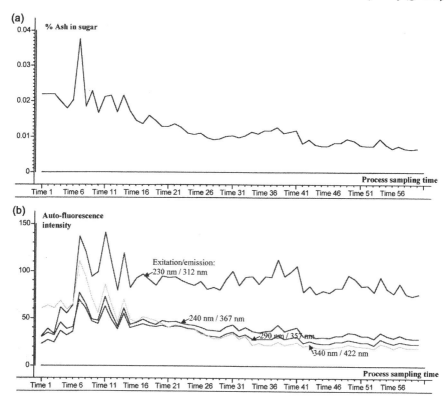

**Figure 15.1.**    Input data from the start-up of a sugar refinery process. Data are plotted as time series for the first $N = 60$ sampling *Times* (44 h). (a) **y** = quality criterion: *Ash* percentage (measured by a traditional, slow laboratory method). (b) Four (out of 200) X-variables: high-speed light *auto-fluorescence* at the excitation/emission wavelengths 230/312, 240/367, 290/357 and 340/422 nm.

2. To what extent does the **X** = *auto-fluorescence* data have *other* useful information about the quality and stability of the sugar refining process?

Figure 15.2a shows the full 200-channel input spectra in the 60 calibration samples. The four excitation-wavelengths are easily distinguishable. Moreover, the spectra appear to contain systematic variation patterns. No gross outlier samples can be seen in the spectra.

## V. Data analysis

### 15.2.8. Weighting of the Variables

*A priori* the noise level is expected to be about the same at all 200 wavelength

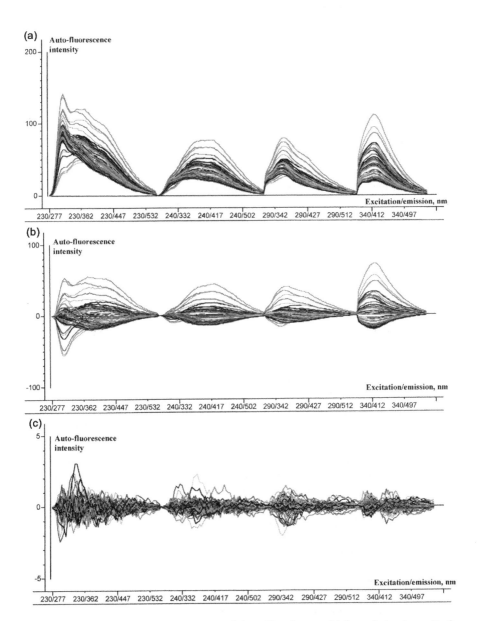

**Figure 15.2.** Auto-fluorescence spectra of the calibration set. Light emission intensity for the first 60 samples at excitation wavelengths 230, 240, 290 and 340 nm/emission wavelengths 280–560 nm. (a) Input spectra $\mathbf{X}$ (60 × 200) (b) Centred input spectra $\mathbf{E}_0 = \mathbf{X} - \bar{\mathbf{x}}$. (c) Residuals $\mathbf{E}_4$ after 4 PCs, amplified 20 times.

channels. Hence, we decide to use unweighted modelling (i.e. weights = 1 for every X-variable). Even **y** is left unweighted.

### 15.2.9. Calibration Modelling of y vs. X

The Y-variable *Ash* was regressed on the 200 X-variables by the bi-linear modelling method (PLSR). Figure 15.2b shows the mean-centred X-data, $E_0 = X - \bar{x}$. This is the variability in the calibration data that could possibly be used for modelling the variations in $Y = Ash$.

Figure 15.2c shows the corresponding residuals $E_4$ after 4 PCs. Note that the ordinate scale is only a 20th of the previous one. This means that most of the variation in the X-data has been picked up by the first 4 PCs. The shape of the residuals indicate that what remains is mostly random measurement noise.

In order to have a reasonably conservative model assessment, a cross-validation was applied with $M = 20$ segments systematically chosen with three consecutive samples in each, i.e. in the time-segmentation pattern $m = 1,1,1,2,2,2,3, 3,3,\ldots,20,20,20$ (cf. Appendix A10.1).

The percent explained variance is shown in Figure 15.3 from the cross-validation (black curves) as well as from the unvalidated model fit (grey curves). Figure 15.3a shows that most of the X-variation (98.6%) is accounted for by the first 2 PCs. The cross-validated and non-validated curves are almost identical, which is good. The cross-validated residual level in **X**, RMSEP(X), was correspondingly reduced from the initial value of 10.6 to 4.2, 1.4, 0.66 and 0.45 optical units by the first 4 PCs, respectively.

Figure 15.3b shows similar results also for **y**, although not quite as clear. Four PCs are seen to give predictive contributions for **y**. The four-factor solution explained 89% of the cross-validated variance in **y**. The estimated prediction error in **y**, RMSEP(Y) changed from 0.60 to 0.33, 0.26, 0.22 and 0.20 ( $\times 10^{-2}$) for models with $A = 0, 1, 2, 3$ and 4 PCs, respectively.

Since Figure 15.3a showed PCs 3 and 4 to be based on less than 1.4% of total initial variance in **X**, they probably also amplify some noise in **X**. Hence, there is reason to be hesitant about using more than 2 PCs in the prediction model for **y**. The 2-PC model explains 81% of the cross-validated **y**-variance. That is considered good enough for getting started.

### 15.2.10. Main Model Overviews

*Score plot*: Figure 15.4a shows the configuration of 60 samples for the first 2 PCs, in terms of score vector $t_1$ (abscissa) and score vector $t_2$ (ordinate). Consecutively numbered sampling times are connected by line segments.

The figure shows PC1 ($t_1$) to describe how the process gradually changes from *Time 7* via, e.g. *13* and *31* to *Time 60* (near *Time 35*). The second PC (or rather, the north-west/south-east diagonal $t_2$ $t_1$) mainly describes the variation from *Time 1* to *Time 7*. But other variations, e.g. between *Time 11–15* and *28–31* and beyond *Time 38* are also picked up in this more or less diagonal direction.

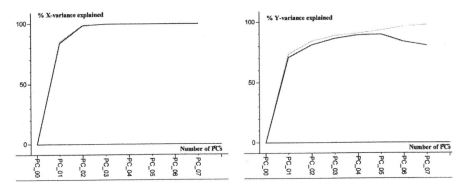

**Figure 15.3.** How the model fits the X- and y-data. Percent explained variance $= 100\% \times$ $(MSE_0 - MSE_A)/MSE_0$ vs. number of PCs, $A$. Black curves, cross-validation; grey curves, calibration fit. (a) *Auto-fluorescence*, **X**, (b) *Ash*, **y**.

This score plot shows that the process has been changed in several disjointed stages, illustrating how the process operators have worked hard to adjust the sugar refinery from its start (*Time 1*) to an acceptable set of stable processing conditions (presumably, at *Time 60*).

This type of two-way map of the process gives a compact representation of the state of the process. Later on, when X-data of new samples arrive, their corresponding scores may be computed and plotted, e.g. on top of the calibration samples.

*Loading plot:* Figure 15.4b shows the corresponding pattern of X-variables in terms of X-loading vector $\mathbf{p}_1$ (abscissa) and $\mathbf{p}_2$ (ordinate). Consecutive X-variables are connected by line segments, and some peak maxima are named explicitly. The straight line points in the direction of the Q-loadings of **y**, i.e. increasing *Ash* content.

An emission around 312 nm is seen clearly both at 230 and 240 nm excitation. An emission around 372 nm is seen clearly at 230 and 290 nm excitation; (at 240 nm excitation it may be present, but does not form a distinct peak). An emission around 417 nm is clearly evident at all four excitations.

This shows that short-term process changes towards the north-west direction in Figure 15.4a (e.g. sample sequences *11–13*) corresponds primarily to decreased emissions around 312 and 372 nm. In contrast, the longer-term process changes towards south-west, e.g. samples *9–10–26–50*, corresponds to a decreased emission around 417 nm.

### 15.2.11. One-way Loading Plots

Figure 15.5a shows another way to understand the same two X-loadings, as one-way spectra. The first PC (black) has a loading spectrum $\mathbf{p}_1$ with positive elements for all of the 200 X-variables, with peak maxima at 372 and 417 nm

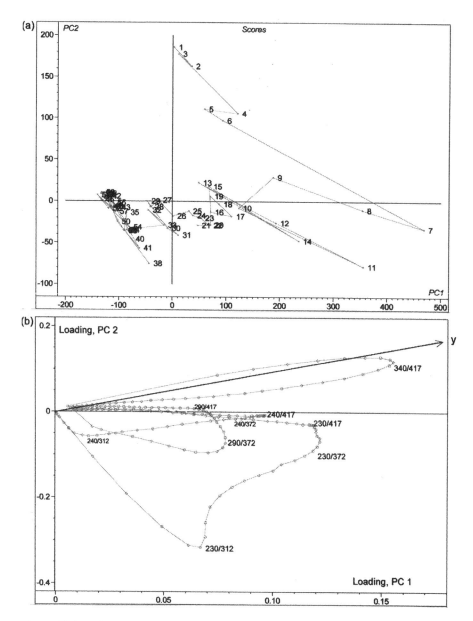

**Figure 15.4.** The main variations at the start-up of the production. (a) Score plot of the $N = 60$ sampling *times* (numbered) used as the calibration set for the initial model, shown for the first 2 PCs. Abscissa, $t_1$; ordinate, $t_2$. (b) Loading plot for the 200 X-variables, some of which are named by their excitation/emission wavelengths (nm). Abscissa, $p_1$; ordinate, $p_2$.

at all four excitations. The second PC's loading spectrum $p_2$ (grey) shows a contrast between the first three excitations (230, 240 and 290 nm) and the last excitation (340 nm), with the most extreme values near 312 nm and/or 372 nm at the former excitations and near 417 nm at the latter.

The loading spectra of the third and fourth PC are shown in Figure 15.5b. Together with $p_1$ and $p_2$ in Figure 15.5a, the structures of $p_3$ and $p_4$ indicate that several constituents are giving overlapping signals, e.g. in the first and the fourth excitation band. But some of these effects could also be due to response curvature in the emission measurements. With more background knowledge one might deduce the nature of these impurity variations from the auto-fluorescence data. More detailed studies by Bro (1999) have revealed that some major sources of auto-fluorescence in the sugar are phenolic substances plus the amino acid tryptophan.

Attempts at removing outliers or useless variables did not appreciably improve the modelling. Thus, we accept the above model as a description of the main variation types in the auto-fluorescence X-data, and how these relate to the Y-variable *Ash*.

### 15.2.12. The Prediction Model

Figure 15.6a shows the regression coefficient $\hat{b}_{A=2}$ obtained from the above calibration model, i.e. using the 2-PC model considered to be robust. Some typical excitation/emission wavelength pairs are noted.

The figure shows how **y** is predicted from **X**. The *Ash* level is predicted to be high from high light emission at all excitations, but in particular at 340 nm excitation, around 417 nm emission. However, this is modified by some negative corrections, in particular by the first emissions at 230 nm excitation, near 312 nm emission.

Figure 15.6b shows the cross-validated prediction ability. Considering how extreme sample 7 is in the raw data (Figure 15.1a) and in the score plot (Figure 15.4a), it is interesting to note that the *auto-fluorescence* data predict the *Ash* reasonably well even for this outlier.

***VI. Conclusion calibration part.*** A calibration model for predicting sugar quality, *Ash*, from auto-fluorescence has successfully been established.

### 15.3. MSPC AND UPDATING

***I. Purpose of MSPC.*** It is cognitively impossible for process operators to keep an eye on all the channels from modern multichannel process instruments (in this case the 200 X-variables from one single instrument). Fortunately, the multivariate model provides some nice summary plots that provide informative and sensitive overviews of the process. The 2D score plot above (Figure 15.4a) is one

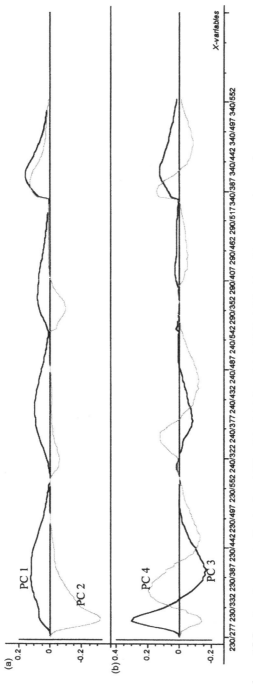

**Figure 15.5.** Loading spectra for the first 4 PCs of the initial model. (a) PCs 1 (black) and 2 (grey): $\mathbf{p}_1'$ and $\mathbf{p}_2'$. (b) PCs 3 (black) and 4 (grey): $\mathbf{p}_3'$ and $\mathbf{p}_4'$.

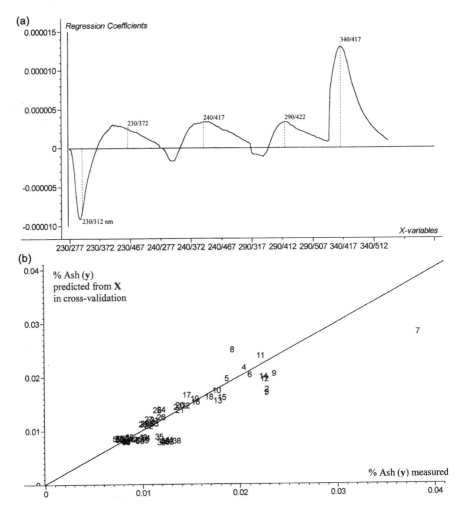

**Figure 15.6.** Assessment of the prediction model. (a) Regression coefficient vector $\hat{\mathbf{b}}_{A=2}$. (b) Measured quality $\mathbf{y}$ (abscissa) for the 60 calibration samples vs. predicted quality $\hat{\mathbf{y}}_{A=2}$ (ordinate, cross-validated). Diagonal: target line, $\mathbf{y} = \hat{\mathbf{y}}_{A=2}$.

of them. Now the obtained calibration model is to be used for multivariate statistical process control purposes, even for new samples.

***II–IV. Experimental planning, work and pre-processing and QC of data.*** These steps were the same as above, but this time $N_{\text{new}} = 46$ new prediction samples were analysed.

### V. Data analysis

#### 15.3.1. Predicted y, With Outlier-sensitive Uncertainty Estimate

Figure 15.7 shows the predicted *Ash* content, both for the 60 initial calibration samples as well as for the subsequent 46 available new sampling times, based on the calibration model using the first two PCs (Figure 15.6a). The predictions $\hat{y}_i$, $i = 1,2,...,106$ are given by the white marks. These predictions were obtained from the model $\hat{y} = \mathbf{X}\hat{b}_{A=2} + \hat{b}_{0,A=2}$. For each point in time, the outlier-sensitive standard prediction uncertainty (eq. (10.6) in Appendix A10) is represented by the grey bars.

The offset is $\hat{b}_{0,A=2}$ equals $0.57 \times 10^{-2}\%$, but as usual, that is not important.

The figure confirms a general decrease in *Ash* content from *Time 1* to *Time 50*, to about $\mathbf{y}_{50} = 0.01\%$. After that, the predicted *Ash* content remains relatively constant for some time.

However, after about *Time 75* the outlier-sensitive uncertainty bars are seen to increase. This shows that for some reason, the X-data of these samples do not fit the calibration model. The process is changing unexpectedly. It also indicates that the predictions $\hat{y}_i$, $i = 75, 76,...,106$ may be unreliable. What could be the problem? Not knowing the true y-values, we have to study the auto-fluorescence data **X** more closely.

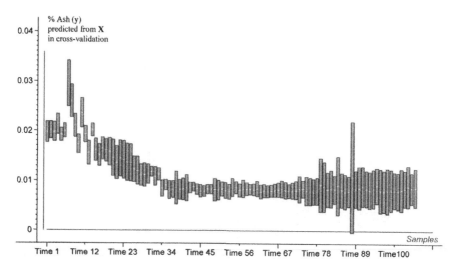

**Figure 15.7.** MSPC: prediction of **y**. Predicted *Ash* percentage $\hat{\mathbf{y}}_{A=2}$ (white line segments) for samples from process *Time 1–106*, with individual outlier-sensitive standard uncertainty estimates (grey bars).

### 15.3.2. Scores as Time Series

An alternative to the 2D score map in Figure 15.4a is to plot even the sample scores as continuous time series. First, we look at the X-information from the process represented as individual PC scores. Then we look at two summary plots that display the X-information in terms of what is picked up *inside* the bi-linear model and what remains *outside* the bi-linear model.

Figure 15.8a shows the scores for the first 2 PCs, $t_1$ and $t_2$ from *Time 1* to *Time 106*. The first 60 sampling times are the same as those plotted in the two-way score plot in Figure 15.4a. These PCs show a rather smooth development, with a couple of new sharp peaks, primarily at *Time 88*. This indicates particular problems at *Time 88* but otherwise nothing dramatic. Even though PCs 3 and 4 were ignored as small and unreliable in the calibration model for **y**, they may still have some interest in **X**. In Figure 15.8b they show increases after about *Time 70*.

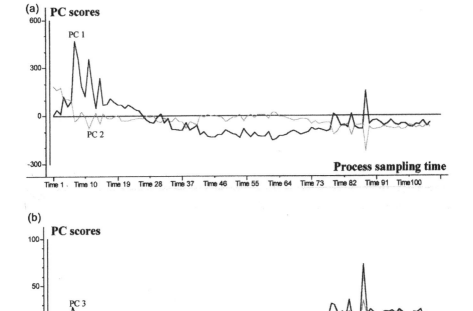

**Figure 15.8.** MSPC: scores time series. Scores (ordinate) for process sampling *Time 1–106* (abscissa). (a) PCs 1 (black) and 2 (grey): $t_1'$ and $t_2'$. (b) PCs 3 (black) and 4 (grey): $t_3'$ and $t_4'$.

Conventional SPC principles (see e.g. Massart et al., 1997, pp. 150–170), with control limits and warning-rules based on trends, could have been used for detecting abnormality in the individual PCs. Instead, we shall here use two ways to summarise how the X-data are modelled inside and outside the bi-linear model:

### 15.3.3. Leverage and X-residual: Two Complementary Summaries of X

❏ LEVERAGE OF THE SAMPLES: WHAT THE MODEL PICKS UP

In Figure 15.9a the leverage $h_i$ (eq. (9.3)) is shown for the calibration samples $i = 1,...,60$ as well as for the new samples $i = 61,...,106$. The leverage vector $\mathbf{h} = [h_i, i = 1,2,...,106]$ sums up the amount of X-information picked up by the $A = 4$ PCs: $\mathbf{t}_1, \mathbf{t}_2, \mathbf{t}_3$ and $\mathbf{t}_4$.

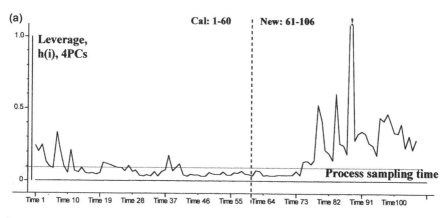

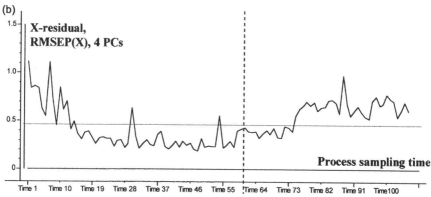

**Figure 15.9.** MSPC: summaries of **X**. Abscissa: process sampling *Time 1–106*. Vertical dashed line: split between calibration sampling *Time 1–60* and new sampling *Time 61–106*. Horizontal dotted lines: error warning limits, based on the calibration samples 45–60. (a) Abnormal fit: leverage $h_{i,A=4}$, summarising X-variations *inside* the bi-linear model. (b) Lack-of-fit: RMSEP(X)$_{i,A=4}$, summarising X-variations *outside* the bi-linear model.

The mean leverage of the 15 last calibration samples (*Time 45* to *Time 60*) is about 0.05. Therefore, at *Time 60*, *a warning limit* for the future leverages of about 0.1 was chosen (horizontal dotted line).

The figure shows that after *Time 70* the leverage starts to increase sharply to values between 0.2 and 0.5, i.e. above the warning limit. Already at *Time 75* the first warnings for unexpected process trouble were detected, and at *Time 85* the problem was obvious. Three extreme peaks may be observed in the leverage beyond *Time 70*, the highest one at $h_{i=88} = 1.7$.

The leverage thus summarises the X-information included *inside* the bi-linear model with the chosen rank (here $A = 4$).

□ RMSEP(X) OF THE SAMPLES: WHAT THE MODEL DOES NOT PICK UP

Figure 15.9b shows a complementary process monitoring tool, namely the lack of fit for each sample to the model. $RMSEP(X)_{i,A=4}$ summarises the X-residuals left unmodelled by the $A = 4$ PCs used.

For the first 60 calibration samples, the X-residual levels $RMSEP(X)_{i,A=4}$ were estimated by cross-validation (eq. (10.5a)). For the new *test set* of samples, *Time 61–106*, $RMSEP(X)_{i,A=4}$ was obtained by fitting their X-data to the 4-PC model.

The mean residual level for the last 15 calibration samples, *Time 45–Time 60*, is about 0.25. Therefore, a warning limit of about 0.5 was chosen. The figure shows that up until *Time 75* the auto-fluorescence X-data fit well to the calibration model. But then the lack-of-fit increases abruptly, and remains high after *Time 78*. A particularly abnormal X-residual is evident at *Time 88*.

$RMSEP(X)_i$ thus summarises the X-information *outside* the bi-linear model.

In the previous application example, Figure 15.4 illustrated how leverage $h_i$ and residual $RMSEP(X)_i$ together give an even more sensitive outlier detection.

□ THE PROCESS SHOWS UNEXPECTED PATTERNS

All the process-monitoring curves in Figs. 15.7–15.9 indicate that the process settles down to a relatively steady state between *Time 45* and *Time 56*. At about *Time 58* the residual level $RMSEP(X)_i$ increases just a little. After *Time 75* the increase in $RMSEP(X)_i$ together with the increase in leverage $h_i$, causes a clear increase in the outlier-sensitive uncertainty estimate in the y-predictions.

Hence, beyond *Time 75* something has happened, either with the process or with the instrument, that makes the calibration model obsolete and the predictions unreliable.

### 15.3.4. Checking the Process Data

*The new input spectra:* Let us now try to find out what has happened, and if possible update the calibration model. Figure 15.10a shows the X-data of the new 46 samples. At first glance, there is nothing wrong with these data. However, the residuals $E_4$ in Figure 15.10b show clear non-random spectral signatures remaining after being fitted to the 4-PC calibration model. This is particularly evident when compared with the same residuals of the calibration samples (Figure 15.2c).

Let us now see how well the reference values *Ash* in the new samples are predicted.

*New Ash data:* The *Ash* content was now measured even for the new samples saved from *Time 61* to *Time 106*, by the original, slow and cumbersome reference method. Figure 15.11a shows the measured $y$ (abscissa) vs. predicted $\hat{y}_{A=2}$ (ordinate, based on the old 2-PC calibration model) for these 46 new samples. The diagonal line represents the *ideal* target.

The figure shows that the predictions are systematically wrong and completely unsatisfactory. While the reference measurements $y$ indicate a variation between 0.6 and $23 \times 10^{-2}\%$, the predictions $\hat{y}$ only range between 0.8 and $1.0 \times 10^{-2}\%$. The prediction for *Time 88* is totally wrong. This is in stark contrast to how well

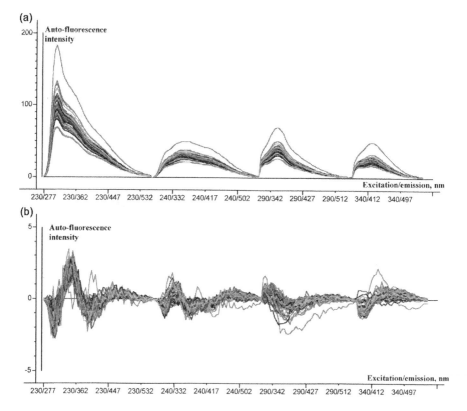

**Figure 15.10.** *Auto-fluorescence* spectra of later samples. Light emission intensity for the 46 new samples. (a) Input spectra $X$. (b) Residuals $E_4$ after 4 PCs, amplified 20 times.

the model performed for the calibration samples, which even in the cross-validation showed very good predictive fit. It seems that the system has changed in such a way that the old calibration model has become obsolete.

### 15.3.5. Updating the Calibration Model

*Learning something new, forgetting something old:* There are different ways to update a bi-linear model over time, after new X- and y-data have arrived. One alternative is simply to append the new samples to the old calibration sample set. An extension of this is to let the system *forget* the least relevant information, by leaving out the oldest calibration samples.

An alternative way to update the bi-linear models is to combine the old model's X- and y-loadings with the new X- and y-data, as described by Helland et al. (1991). Old, irrelevant samples may be *forgotten* asymptotically by the use of a sample-weighting function (Dayal and MacGregor, 1997).

A more primitive way to update the model would be to change the regression coefficient (Figure 15.6a) by a slope – and offset correction based on the systematic errors observed Figure 15.11a.

A simplified *Bayesian* approach to the calibration updating would be to use a weighted average of the models from the old and new data, with weights reflecting the relevance and reliability of the two models.

The first 15 sampling times on the very first day of the sugar production represent extreme start-up problems, with little or no relevance later in the production season. Hence, we leave them out from the next calibration modelling.

A new, updating calibration sample set was thus defined from *Time 16* to *Time 106*, i.e. $N = 91$ samples. A new bi-linear regression model was then estimated, with $y = Ash$ and $X = 200$ *auto-fluorescence* channels like before, and again with cross-validation based on the segmentation of three consecutive readings in each segment: 1,1,1,2,2,2,3,3,3,....,30,30,30.

The cross-validation showed that the first two PCs in the updated model explained 84% of the CV-variance in $y$ and 99% of the variance in $X$. The subsequent 3 PCs improved the prediction of $y$ up to 88.5%, but this improvement was considered too small to be reliable for such an unstable process. Attempts at improving the predictive ability by removing outliers or useless X-variables did not affect the modelling appreciably, and was therefore stopped. Hence, the 2-PC solution was pragmatically chosen as the optimal updated model.

Figure 15.11b shows the updated regression coefficients $\hat{b}$, based on $A = 2$ PC. Compared with the original calibration model in Figure 15.6a, the updated model has a simpler structure, with smaller values and no negative contributions.

This indicates that the updated model may be more stable than the first calibration model.

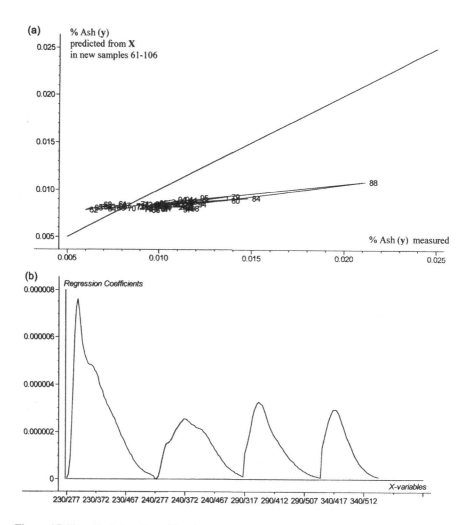

**Figure 15.11.**    Updating the calibration model. (a) Before the updating: measured quality **y** (abscissa) vs. predicted $\hat{y}_{A=2}$ (ordinate, Figure 15.7), obtained by applying the initial model (Figure 15.6a) to the X-data of new samples at *Time 61–106* (Figure 15.8a). Consecutive sampling times are connected. (b) The updated model: regression coefficient vector $\hat{b}_{A=2}$ from the updated model based on *Time 16–106*. (c) After the updating: measured quality **y** (abscissa) vs. predicted $\hat{y}_{A=2}$ (ordinate, cross-validated), obtained by applying the updated model to the X-data of all the samples in the updating calibration set *Time 16–106*. Consecutive sampling times are connected. (d) Measured and predicted *%Ash*: black line, **y** measured; grey line, $\hat{y}_{A=2}$ predicted from the updated model, by cross-validation (*Time 16–106*) or by prediction (*Time 1–15*).

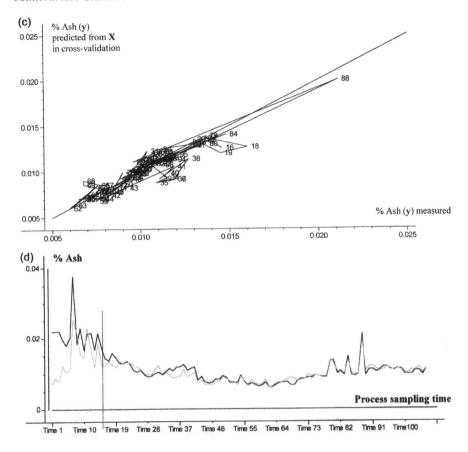

**Figure 15.11.** (*continued*)

Figure 15.11c shows the corresponding measured *Ash* content **y** (abscissa) vs. predicted **ŷ** (ordinate, cross-validated). The estimated prediction error RMSE-$P(Y)_{A=2}$ equals $0.10 \times 10^{-2}\%$. The results for all 91 samples in the updating calibration set are plotted. A comparison with Figure 15.11a shows the updated model to have improved the predictive ability greatly.

The figure shows that the sample at *Time 88* is extreme. However, since it is well predicted in the cross-validated modelling, this means that it is extreme with respect to a phenomenon also spanned by other samples and hence included in the updated calibration model. That explains why removal of this outlier did not affect the modelling very much.

After the updated 2-PC model has been applied even to the *forgotten* samples (*Time 1–15*), the full time series may now be displayed. Figure 15.11d shows the measured (black) and the CV-predicted (grey) levels of *%Ash*. The two curves follow each other quite well within the updating range, particularly for the last half of the sampling period. As expected, the predictions are not good for the first *15 forgotten* samples.

This serves to show that the updated model seems to be reliable so far, and that it made sense to *forget* the *15* initial samples. But it also serves to remind us that we are here dealing with an *indirect* measurement of **y** from **X** of *Ash* from *auto-fluorescence*, and the y-predictions are sensitive to unexpected phenomena in **X**.

*Further work:* The logical next steps would now be:

1. To test the updated model over a longer production period and update it again if necessary.
2. To try to develop a simplified *auto-fluorescence* analyzer that would work on-line and more or less in real-time.
3. Another potential development would be to introduce proper dynamic (time-variant) information in the modelling, in order to *forecast* the unknown quality at future points in time from measurements available at the present and past points in time. This is briefly discussed in Appendix A15.

It should be noted, that cybernetical control theory/systems identification, statistical process control and quality control represent three highly specialised fields (cf. Appendix A1), beyond the scope of this introductory book. But the present example demonstrates how the same BLM-based *soft modelling* previously used for many other types of data analysis, may also be applied to the modelling and monitoring of dynamical processes.

***VI. Conclusion for MSCP and updating.*** Using the previous calibration model for prediction of the new samples, gave warnings for unexpected process trouble in time. A successful updating of the calibration model was performed.

## 15.4. CONCLUSIONS

The multivariate soft modelling was used as MSPC to provide information about the product quality (**y** = *Ash*) from a multichannel instrument (**X** = *auto-fluor-escence*). Moreover, the soft modelling provided informative maps and time series plots of how the process in general developed over time. These plots summarise the X-information *inside* and *outside* the bi-linear model of **X**.

When the process was found to have drifted away from the states originally calibrated for, the bi-linear calibration model was updated, based on a new *time window* of samples.

For simplicity, data were only presented for the first few days of the sugar production season. But already these data shed new light on the nature of the impurities that have to be removed in the sugar production. Moreover, the present project has indicated that *auto-fluorescence* measurements have a poten-tial for process monitoring, not only for predicting the traditional quality criterion *Ash*, but also as a more sensitive, *global fingerprinting* method (Munck et al., 1998).

### 15.4.1. Rhythm and Blues in Research

The technical monitoring of an industrial process (Kettaneh-Wold et al., 1994; Wise and Gallagher, 1996) is similar to how we as humans sense the world. In fact, experienced process operators in industry often check the process by listening to the humming and hissing of the equipment.

Modern multichannel analytical instruments are like grand pianos, and deserve to be played with more than one finger at a time. A multivariate information space is so much larger than a univariate one, and makes it so much easier to distinguish useful information from useless noise.

The present example has indicated how we can decompose the *cacophony* of data from multichannel process instruments into *music* and *noise*. The *music* is seen in terms of underlying *harmonies* (latent variables' loadings), *melodies* and *rhythms* (time-variant scores and y-predictions), while the *noise* is seen as residuals and resulting uncertainty estimates. Abnormal phenomena are detected as disharmony and lack of rhythm.

If the system under study is very complex or changes systematically over time, then it is important not to try to model all of it at once: As mentioned in Chapter 1.1, high-dimensional heterogeneous, non-linear dynamic mathematical models may be formulated theoretically. But they are statistically difficult to handle, and even more difficult to understand. This example has shown how a complex, non-stationary system may be studied visually with the help of local, stationary bi-linear models. Thus, the underlying harmonies were revealed!

### 15.5. TEST QUESTIONS

1. What are the benefits of predicting product quality from high-speed process measurements?
2. What are the risks involved?
3. How was the calibration sample set designed in the present example?
4. Was this design satisfactory?
5. What are the main plots when the *soft* bi-linear regression modelling is used for MSPC?
6. How do you detect a need for updating the model?
7. How do you update the model?
8. What is the difference between checking for non-stationary processes and checking the predictive ability of a model by the use of a test set?
9. Which other possible uses of the same data analytical approach can you think of?

## 15.6. ANSWERS

1. The benefits of predicting product quality from high-speed process measurements include: (a) getting relevant product information fast enough, (b) maintaining an overview of the developing process, and (c) detecting instrument problems or process drift.
2. Alienating the operators, becoming vulnerable to errors in yet another level of technical gadgets.
3. The calibration set was simply the samples taken consecutively as they come during the start-up period.
4. The present design was OK for an initial proof of the concept of predicting *Ash* from *auto-fluorescence*. But for developing a reliable, final process calibration model, it would be necessary to perturb the production process consciously, to ensure sufficiently informative data.
5. (a) Input data plots, (b) plots of loadings, scores, Y-predictions and model summaries such as leverage and RMSEP(X). These plots may be one-way time-series plots, or two-way *maps*.
6. When, e.g. the time-series plots of leverage, RMSEP(X) and/or Y-predictions indicate consistent and increasing levels of abnormality relative to the calibration sample set.
7. By recalibrating it once in a while, based on an updated set of calibration samples, as shown here.
8. Checking and correcting for drift by the use of X- and y-data from *a future set of new samples* is important when dealing with *non-stationary* processes, where the general type of samples (or the data-generating instrument) *changes over time*. On the other hand, the validation of a model of a *stationary* process by the use of X- and y-data from *a new, independent test set of samples* is just intended to check against over-fitting the model. CV is then often a better alternative (Martens and Dardenne, 1998).
9. The present way to analyse data has already proven very useful for monitoring and controlling a number of industrial process types. However, it may also be applied, e.g. to other dynamic processes.

One could envision it being used for continuous monitoring of water quality, with $X =$ several cheap, high-speed but non-selective multichannel sensors (e.g. auto-fluorescence spectra, conductivity, etc.) and $y =$ some purity index, e.g. from human sensory analysis. Another possible alternative is to use it to monitor the growth of agricultural produce, with $X =$ multichannel cameras (fixed, tractor-mounted, air-borne or satellite) as well as other measured inputs concerning the plants and the local climate, and $Y =$ the health status of the plants, as judged by experts.

The same approach could be used for monitoring and forecasting car traffic, with $X =$ traffic pattern on various streets and $Y =$ traffic pattern at other streets or at later points in time. It could also be used for monitoring and forecasting computer network load at different points in time, with e.g. $X =$ load on different computer resources, with quality $\mathbf{y}$ = total throughput. It could even be used for monitoring the performance of a computer simulation process.

Moreover, this *soft modelling* approach to MSPC could be used in hospitals, making more effective monitoring of patients in intensive-care units, e.g. with the X-variables being the numerous sensors hooked up to the patient and $\mathbf{Y}$ being various medical conditions of the patient; the calibration model could be started with data from previous patients, and it could even be trained to forecast the conditions of the patient in the immediate future.

Finally, this approach could be implemented in motorised vehicles (cars, ships, etc.) for slowly developing models that could improve fuel efficiency. A variety of sensor inputs $\mathbf{X}$ from driver's control, from the motor and from the surroundings could be related to $\mathbf{y}$ = amount of fuel consumed per distance travelled.

# Chapter 16

# Design and Analysis of Controlled Experiments: Reducing Loss of Quality in Stored Food

*This final example shows how to use the BLM method for explorative and confirmative analysis of designed experiments. First, a small factorial screening design is developed. Based on the results, a more ambitious and expensive experiment is performed, using a response surface design.*

*In each case, the power of several alternative designs is first computed, and a cost-effective design is chosen. Then the experiment is performed, and the data analysed. The statistical reliability is checked by cross-validation/jack-knifing, as well as by visual interpretation. For comparison, the BLM results are shown to be very similar to those from traditional ANOVA.*

*The data come from a research project in predictive microbiology, concerning the quality of food: how to reduce the growth of harmful microorganisms.*

## 16.1. INTRODUCTION

### 16.1.1. Purpose of the Project

The project concerns food safety with the aim to explore optimal conditions for preventing quality loss in foods due to mould growth. Some food mould species can generate potent toxic substances, like aflatoxin from *Aspergillus flavus* (*A. fla*) in peanuts, while others are benign and even desirable in some product, e.g. *Penicillium roqueforti* (*P. roq*) in blue cheese.

Four general mould-related experimental factors are to be investigated: water activity ($a_W$) and acidity (pH) in the samples, and oxygen level ($O_2$) and carbon dioxide level ($CO_2$) in the air in the food package. The effects of these factors on the growth of various mould species in a controlled, food-like sample type will be determined.

The data (file: MOULD-DESIGN-I, -II) are taken from Haasum and Nielsen (1998), and adapted didactically in order to illustrate one way to approach the design and analysis of laboratory experiments.


### 16.1.2. Purpose of the Data Analysis

In a *controlled* experiment, the design factors **X** are consciously changed in order to study their effects on certain response variables **Y**. The present example is based on *factorial designs*. In particular, the possibility of non-linear effects (curvatures and interactions) will be investigated.

First, a small screening-experiment will be used for selecting reasonable experimental conditions. Secondly, in the main experiment this is expanded with more experimental conditions and more response variables, in order to study the effects in more detail.

In each case, the planning of each experiment starts with a definition of the purpose and an identification of the research opportunities and resource limitations. Then several alternative experimental designs are defined and their statistical power is checked. The experiment is performed, and the obtained response data analysed according to the design used.

As outlined in Figure 3.1, the experimental planning requires a sensible choice of variables and samples. Figure 3.2 outlined the role of experimental planning in the research project cycle. Figures 11.1 and 11.2 showed various types of experimental design. An example of mixture design was illustrated for cocoa/sugar/milk mixtures in Chapters 8–11. The statistical power of that design was assessed in Chapter 11.4. The reader is invited to see the present illustration as one example of how his or her own experiments might be planned, optimised and analysed with low risk, and yet with low cost.


### 16.2. SCREENING EXPERIMENT

*I. Purpose of screening experiment.* The purpose is, at first, to gain preliminary insight: the conclusions are going to be used in planning a more expensive and ambitious experiment.

This initial screening phase is intended to check that the chosen levels of the design factors are relevant, to look for possible interactions between these factors, and to see if the main experiment may be simplified in some way.

*II. Experimental planning*


### 16.2.1. Selection of Variables

*Response variable*: The mould species *(A. fla)* is chosen as model organism. Its growth, measured as the diameter of the mould colony after 7 days' incubation, will be used as the Y-variable. Hence it is here called $\mathbf{y} = Diam.\_7days$. This measurement is expected to lie between 0 mm (no growth) and 70 mm.

Seventy millimetres is the maximum diameter on an 11 cm Petri dish, when the fungus has been *seeded* (inoculated) at three points, 2 cm apart.

*Expected noise level in response variable*: Based on earlier experience with this type of measurement, a standard uncertainty of the measured diameters about 2.5 mm is expected. This includes repeatability errors from the way the mould is applied to the sample surface, the way the mould is grown and the way the colony diameter is measured. It does not include production variations in the material (substrate) on which the moulds are grown: for the present experiment that is considered fixed.

*Design factors*: The controlled design requires concise definition of which experimental factors to be varied, as well as their actual levels. The effect of four design factors are tested: the amount of biologically available water in the samples, i.e. the water activity ($a_W$), the sample pH, and the $O_2$ and $CO_2$ content of the atmosphere around the samples.

### 16.2.2. Selection of Samples

Within the given resources, an experiment with about 20 samples was considered to be practically feasible.

☐ DESIGN TYPE

In order to be able to see possible *interactions* between the design factors, a factorial screening design called $2^K$ is chosen with $K = 4$. This implies 16 experimental conditions. As illustrated in Figure 11.2b, a centre point (here named $c$), is added to the design in order to establish a *reference condition* and to get a first indication of possible *curvatures* in the mould response to these factors. This centre point is repeated four times in order to get a rough check on the repeatability of the actual experiment.

We could have used an even simpler design, the so-called *fractional factorial* $2^{K-L}$ *design* (Figure 11.2a). The design $2^{4-1}$ requires only eight ($2^3$) samples. But that would have made the experiment less transparent (by causing confounding of some interactions), and this simplification was not presently considered necessary.

Hence, the total number of experiments in this planned screening design may be termed $2^4 + 4$, and has $N = 20$ samples, which is considered to be within the resource limitations.

**Table 16.1.** Initial screening design: the four design factors, at their chosen levels.[a]

|        |    | $aW$ $1'$ | pH $2'$ | $O2$ $3'$ | $CO2$ $4'$ |
|--------|----|-----------|---------|-----------|------------|
| *wpoc*  | 1  | 0.96 | 5 | 4  | 5  |
| *wpoC*  | 2  | 0.96 | 5 | 4  | 25 |
| *wpOc*  | 3  | 0.96 | 5 | 16 | 5  |
| *wpOC*  | 4  | 0.96 | 5 | 16 | 25 |
| *wPoc*  | 5  | 0.96 | 7 | 4  | 5  |
| *wPoC*  | 6  | 0.96 | 7 | 4  | 25 |
| *wPOc*  | 7  | 0.96 | 7 | 16 | 5  |
| *wPOC*  | 8  | 0.96 | 7 | 16 | 25 |
| *Wpoc*  | 9  | 0.98 | 5 | 4  | 5  |
| *WpoC*  | 10 | 0.98 | 5 | 4  | 25 |
| *WpOc*  | 11 | 0.98 | 5 | 16 | 5  |
| *WpOC*  | 12 | 0.98 | 5 | 16 | 25 |
| *WPoc*  | 13 | 0.98 | 7 | 4  | 5  |
| *WPoC*  | 14 | 0.98 | 7 | 4  | 25 |
| *WPOc*  | 15 | 0.98 | 7 | 16 | 5  |
| *WPOC*  | 16 | 0.98 | 7 | 16 | 25 |
| *c1*    | 17 | 0.97 | 6 | 10 | 15 |
| *c2*    | 18 | 0.97 | 6 | 10 | 15 |
| *c3*    | 19 | 0.97 | 6 | 10 | 15 |
| *c4*    | 20 | 0.97 | 6 | 10 | 15 |

[a] $aW$, water activity (low and high levels named $w$ and $W$; pH, degree of alkalinity (low and high levels named $p$ and $P$); $O2$, % $O_2$ (low and high oxygen levels named $o$ and $O$); $CO2$, % $CO_2$ (low and high carbon dioxide levels named $c$ and $C$); Centre-points: $c1$, $c2$, $c3$, $c4$.

❐ DESIGN FACTOR VALUES

Table 16.1 shows the chosen levels of the four experimental factors. In order to be compatible with other computer outputs, the naming notation has been simplified: the water activity $a_W$ is sometimes written $aW$; $O_2$ is written $O2$ and $CO_2$ is written $CO2$.

In the sample names (rows), the four factors $aW$, pH, $O2$ and $CO2$ are represented by the letters $w, p, o$ or $c$, at the low level and $W, P, O$ or $C$ at the high level. So, for instance, sample 1 (*wpoc*) has low levels of all the four design factors, while sample 16 (*WPOC*) has high levels of all the factors. Samples *c1–c4* represent the four replicates of the centre point: they have an average level of each design factor.

In order to comply with the quality criteria for designs listed in Chapter 11.3.1, the high and low levels of the four design factors have been chosen with sufficient span to be able to give appreciable effects, but still to be within the range relevant for storage of foods: 0.96–0.98 $a_W$ units, 5.0–7.0

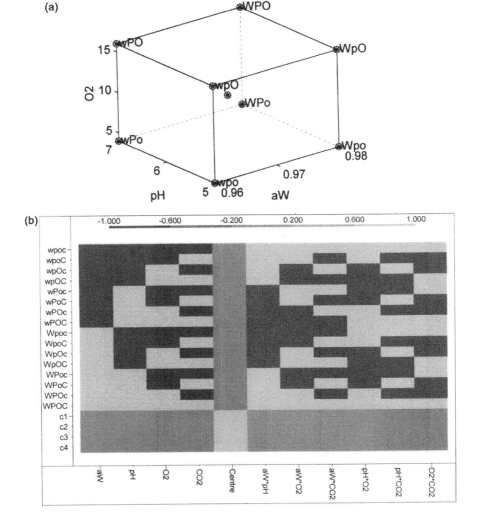

**Figure 16.1.** Basic screening design. (a) 3D spatial representation of three of the four design factors, *aW* (levels named *w,W*), pH (*p,P*) and *O2* (*o,O*). This is repeated at low and at high $CO_2$. (b) The design variables **X**. Light grey, +1; dark grey, 0; black, −1.

pH units, 4–16% $O_2$ and 5–25% $CO_2$. The error in attaining these levels of the design factors in the actual experiment is considered small enough to be ignored.

A four-factor design like $2^4$ is difficult to visualise. In order to illustrate the naming convention of the corner points, Figure 16.1a shows one part of the

**Table 16.2.** Initial screening design: design variables **X**: coded design factors, with interactions.[a,b]

|  |  | aW 1' | pH 2' | O2 3' | CO2 4' | Centre 5' | aW*pH 6' | aW*O2 7' | aW*CO2 8' | pH*O2 9' | pH*CO2 10' | O2*CO2 11' |
|---|---|---|---|---|---|---|---|---|---|---|---|---|
| wpoc | 1 | -1 | -1 | -1 | -1 | 0 | 1 | 1 | 1 | 1 | 1 | 1 |
| wpoC | 2 | -1 | -1 | -1 | 1 | 0 | 1 | 1 | -1 | 1 | -1 | -1 |
| wpOc | 3 | -1 | -1 | 1 | -1 | 0 | 1 | -1 | 1 | -1 | 1 | -1 |
| wpOC | 4 | -1 | -1 | 1 | 1 | 0 | 1 | -1 | -1 | -1 | -1 | 1 |
| wPoc | 5 | -1 | 1 | -1 | -1 | 0 | -1 | 1 | 1 | -1 | -1 | 1 |
| wPoC | 6 | -1 | 1 | -1 | 1 | 0 | -1 | 1 | -1 | -1 | 1 | -1 |
| wPOc | 7 | -1 | 1 | 1 | -1 | 0 | -1 | -1 | 1 | 1 | -1 | -1 |
| wPOC | 8 | -1 | 1 | 1 | 1 | 0 | -1 | -1 | -1 | 1 | 1 | 1 |
| Wpoc | 9 | 1 | -1 | -1 | -1 | 0 | -1 | -1 | -1 | 1 | 1 | 1 |
| WpoC | 10 | 1 | -1 | -1 | 1 | 0 | -1 | -1 | 1 | 1 | -1 | -1 |
| WpOc | 11 | 1 | -1 | 1 | -1 | 0 | -1 | 1 | -1 | -1 | 1 | -1 |
| WpOC | 12 | 1 | -1 | 1 | 1 | 0 | -1 | 1 | 1 | -1 | -1 | 1 |
| WPoc | 13 | 1 | 1 | -1 | -1 | 0 | 1 | -1 | -1 | -1 | -1 | 1 |
| WPoC | 14 | 1 | 1 | -1 | 1 | 0 | 1 | -1 | 1 | -1 | 1 | -1 |
| WPOc | 15 | 1 | 1 | 1 | -1 | 0 | 1 | 1 | -1 | 1 | -1 | -1 |
| WPOC | 16 | 1 | 1 | 1 | 1 | 0 | 1 | 1 | 1 | 1 | 1 | 1 |
| c1 | 17 | 0 | 0 | 0 | 0 | 1 | 0 | 0 | 0 | 0 | 0 | 0 |
| c2 | 18 | 0 | 0 | 0 | 0 | 1 | 0 | 0 | 0 | 0 | 0 | 0 |
| c3 | 19 | 0 | 0 | 0 | 0 | 1 | 0 | 0 | 0 | 0 | 0 | 0 |
| c4 | 20 | 0 | 0 | 0 | 0 | 1 | 0 | 0 | 0 | 0 | 0 | 0 |

[a] Low, intermediate and high levels have been coded by $-1$, 0 and $+1$, respectively.

[b] $aW$, water activity (low and high levels named $w$ and $W$); pH, degree of alkalinity (low and high levels named $p$ and $P$); $O2$, % $O_2$ (low and high oxygen levels named $o$ and $O$); $CO2$, % $CO_2$ (low and high carbon dioxide levels named $c$ and $C$); Centre-points: $c1$, $c2$, $c3$, $c4$.

design, the $2^3$ factorial design of $aW \times pH \times O2$, in analogy to Figure 11.2b. This is repeated at each of the two levels of $CO2$.

The optimal number of replicates, and hence the total number of samples and the cost of the experiment, will be determined below.

◻ DESIGN FACTOR REPRESENTATION AS DESIGN VARIABLES X

Figure 16.1b and Table 16.2 show the data to be used as X-matrix in the subsequent data analysis. In columns 1–4 the levels of the four experimental factors have been coded −1 or +1 for the low and high levels, respectively. These are the linear design variables, intended to reveal *the main effects* of the four design factors. The fifth column codes for the centre point $c$, and is intended to reveal possible *curvature* effects. Columns 6–11 represent all the so-called *two-factor interactions*, defined as the product of two and two design factors. Three- and four-factor interactions are to be ignored in the modelling because they complicate the modelling and are difficult to interpret. Instead, we shall look for outliers and extreme values in resulting plots.

### 16.2.3. Model and Estimation Method to be Used

The bi-linear model with usual linear regression model summary, $y = b_0 + \mathbf{X}b + \mathbf{f}$ (eq. (6.1d)), will be used, with $\mathbf{X}$ defined by Table 16.2 and illustrated by Figure 16.1b, and $\mathbf{y}$ is the *Diam._7days* of *A. fla*. Hence, there are 11 X-variable and one Y-variable. The total number of samples, $N$, is the number of treatments (20) × the number of replicates.

The model parameters $b_0$ and $\mathbf{b}$ are to be estimated by APLSR, and their reliability checked by leave-one-sample-out cross-validation (*jack-knifing*). There are 11 X-variables, and they all vary independently, so we may extract $A_{\text{Max}} = 11$ non-zero PCs from $\mathbf{X}$.

Since $\mathbf{X}$ from orthogonal designs has very little covariance information, we expect the bi-linear model to become independent of the number of PCs after very few PCs: the estimates for $b_0$ and $\mathbf{b}$ will be constant beyond the first few PCs. However, in order to retain compatibility with traditional linear modelling, we use all 11 PCs ($A_{\text{Opt}}$), estimated by stabilised PLSR (Appendix A6.8).

### 16.2.4. How Many Replicates Needed?

Is this experimental design in Table 16.2 and Figure 16.1b, with its 20 different *experimental conditions*, powerful enough to ensure that interesting effects will be revealed, given the expected measurement noise level? Or do we need to take replicates?

Many researchers seem to be able to give good intuitive answers to this type of questions, based on experience. Readers of this category may skip this subsection. However, if more explicit answers are desirable, we may estimate the *power* of the experimental designs with different number of replicates.

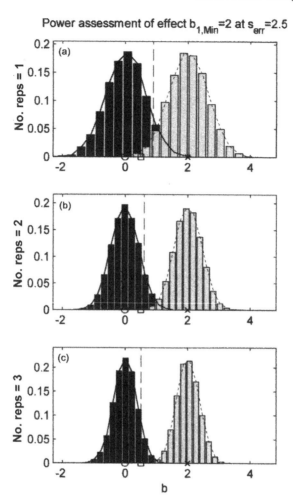

**Figure 16.2.** Pre-experiment comparison of three alternative screening designs: results from Monte Carlo estimation of statistical power. Distributions of regression coefficient $\hat{b}_1$ (the main effect of $x_1 = aW$) estimated in 10 000 *future experiments*. Left (black) histograms, null-hypothesis H0: the true effect $b_{1,true} = 0$. Right (grey) histograms, experimental hypothesis H1: the true effect $b_{1,true} = b_{1,Min} = 2$ mm per coding step in $x_1$. (a) one replicate, $N = 20$ samples (i.e. no replication). (b) Two replicates, $N = 40$ samples. (c) Three replicates, $N = 60$ samples.

◻ MONTE CARLO POWER ESTIMATION

Figure 16.2 shows the outcome of Monte Carlo simulation of power (Chapter 11.4) for three alternative designs, using one, two or three full replicates (i.e. $N = 20$, 40 or 60). The assumptions behind these simulation results are: the measurement noise in the future y-responses is assumed to be random and normally distributed with a standard uncertainty of $s_{err} = 2.5$ mm. The data from the experiment are intended to be analysed by bi-linear regression. The minimum main effect that would make the experiment worth while is considered to be $b_{Min} = 2$ mm, i.e. a change in **y** of 4 mm when a design factor increases 2 units, from $-1$ to $+1$.

The figure shows the distribution of how the estimate of a main effect, e.g. $\hat{b}_1$ (effect of $aW$) would be expected to vary, due to experimental errors in the response data, if the "true" value of $b_1$ were either 0 (black) or 2 (light grey). As the present design is set up, the results are equally applicable for all the other main effects and two-factor interactions.

Results are given for three alternative designs, each based on the design in Table 16.2, but with one, two or three replicates, in (a), (b) and (c), respectively. Here *one replicate* is the same as *no replicates*. Each histogram shows the results from analysing 10 000 simulated, hypothetical *experiments*.

In each subplot, the light *grey* histogram represents the experimental hypothesis H1, that the *true* effect of one of the coded design factors **x** on the response **y** is just large enough to be *interesting* for us; $b_{true} = b_{Min} = 2$. The *black* histogram shows the corresponding null-hypothesis H0, that **x** has *no effect* on **y**; $b_{true} = 0$.

The histograms show that when the number of replicates increases from one to three, the resolving power of the design increases. But already in Figure 16.2a, with only one replicate, there is a clear distinction between the two probability distributions of the outcomes, although some overlap is evident around the crossover value near $b = 1$.

When the actual response data **y** arrive in the future, we shall be able to estimate the vector of effects $\hat{\mathbf{b}}$. We intend then to judge the reliability of each estimated effect $\hat{b}_k$, based on whether or not the value $b_k = 0$ lies within the cross-validated reliability range $\hat{b}_k \pm 2\hat{s}(\hat{b}_k)$. As discussed in Chapter 11.4, this corresponds, roughly, to a 5% risk of committing Type I error (being fooled by just noise), i.e. $\alpha \approx 0.05$.

The vertical dashed lines in Figure 16.2 represent the critical $b$-value, found in the MC simulation, for the chosen level of confidence $\alpha = 0.05$. For the smallest design, without any replication (Figure 16.2a), the risk of Type II error, overlooking a *true* effect of $b_{true} = 2$ due to experimental errors in **y**, is about 4% ($\beta \approx 0.04, power = 0.96$). When using two replicates, this risk reduces to about 0.2%, and with three replicates it reduces further to less than 0.01%.

A probability of about 4% for overlooking a *true* main effect of size $b_{true} = 2$ is considered an adequately low risk for this screening experiment. Therefore, the experimental design in Tables 16.1 and 16.2, without any replication, is chosen to be the one used in practice. Now we are ready to conduct the actual experiment.

The same power was found for the two-factor interactions. That is not unexpected, since in this design, there is the same support for each interaction parameter as for the main effect parameters. In contrast, the estimation of the *curvature* effect, $x_5$ (=*Centre*, column 9, Table 16.2) is more difficult. With only one replicate of the 20 treatments, the probability was about $\beta \approx 0.60$, i.e. about 60% risk of overlooking a curvature effect of size $b_{5,\text{true}} = 2$. The reason is that the curvature-effect is supported by fewer observations than the main effects and interactions.

***III. Experimental work***. The experimental work in this case consists of making $N = 20$ artificial, food-like samples (*substrates*) with different controlled chemical compositions, inoculating these samples by placing a few cells of mould on the surface of each of them, and storing the samples at room temperature (20°C), in different controlled atmospheres.

The response variable, the diameter of the mould colonies, is measured after 7 days of storage.

## *IV. Pre-processing and QC of data*

**Table 16.3.** Measured response, **y**

|        |    | Diam._7days (mm)[a] |
|        |    | 1$'$              |
|--------|----|--------------------|
| *wpoc* | 1  | 49 |
| *wpoC* | 2  | 34 |
| *wpOc* | 3  | 50 |
| *wpOC* | 4  | 22 |
| *wPoc* | 5  | 44 |
| *wPoC* | 6  | 30 |
| *wPOc* | 7  | 44 |
| *wPOC* | 8  | 16 |
| *Wpoc* | 9  | 51 |
| *WpoC* | 10 | 35 |
| *WpOc* | 11 | 57 |
| *WpOC* | 12 | 37 |
| *WPoc* | 13 | 50 |
| *WPoC* | 14 | 33 |
| *WPOc* | 15 | 52 |
| *WPOC* | 16 | 30 |
| *c*1   | 17 | 41 |
| *c*2   | 18 | 46 |
| *c*3   | 19 | 41 |
| *c*4   | 20 | 44 |

[a] Diameter of the mould *Aspergillus flavus* (*A. fla*), measured after 7 days of growth under the experimental conditions in Table 16.1.

### 16.2.5. Preliminary Inspection of Input Data

Table 16.3 shows the measured response **y** for the $N = 20$ samples. The measurements fall nicely inside the 0–70 mm range with safe margins to the 0 or 70 mm extremes. No obvious outliers are evident. The mean diameter is 40.3 mm and the total initial standard deviation is 10.6 mm.

The measurement error at the repeatability level may be roughly estimated from the four centre points *c1–c4*. Their mean is 43 mm and their standard uncertainty is 2.4 mm. Since this estimated standard uncertainty is based on only four numbers, it is rather imprecise. But at least it is compatible with the standard uncertainty $s_{err} = 2.5$ mm noise assumed in the planning prior to the experiment.

Hence, we conclude that it makes sense to proceed with the PLSR modelling of the measured response data (MOULD-DESIGN-I).

### V. Data analysis

### 16.2.6. Modelling the Response Data

The data for 11 design variables **X** (Table 16.2) and response **y** (Table 16.3) in the $N = 20$ samples were submitted to bi-linear regression. Full leave-one-treat-

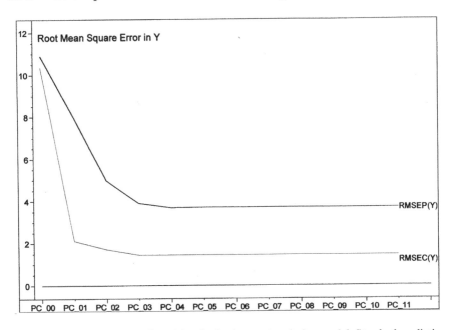

**Figure 16.3.** Predictive ability of the obtained screening design model. Standard prediction uncertainty (RMSEP(Y)) and calibration fit (RMSEC(Y)), given as functions of the number of PCs, $A = 1, 2, \ldots, 11$, estimated from the measured responses of the $N = 20$ samples.

ment-out cross-validation ($M = 20$ segments) was used for estimating the prediction error RMSEP(Y): hence the validation level is at the interpolation ability level.

## ❐ CHECKING THE PREDICTIVE ABILITY

Before looking closer at the estimated effects, we first check that the obtained model in fact has predictive ability at all. Figure 16.3 shows how the model's predictive performance, in terms of the root-mean-square error in **y** as function of the number of PCs, $A = 0,1,2,\ldots,11$.

The bottom curve (grey) shows the unvalidated fit, RMSEC(Y) (eq. (10.3c)). This is included here in order to show that when all 20 samples are used, the fit to the data is constant beyond $A = 2$ PCs. RMSEC(Y) = 1.4 mm for $A \geq 2$. But, as usual, this must not be taken as an estimate of the actual prediction error. The top curve (black) shows the estimated error of prediction at the interpolation level, RMSEP(Y) (eq. (10.5e)). With each of the 20 samples in turn being left out of the modelling, the covariance structure in **X** changes slightly, so the PLSR needs $A_{Opt} = 3$ PCs to attain minimum RMSEP(Y).

Beyond three PCs the model is stable also in the cross-validation, with an estimated between-treatments RMSEP(Y)$_{A=11} = 3.7$ mm. This corresponds to 88% of the cross-validation variance in **y** being correctly explained from **X**. The estimated approximate confidence range of RMSEP(Y) is 3.0–4.3 mm (eq. (10.5i)).

This shows that the obtained regression model might, at least in principle, be used for predicting how *A. fla* would be able to grow in future unknown samples, from their known values of *aW*, pH, *O2* and *CO2* in their relevant ranges.

Inspection of the individual prediction errors, $\hat{\mathbf{f}} = \mathbf{y} - \hat{\mathbf{y}}$, did not reveal any outliers. However, the error estimate RMSEP(Y) is somewhat higher than the noise standard deviation assumed during the experimental planning ($s_{err} = 2.5$ mm). This is not unexpected: in addition to the contribution from the actual repeatability noise in the y-data $s_{err}$, RMSEP(Y) contains model specification errors, as well as model estimation errors in $\hat{\mathbf{y}}$ due to estimation errors in the regression coefficients $\hat{\mathbf{b}}$.

An alternative estimate of the standard uncertainty of the noise in **y** may in the case of full-rank PLSR modelling of orthogonal X-variables be obtained from

$$s_{err} = \text{RMSEC(Y)} \times \sqrt{N/(N - K - 1)} \qquad (16.1)$$

$$= 1.44 \times \sqrt{20/(20 - 11 - 1)} = 2.3 \text{ mm}$$

Having inspected the data and ensured that the model has predictive ability, we are now ready to study the estimated effects of the individual design factors.

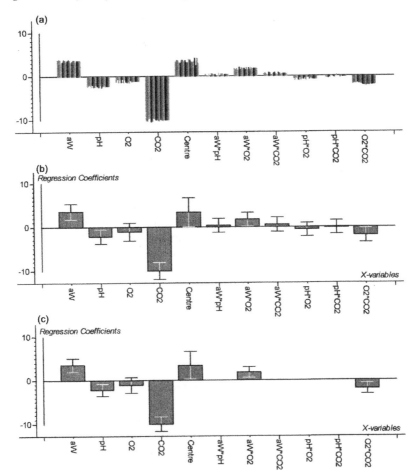

**Figure 16.4.** Screening design model: estimated effects and their reliability. (a) Cross-validation: the partially perturbed regression coefficient vector $\hat{\mathbf{b}}_{-m}$ for the 11 X-variables (abscissa), estimated $M = 20$ times, each time keeping one sample out: $m = 1, 2, \ldots, M$. (b) Full model: the regression coefficient vector $\hat{\mathbf{b}}$ based on all the samples and all the variables, with reliability range $\hat{\mathbf{b}} \pm 2\hat{s}(\hat{\mathbf{b}})$. (c) Optimised full model: the regression coefficient vector $\hat{\mathbf{b}}$ and its reliability range, based on all the samples, after making the non-significant interactions passive.

☐ RELIABILITY ASSESSMENT OF THE ESTIMATED EFFECTS

During cross-validation the effect vector **b** is re-estimated $M = 20$ times. Figure 16.4a shows the 20 partially perturbed estimates, $\hat{\mathbf{b}}_{-m}$, $m = 1, 2, \ldots, M$. Again, no obvious outliers are evident, so it makes sense to summarise these results statistically.

Figure 16.4b shows the full model estimate, $\hat{\mathbf{b}}$ with its resulting reliability range ($\pm 2\hat{s}(\hat{\mathbf{b}})$), as estimated from the partial perturbations between, $\hat{\mathbf{b}}_{-m}$ and $\hat{\mathbf{b}}$ (eq. (10.10)).

The figure shows that the main effects of $aW$, pH and particularly $CO2$ are clearly significant, in the sense that the value zero is obviously not included inside the reliability range. The effects of *Centre*, $aW*O2$ and $O2*CO2$ are also probably real, although this is a little more uncertain, since zero lies just outside the reliability range.

But the main effect of $O2$ and of the interactions $aW*$pH, $aW*CO2$, pH$*O2$ and pH$*CO2$ are not significantly different from zero, at least not at the level of validity chosen here (ability to predict each treatment from the other treatments). The value zero is well inside the reliability range of these estimated effects, and is hence a plausible value for the true, but unknown effects. In other words, their non-zero $\mathbf{b}$-values may well have been caused just by incidental errors in $\mathbf{y}$.

More detailed significance testing is possible, if we are willing to make more detailed statistical assumptions about the distribution of the different types of data and model errors involved. This is done in Appendix A16.

◻   WHAT CAN BE SAID ABOUT THE NON-SIGNIFICANT EFFECTS?

To report that the effects of some design factors are *non-significant* (n.s.) may not be informative enough if it concerns an important effect (Chapter 2.5.2). There may be a huge difference between lack of significance due to a small effect $\hat{b}$ and due to a large standard uncertainty $\hat{s}(\hat{b})$.

In the present case, the estimated main effects of $O2$ was $\hat{b}_3 = -1.11$ with a standard uncertainty of $\hat{s}(\hat{b}_3) = 0.95$. This effect estimate is small, and clearly not significantly different from zero. But with high confidence (roughly 95%) we can state that for coded X-variables, the true effect of $x_3 = O2$ on the diameter $\mathbf{y}$ is expected to be within the uncertainty region $\hat{b}_3 \pm 2\hat{s}(\hat{b}_3)$, i.e. between $+0.8$ and $-3.0$, as shown in Figure 16.4b.

It may sometimes be practical to express the effect-estimates in a more physical meaningful unit. Here, the coding step-size for $x_3$ equals 6% $O_2$ (16–10 and 10–4% $O_2$ for steps 1 to 0 and 0 to $-1$, respectively). By dividing the estimated main effects by the coding step size, we obtain them expressed in the physical unit, in this case: $\hat{b}_{3,\text{phys}} = -0.18 \pm 0.32$ mm/% $O_2$. This means that true main effect of $O2$ is probably not outside the range $-0.14$ and $+0.50$ mm/% $O_2$. It is more informative to report this, than just reporting the effect of $O2$ as *n.s.* On the other hand, it requires more information to convey this reliability region. For unimportant variables, the more parsimonious *n.s.* is satisfactory. More details are given in Appendix A10.

### 16.2.7. Elimination of Useless Variables

The four interaction effects found to be clearly unreliable by the cross-validation, were eliminated from $\mathbf{X}$, and the cross-validated PLSR modelling repeated.

With only seven remaining X-variables, the maximum number of PCs is now seven. The resulting effect-estimates are given in Figure. 16.4c, using $A = 7$. It shows the estimates $\hat{\mathbf{b}}$ to have changed very little, but the estimated reliability ranges of $\hat{\mathbf{b}}$ have become somewhat smaller. The main effect of *O2* remains non-significant.

By the above elimination of the four non-significant interactions, the between-treatments prediction error RMSEP(Y) was now reduced from 3.7 mm in the first PLSR model to 2.9 mm in the second PLSR model. The main reason is that the simplified regression model has fewer parameters to be estimated, and is therefore less prone to being contaminated by measurement noise and other errors. Thus, the elimination of useless variables improved the modelling.

### 16.2.8. Checking the Model Interpretation in the Raw Data

❑ THE PREDICTIVE ABILITY OF THE SIMPLIFIED SCREENING DESIGN MODEL

Figure 16.5a shows the predicted $\hat{\mathbf{y}}$ (ordinate) vs. the measured $\mathbf{y}$ (abscissa) for this second, simplified PLSR model. It confirms that the model is quite successful: the correlation coefficient between the measured and cross-validation predicted diameter of *A. fla* is as high as 0.96 (i.e. 93% correctly explained cross-validation variance). There are no obvious outliers in the figure, and closer inspection of the residuals did not reveal any other model failures. So we do not attempt to refine the screening model any further.

❑ LOOKING FOR THE MOST EXTREME EXPERIMENTAL CONDITIONS

As Figure 16.5a shows, the least mould growth, both measured and predicted, is obtained in samples *wPOC* and *wpOC*, both of which have low *aW* and high *O2* and *CO2*. This appears to be the most promising of the tested storage conditions, if we want to avoid quality losses due to *A. fla* moulding.

Conversely, the most growth is found in samples *WpOc* and *WPOc*, i.e. with high *aW* and *O2* and low *CO2*. This appears to be the most dangerous storage condition with respect to growth of *A. fla*. But these detailed results need to be verified.

❑ A CLEARLY NEGATIVE MAIN EFFECT

The effect of $CO_2$ on $\mathbf{y}$ was seen to be strongly negative and highly significant. In the second model the effect was estimated to be of $\hat{\mathbf{b}}_4 = -10.0 \pm 1.4$. What does that mean?

Figure 16.5b shows **y**-values plotted against *CO2* ($\mathbf{x}_4$). The latter is shown in its un-coded form, with values 5, 15 and 25% $CO_2$, respectively ( = coded steps $-1$, 0 and 1, as also shown). Since the experimental design in this case is

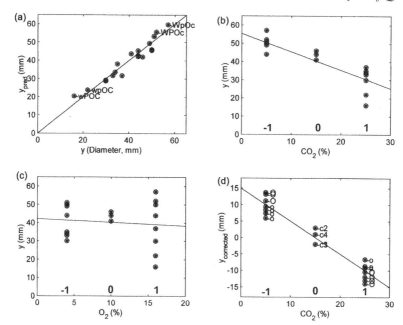

**Figure 16.5.** Checking the screening design model by inspection of input data. (a) Predicted $\hat{y}$ from the optimised model with 2 PCs (ordinate), plotted against the measured **y**. The four most extreme samples are named explicitly. (b) The measured **y** (ordinate) plotted against the coded level of *CO2*. (c) The measured **y** (ordinate) plotted against the coded level of *O2*. (d) The measured **y** (ordinate), corrected for all the effects except *CO2* and *O2\*CO2*, plotted against the coded level of *CO2*. The *O2* level (*o* or *O*) is printed (together with the centre point names *c1–c4*).

orthogonal, the univariate regression model (the straight line in the figure) is compatible with the multivariate regression model.

The diameter decreases, on the average, from about 50 to about 30 mm, i.e. 20 ($10 \times 2$) mm when the $CO_2$ level changes two steps, from $-1$ to 1. Hence, the slope of the line is $\hat{b}_4 = -20/2 = -10$. The figure confirms that it is highly unlikely that the observed effect of *CO2*, $\hat{b}_4 = -10.0 \pm 1.4$, could have been caused by random errors alone.

◻ A NON-SIGNIFICANT EFFECT

The main effect of *O2* was seen to be not significantly different from zero. In the second model the effect was estimated to be of $\hat{b}_4 = -1.1 \pm 1.4$ (Figure 16.4c). Figure 16.5c shows **y** plotted against the *O2*, with coded values $-1$, 0 and 1 corresponding to 4, 10 and 16% $O_2$, respectively. With $\hat{b}_4 = -1.1$, the line is almost flat. This small observed effect could well have been caused just by random errors.

## ◻ A NEGATIVE TWO-FACTOR INTERACTION

The interaction term $x_{11} = O2*CO2$ was found to be small, but significantly negative: $\hat{b}_{11} = -2.3 \pm 1.4$. What does this mean in practice?

A negative two-factor interaction means that when both of the involved design factors are at the *same* coded level ($O2$ and $CO2$ both $-1$ or both $+1$), this is expected to give a *negative* contribution to **y**, compared to what is expected when they have opposite levels. Conversely, a positive two-factor interaction means that when both of the involved design factors are at the *same* coded level ($O2$ and $CO2$ both $-1$ or both $+1$), this is expected to give a *positive* contribution to **y**, compared to what is expected when they have opposite levels.

This is illustrated in Figure 16.5d for the negative $O2*CO2$ effect: here, for illustration purposes only, the response data have been corrected for the effects of the other X-variables. For increased clarity, only the samples' symbol for low or high levels of $O_2$ (*o*, *O*) is printed. The correction of **y** (cf. Appendix A4.5) was done by forming a new regression model, this time with only nine X-variables; the main effect $x_4 = CO2$ and its three interactions were excluded from the model. Then the y-residual **f** from this model is plotted (ordinate) vs. $CO2$ ($x_4 =$ abscissa).

We now expect to see the unmodelled main effect of $CO2$ in the negative slope of the regression line. In addition, the negative $O2*CO2$ effect makes us expect to find the samples with equal coded levels of $O2$ and $CO2$ below the regression line and with oppositely coded levels above the line.

This is indeed what we find: the figure shows that at low $CO2$, low O2 (*o*) is found below and high $O2$ (*O*) above the regression line. At high $CO2$, *o* is found above and *O* below the regression line. What this small interaction effect *means* is not clear, so we regard it with considerable scepticism at this stage.

***VI. Conclusions from the screening experiment.*** During the planning of the screening experiment, the MC power estimation indicated that an experimental design $2^4$ with four centre points had sufficient power to reveal the main effects and two-factor interactions of interest, given the expected measurement noise level. No extra replicate appeared to be necessary; the factorial design had enough redundancy to allow the different treatments to serve as *replicates* of each other around the linear model.

The experiment was then performed, and the diameter of *A. fla* colonies was measured after 7 days of growth in samples produced according to the chosen design.

When analysed by cross-validated BLM, the data showed that the diameter was clearly reduced by $CO_2$, as seen from the clearly negative value of $\hat{b}_4$, the main effect of $CO_2$. A clearly positive main effect of $a_W$ and a clearly negative main effect of pH were also observed.

$O_2$ by itself did not yield any significant effect. But in combination with $CO_2$ it gave a small and relatively clear negative interaction effect. In combination with $a_W$, $O_2$ gave a small and somewhat less reliable positive interaction effect.

Moreover, the centre points gave a positive effect, indicating a curvature in response to one or more of the design factors. However, it is not possible from the present design to conclude which of the design factors cause this curvature.

Finally, the experiment showed that the root mean square error of prediction of individual treatments from other treatments in the experiment was about 3 mm.

On the basis of the conclusions from this preliminary screening experiment with only 20 samples, we are ready to expand the experiment.


## 16.3. MAIN EXPERIMENT: RESPONSE SURFACE ANALYSIS


*I. Purpose of main experiment.* In the previous study we found that *CO2* had strong inhibitory effect on *A. fla*, while *aW* enhanced it. We would like to verify this and to check if this is also the case for some other mould species.

Moreover, we would like to find the reason for the marked effect of the *centre* point: Which design factor is responsible for the curvature? We would also like to check the interactions found.

## II. Experimental planning

### 16.3.1. Selection of Variables

❏   RESPONSE VARIABLES

In addition to the mould from the screening experiment, (*A. fla*), four other moulds are also to be studied: *Geotricum candidum* (*G. can*), *Penicillium roqueforti* (*P. roq*), *P. nalgiovense* (*P. nal*) and *P. verrucosum* (*P. ver*). Their colony diameter after 7 days of growth are to be used as Y-variables in the PLSR modelling.

From microbiological background knowledge we expect *G. can* ($y_1$) and *P. roq* ($y_2$) to form one group and *P. nal* ($y_4$) and *P. ver* ($y_5$) to form another, with *A. fla* ($y_3$) in between.

❏   EXPECTED NOISE LEVEL IN RESPONSE VARIABLES

In the previous screening design we found the pure repeatability error in measuring the *A. fla* diameter to have a standard deviation of about 2.5 mm. Since we are now going to probe a wider experimental range, the model error as well as the sampling and measurement noise may now be somewhat higher. The different moulds may be more or less easy to investigate. For lack of more knowledge, we assume a future error with a repeatability standard deviation of 4 mm, common to all five moulds. Again, this standard uncertainty is supposed to include errors from the way the mould is applied to the sample surface, the way the mould is grown and the way the colony diameter is measured, but not variation in the growth medium.

◻ DESIGN FACTORS

The experimental factors are the same as during the screening experiment: *aW,* pH, *O2* and *CO2.*

## 16.3.2. Selection of Samples

◻ CHOICE OF DESIGN TYPE

A factorial *response surface design* is to be developed, to allow us to study the two-factor interactions and curvatures in more detail: a higher number of distinct levels of *aW,* pH and *CO2* are to be tested.

The results for *A. fla* already available from the screening experiment are to be used as part of the extended design. Therefore the 20 original screening conditions are going to be tested also for the four other moulds, as part of the new, extended design.

Since *O2* in the screening design showed no clear main effect, and only some very small interaction effects, we do not expect *O2* to be responsible for the curvature. Hence we keep this factor constant at 10% for the *new* treatments, in order to reduce the cost of the experiment.

◻ CHOICE OF DESIGN FACTOR VALUES

A *central composite design* (Figure 11.2d) of the sample composition *aW*\*pH is set up with five different levels of each, with the *aW* levels 0.95, 0.96, 0.97, 0.98 and 0.99 and the pH levels 4, 5, 6, 7 and 8. This central composite design is shown in Figure 16.6a. It consists of the *aW*\*pH combinations from the previous screening design (⊛) as well as four new combinations.

These four new combinations were tested at two different *CO2* levels: 0.9% and 25%, with four replicates of the condition *aW* = 0.97, pH = 6. Together the old and new experimental conditions generated a total $N = 43$ treatment samples.

Due to practical considerations, a few of the treatments from the original screening design had to be modified slightly.

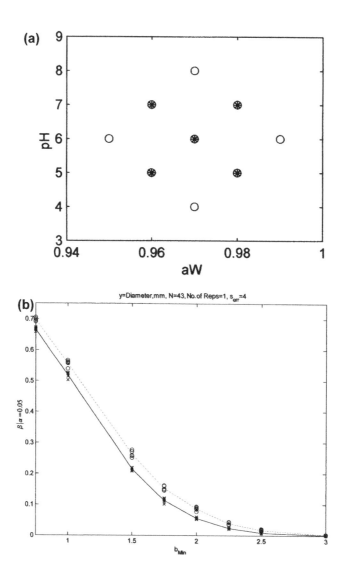

**Figure 16.6.** Response surface design. (a) The central composite design plotted for design factors $\mathbf{x}_1 = aW$ and $\mathbf{x}_2 = pH$. (b) Power-assessment of the response surface design ($N = 43$) for $aW$. The lack of power, $\beta_{|\alpha=0.05}$ (ordinate), vs. the minimum effect to be detected, $b_{\text{Min}}$ (abscissa). This shows the risk $\beta$ of overlooking various true real effect-sizes $b_{\text{Min}} = b_{\text{true}}$ as non-significant at the 95% significance level, as estimated by MC simulation. Lower curve (×): the main-effect of $\mathbf{x}_1 = aW$. Upper curve (○): the curvature-effect of $\mathbf{x}_{11} = aW^{**}2$.

❏ CHOICE OF DESIGN FACTOR REPRESENTATION IN DESIGN VARIABLES X

Like in the screening design, the four main effects $aW$, $pH$, $O2$ and $CO2$ are to be used as $x_1,...,x_4$. In order to study the effects that the four design factors have on each other's impact on the mould growth, the six two-factor interactions $aW*pH,...,O2*CO2$ (eq. (8.4)) are to be used as $x_5,...,x_{10}$, computed after centring, e.g. as $x_5 = (x_1 - \bar{x}_1)(x_2 - \bar{x}_2)$.

In addition, to determine which of the design factors are responsible for the apparent curvature, seen in the screening design as a significant *Centre* point effect, the four squared effects (eq. (8.4)) of $aW$, $pH$, $O2$ and $CO2$ are to be used as $x_{11},...,x_{14}$, computed after centring, e.g. as $x_{11} = (x_1 - \bar{x}_1)^2$.

### 16.3.3. Model and Estimation Method to be Used

Thus, $X$ now has 14 variables (four main effects, six two-factor interactions and four square terms), and $Y$ is going to have five variables. The data analysis will be based on the multiresponse linear model (eq. (6.2d)), $Y = b_0 + XB + F$.

This corresponds to a second-degree polynomial regression. For analysis of designed data it is often called a response surface model: the reason will be illustrated below. Its parameters $b_0$ and $B$ and the error level in $F$ are to be estimated by cross-validated bi-linear modelling, as shown previously.

### 16.3.4. Does the Extended Design Have the Necessary and Sufficient Power?

We check the statistical power of the response surface design before we use it for actual experimental work. There will not be capacity for any replication of the present surface design experiment; the number of samples is limited to about $N = 43$. But we would like to know how small effects the present design will be able to reveal with acceptably low risk. Therefore, the design is submitted to MC estimation of power (Chapter 11.4).

Figure 16.6b shows the estimated *lack-of-power* ($\beta$) (ordinate), the risk of committing Type II error, as functions of different *true*, but unknown sizes of these effects (abscissa). The ordinate represents the estimated probability of overlooking, e.g. the *main effect* of $aW$ ($x_1$, solid curve) or its *square effect* $aW**2$ ($x_{11}$, dotted curve) as *significant* at $\alpha = 0.05$, when using the test of whether or not $\hat{b} \pm 2\hat{s}(\hat{b})$ includes $b = 0$. Each Monte Carlo simulation is repeated five times, with 10 000 hypothetical *experiments* in each. The basic assumption is that the standard uncertainty in the Y-data is $s_{err} = 4$ mm.

The size of $b$ on the abscissa is here given in terms of *standardised* X-variables, but unweighted Y-variables. It thus represents the change (mm) in diameter when design factor $aW$ or its square changes *one standard devia-tion*.

The figure shows, for the $aW$ main effect ($\times$), that there would only be about 5% risk ($\beta \approx 0.05$) of overlooking a *true* effect of $b_{1,true} = 2$ (i.e. a *true* response change in $y$ of 2 mm in diameter when $x_1 = aW$ is changed by one standard deviation). Since the total initial standard deviation of $aW$ is 0.01 in this design, this corresponds to a diameter-increase of 8 mm when $aW$ increases over the full range chosen, from 0.95 to 0.99 (i.e. four standard deviations).

If, instead, the *true* effect $b_{1,true}$ is as high as 3, the figure shows that the risk of over-looking it is virtually nil. On the other hand, if the true effect is only 1.5, the risk increases, and if it is as low as 1.0, the chance of distinguishing it from noise is only 50/50.

The corresponding risk for the *square* term ($x_{11}$, ○) is seen to be only slightly higher. The reasons for this is that, in contrast to our previous screening design, our present design has values at five different $aW$ levels and is hence good for revealing response curvatures; this is why it is called a response surface design.

Hence, according to these simulation results, the present design seems to have the necessary and sufficient power, and may be put to use: The actual response surface measurements are now started.

***III. Experimental work***. The 43 experimental conditions were generated according to the response surface design, and tested in a more or less randomised order for all five mould species (MOULD-DESIGN-II).

***IV. Pre-processing and QC of data***

### 16.3.5. Preliminary Inspection of Input Data

Figure 16.7 shows the distribution histograms of the raw data from the $N = 43$ measured colony diameters.

The figure shows that the five mould species appear to grow more differently than expected: the mean diameters (termed $m$ in the figure) vary from 43 mm ($y_1$, *G. can*) to 15 mm ($y_5$, *P. ver*). The largest diameters of *G. can* approach the maximum physical colony size possible (70 mm). This may create artefacts that could possibly affect the modelling, so care must be taken not to over-interpret curvature and interaction effects for this mould. Other than that, we cannot see any obvious errors in the data.

The *total* initial standard deviation of the Y-variables (termed $s$ in Figure 16.7) varies quite a lot, from 16 mm (*G. can*) to 5 mm (*P. ver*). Although we do not know this for sure, there is reason to expect the standard uncertainties in the Y-variables to vary correspondingly. In order to ensure reasonably similar noise levels and to simplify the comparison of estimated effects, each Y-variable is therefore *standardised* to zero mean and total initial standard deviation of one (eq. (4.7)). To simplify the interpretation of the estimated effects, we also standardise the X-variables, as mentioned above.

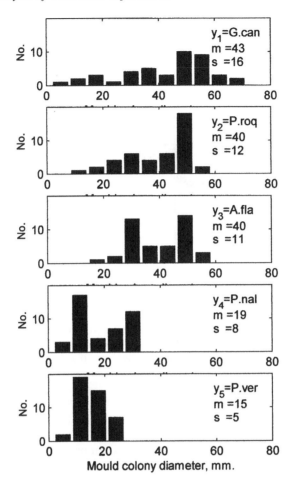

**Figure 16.7.** Response surface experiment: responses $\mathbf{Y} = [\mathbf{y}_1, \mathbf{y}_2, ..., \mathbf{y}_5]$ summarised. Histograms of the frequency distribution of colony diameters for the five mould species *G. can*, *P. roq*, *A. fla*, *P. nal* and *P. ver*.

## *V. Data analysis*.

### 16.3.6. Modelling the Response Data

For simplicity, all five moulds were modelled simultaneously. Full leave-one-sample-out cross-validation ($M = 43$ segments) is to be used in the PLSR modelling; this represents validation at the interpolation ability level.

◻ CHECKING THE PREDICTIVE ABILITY

Two PCs were needed to acquire satisfactory modelling (75–79% explained

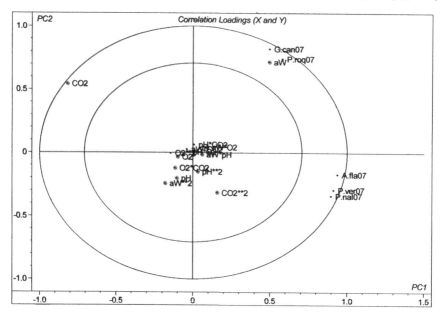

**Figure 16.8.** Bi-linear model of response surface experiment: correlation loadings. Correlation coefficients $\mathbf{r}_1$ and $\mathbf{r}_2$ between the input X- and Y-variables and the first 2 PCs $\mathbf{t}_1$ (abscissa) and $\mathbf{t}_2$ (ordinate). The two ellipses represent 50 and 100% explained variance. X-variables found to be significant for one or more Y-variables are circled.

cross-validation variance for the five Y-variables). After six PCs the predictive ability was more or less constant, with an average predictive ability of 86% explained cross-validation variance (max: 95% for $\mathbf{y}_1 = P.\ nal$ and min: 77% for ($\mathbf{y}_2 = P.\ roq$). This corresponds to unstandardised prediction errors RMSEP(Y) of 7.2, 5.9, 3.2, 1.9 and 1.7 mm for *G. can*, *P. roq*, *A. fla*, *P. nal* and *P. ver*, respectively.

A couple of samples showed particularly high Y-residuals. But re-modelling without these did not improve the modelling very much. Neither did it help much to make non-significant X-variables passive. So these attempts at model improvement were not pursued.

❏  BI-LINEAR MODEL INTERPRETATION

Figure 16.8 gives the correlation loadings for the two most important PCs that together explained about 77% of the total initial cross-validation variance in **Y**. It reveals that the moulds *G. can* and *P. roq* were primarily correlated to water activity level *aW*, while the two moulds *P. ver* and *P. nal* were primarily anti-correlated with *CO2*. *A. fla* falls in between these two groups.

The remaining PCs are very small and do not change the model very much; hence they are not shown here. A number of effects are in Figure 16.8 marked

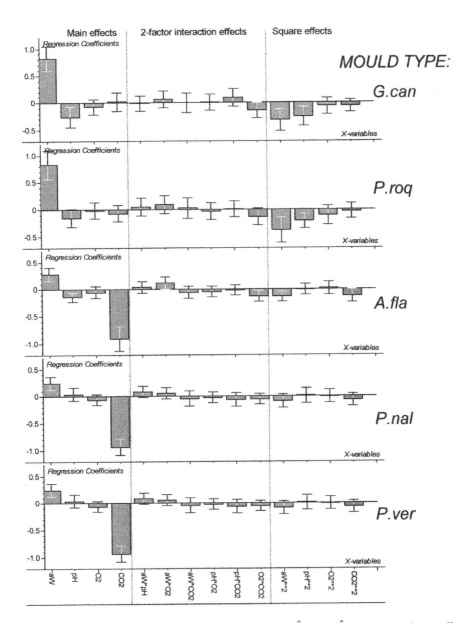

**Figure 16.9.** Linear model summary: estimated effects $\hat{b}_{jk} \pm 2\hat{s}(\hat{b}_{jk})$, the regression coefficients with reliability ranges. Columns, X-variables. $k$: 1–4, main effects; 5–10, two-factor interactions; 11–14, square effects. Rows, Y-variables. $j = 1$–5: (a) *G. can* ($\mathbf{y}_1$), (b) *P. roq* ($\mathbf{y}_2$), (c) *A. fla* ($\mathbf{y}_3$), (d) *P. nal* ($\mathbf{y}_4$), (e) *P. ver* ($\mathbf{y}_5$).

(small circles) as significant for the prediction of at least one Y-variable. This is scrutinised in more detail in the next figure.

## ☐ ANALYSIS OF THE EFFECTS AND THEIR RELIABILITY

Figure 16.9 shows the corresponding standardised regression coefficients for all five moulds, with reliability ranges, $\hat{b} \pm 2\hat{s}(\hat{b})$.

*Main effects*: The pattern of the moulds from Figure 16.8 is clearly illustrated: *G. can* and *P. roq* form one group, *P. nal* and *P. ver* another group, with *A. fla* in between. Here we see that all five moulds show clear positive $aW$ main effect, but the first two show stronger effects than the last three. The first three moulds show negative main effect of pH, while the last two do not. None of the moulds show an appreciable effect of $O2$. The three last moulds (*A. fla*, *P. nal* and *P. ver*) are dominated by the negative $CO2$ main effect.

*Interactions*: The three first moulds show small, but reasonably clear $O2*CO2$ interaction effect. *A. fla* shows a small positive and relatively unclear interaction $aW*O2$.

*Curvatures*: The first two moulds show very clear $aW**2$ and $pH**2$ curvatures. The third, *A. fla*, shows possible curvature effects $aW**2$ and $CO2**2$. The sign of the square effects is *negative*, which indicates a *maximum* diameter at some level of these factors (the second derivative of the X–Y relationship is negative).

Hence, the experiment has revealed a number of interesting effects. That was the goal of the experiment, so in that respect we are satisfied; the experimental design appeared to behave the way we anticipated from the Monte Carlo simulations.

## ☐ INCREASING CONFIDENCE BY SEEING THE SAME EFFECT IN DIFFERENT RESPONSES

The five moulds in Figure 16.9 show many similarities, even for effects that are not quite statistically significant in each individual mould. The small interaction effects of $x_6 = aW*O2$ and $x_{10} = O2*CO2$ are such examples: for all five moulds the estimates are positive for $\hat{b}_6$ and negative for $\hat{b}_{10}$. This indicates small, but real interaction effects in all the moulds, not only in *A. fla* where they are more or less statistically significant. Of course, such patterns, observed in several response-variables, may reflect the same systematic error in all these Y-variables. But given the present factorial, randomised experimental design, we are more inclined to see them as independent evidence of small, but interesting effects.

Additional confidence comes from the fact that the observed grouping of the five moulds in Figures 16.8 and 16.9 is compatible with prior knowledge about these moulds (Haasum and Nielsen, 1998). Since we kept this background knowledge *secret* during the mathematical modelling, it also represents independent evidence.

### 16.3.7. Comparison to the Previous Screening Experiment

The main conclusions are the same as in the screening experiment for the mould *A. fla*. But we have now gained insight into what caused the curvature, mainly $aW**2$ and $pH**2$, instead of just the previous general *Centre* effect.

### 16.3.8. Checking the Model Interpretation in the Raw Data

What now remains, is to ensure that we understand the modelling, by checking the main results in the input data, and then drawing the final conclusions and proposals for further work.

❐ THE MEANING OF A RESPONSE SURFACE

The two first design factors, *aW* and *pH*, in Figure 16.9 showed both clear main effects and clear curvature effects, especially for the moulds *G. can* and *P. roq.* What does that mean?

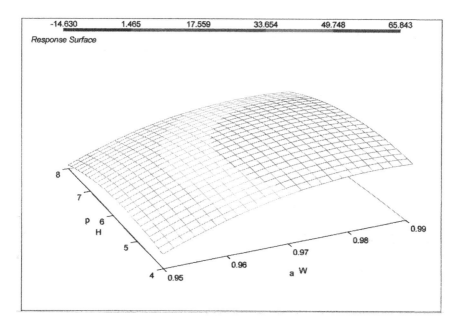

**Figure 16.10.** Estimated response surface. Diameter $\hat{y}_1$ of *G. can* colonies, reconstructed as an interpolated continuum from the PLSR model as a function of design factors $x_1 = aW$ and $x_2 + pH$. The other design factors $x_3 = O2$ and $x_4 = CO2$ were kept constant.

Figure 16.10 shows the estimated *response surface* of one mould, *G. can.* Here, the estimated *G. can* diameter, $\hat{\mathbf{y}}_1$, reconstructed from the model $\hat{\mathbf{b}}_1$ in Figure 16.9a, is plotted as a function of $aW$ ($\mathbf{x}_1$) and pH ($\mathbf{x}_2$). The strong main effect of $aW$ is evident from the slope of the surface. The two clear square terms are also evident from the double curvature of the surface; the growth of *G. can* appears to be maximum around pH 6 and around $aW = 0.98$.

The present response surface is shown for a fixed, average level of the two other design factors, 10% $O_2$ and 14% $CO_2$. Since there were no clear interactions between $aW$ or pH on one hand, and $O2$ or $CO2$ on the other, the estimated response surface will look similar at the other levels of $O2$ and $CO2$.

It is important to realise that the response surface reflects the reconstructed response, based on the estimated model. In itself it says nothing about how *well* this model fits the Y-measurements. It can therefore be deceptive, especially if plotted in striking colours etc.

It is, therefore, also important to check the actual raw data. This is particularly important for the species *G. can*: during our quality control of its input data we have already noticed that some of its values have approached the maximum possible diameter, 70 mm. Could it be that the modelled curvature is just an artefact?

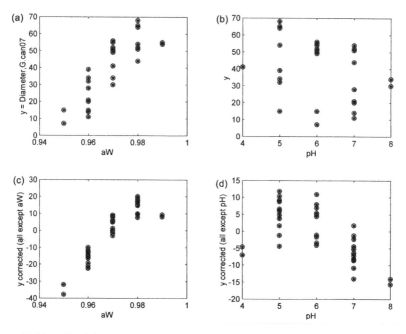

**Figure 16.11.** Checking the response surface curvatures for $\mathbf{y}_1 = G.$ *can* in the raw data. (a) Diameter $\mathbf{y}_1$ vs. $\mathbf{x}_1 = aW$. (b) $\mathbf{y}_1$ vs. $\mathbf{x}_2 =$ pH. (c) Corrected $\mathbf{y}_{1,\text{corrected}*}$ vs. $\mathbf{x}_1 = aW$. (d) $\mathbf{y}_{1,\text{corrected}**}$ vs. $\mathbf{x}_2 =$ pH. *Residual of $\mathbf{y}_1$ after regression on all X-variables except those involving $aW$. **Residual of $\mathbf{y}_1$ after regression on all X-variables except those involving pH.

◻ SEEING THE INDIVIDUAL CURVATURES

The input data behind the response surface in Figure 16.10 are shown in Figure 16.11a,b: the measured diameter of *G. can* ($y_1$) is plotted against $aW$ ($x_1$) and against pH ($x_2$). The strong positive effect of $aW$ is evident. But generally there are too many types of variability present in these input data plots to give a clear interpretation. The detailed effects come out more clearly when the correction method from Figure 16.5d is used again:

Figure 16.11c similar to Figure 16.11a, except that the mean and all the effects of pH, *O2* and *CO2* have been removed (Appendix A4.5) from $y_1$, by a PLSR of $y_1$ on all the X-variables except those involving $aW$. Hence all the $aW$-related effects are left in the corrected y-values. An interesting, but small curvature structure is now more clearly evident: the corrected diameter of *G. can* indicates a maximum at the second highest $aW$ level, 0.98, higher than that at $aW = 0.99$. Hence, it seems that *G. can* is inhibited by too much moisture, as well as (of course) by too little moisture. This was also what the clear negative effect of $aW**2$ in Figure 16.9a and the curvature along the $aW$ axis in Figure 16.10 indicated. This curvature could hardly have been caused by the diameter $y_1$ approaching the 70 mm limit, because the diameter at $aW = 0.99$, which causes the curvature, is only 55 mm.

Likewise, Figure 16.11d is the same as Figure 16.11b, except that the mean and all the effects of $aW$, *O2* and *CO2* have been removed from $y_1$, leaving only the pH-related terms. Even though considerable variability remains, we now see a clear growth maximum between pH 5 and 6. Again, this corresponds to the effect of $pH**2$ seen in Figure 16.9a and the pH-curvature in Figure 16.10. Neither curvature could have been caused just by $y_1$ approaching the 70 mm limit.

◻ LOOKING FOR EXTREME RESPONSES

Detailed inspection shows that the two samples with the very lowest values of $y_1$ in Figure 16.11a,c were the two with the very lowest level of $aW$ (0.95). On the contrary, the two highest ones had the second highest $aW$ (0.98). Thus, the extreme responses here corresponded well to the general model conclusions.

## VI. Conclusions from the response surface experiment

The modelling results in Figures 16.8 and 16.9 are regarded as our final results from the present project. The five moulds were seen to form two distinct groups.

The growth of the first group (*G. can* and *P. roq*) may be strongly reduced by lowering the water content in the samples. They may also be inhibited by low and by high pH.

The growth of the second group (*A. fla*, *P. nal* and *P. ver*) may primarily be reduced by high carbon dioxide levels in the packaging atmosphere. The observed interaction effects were weak, but some of them were consistent for all the moulds.

## 16.4. CONCLUSIONS

This chapter has demonstrated the use of factorial designs to ensure that data from experiments are informative enough. In order to minimise the risk of working under irrelevant conditions, a small screening design was first conducted. Then, certain effects were studied in more detail in a follow-up experiment based on response surface design.

In each case, the intended experimental design was first checked for statistical power, by MC simulation. Thereby, cost-effectiveness could be ensured. Then the experiments were performed, and the data submitted to cross-validated BLM analysis. The model parameters could be interpreted graphically, and cross-validation showed which effects were reliable.

The results showed how the growth of the fungi could be controlled. All five species were inhibited by low water content. Three of them were inhibited by high carbon dioxide concentrations in the packaging atmosphere, two were sensitive to extreme pH levels.

In view of the purpose of the experiment, the results are considered quite interesting for the preservation of food quality. However, they need to be reproduced in independent experiments before we can regard them as proper scientific *knowledge*. Also, they need to be tested in practice in actual food products, instead of the present food *model* system.

This example has shown that data from designed experiments may be analysed just like any other type of data, by cross-validated BLM. For comparisons to more traditional analysis of designed experiments, Appendix A16 shows a more formal significance testing based on the cross-validated PLSR results. The estimated significance levels are shown to be very similar to those from traditional ANOVA, in both designs.

## 16.5. TEST QUESTIONS

1. What are the most important reasons for using designed experiments?
2. What is a factorial design?
3. How does a mixture design differ from a factorial design?
4. What is the difference between a screening design and a response surface design?
5. What are the arguments for splitting the experimental work into two or more small experiments, instead of one big experiment?
6. What is the difference between Type I and Type II errors, and between statistical significance and power?
7. What is the difference between coded and standardised design variables?

8. Is it possible to analyse several Y-variables at the same time?
9. What are the regression coefficients **b** called when **X** = design variables?

## 16.6. ANSWERS

1. Designed experiments, compared to undesigned experiments, have better chance of revealing interesting effects at a given cost of the experiment.
2. A factorial design is an experimental plan where two or more design factors are consciously varied independently of each other.
3. A mixture design differs from a conventional factorial design in that the sum of two or more quantitative design factors is constant (cf. Chapter 11.3).
4. A screening design usually has fewer levels of each design factor than a response surface design. Thereby, one can afford to include a higher number of design factors in a screening design before combinatorial explosion makes it too big.
5. By splitting the experimental work into two or more small experiments, we reduce the risk of wasting work on uninformative conditions, and can advance with relatively small experiments.
6. Statistical significance concerns the risk, $\alpha$, of committing Type I errors: being fooled into believing in an observed effect caused only by incidental errors in the data. The lack of statistical power is the risk, $\beta$, of committing Type II errors: regarding a real effect as *non-significant* and therefore overlooking it (cf. Chapter 11.4).
7. Both concern the numerical representation of the variables. Coded variables represent quantised translations of the original variables, usually with only a few levels, e.g. into 0 or 1, or into 1, 0 or $-1$. Standardised variables have the same resolution as the original variables, but have been mean centred and scaled to a total initial standard deviation of 1.
8. Yes, it is possible to analyse several Y-variables at the same time. The resulting loading plots may simplify the model interpretation.
9. When **X** = design variables, the PLSR is called APLSR, and the regression coefficients **b** the *estimated effects*.

# PART FOUR: APPENDICES (A1–A16)

This part contains appendices for Chapters 1–16, with background inform-ation and technical details.

**Appendices for Part One:**
A1 How the present book relates to some mathematical modelling traditions in science
A2 Sensory science
A3.1 Bi-linear modelling has many applications
A3.2 Common problems and pitfalls in soft modelling

**Appendices for Part Two:**
A4 Mathematical details
A5 PCA details
A6 PLS Regression details
A7 Modelling the unknown
A8 Non-linearity and weighting
A9 Classification and outlier detection
A10 Cross-validation details
A11 Power estimation details

**Appendices for Part Three:**
A12 What makes NIR so information-rich?
A13 Consequences of the working environment survey
A14 Details of the molecule class models
A15 Forecasting the future
A16 Significance testing with cross-validation vs. ANOVA

# Appendix A1

# How the Present Book Relates to Some Mathematical Modelling Traditions in Science

The authors fear that the present book may provoke some traditional members of the field of mathematical statistics: the book makes little or no use of traditional statistical distribution theory, and takes a dim view on hypothesis testing. The book is in good company in respect to the latter (Tukey, 1977; Cohen, 1994; Gigerenzer and Murray, 1987; Thompson, 2000). For instance, Deming (1987) states in his paper *On the statistician's contribution to quality*: 'a process that is stable today, may not be stable tomorrow...test of hypothesis and tests of significance belong to the philosophy of some other world, not this one'.

The field of mathematical statistics seems to hold a virtual world monopoly on teaching students data analysis. That reflects the only traditional way it was possible to deal rationally with uncertainty, before the advent of computers. Now that is changing. Computerised measurements today provide thousands of variables a second, giving us *fat* tables of input data with more variables than objects, instead of the *tall* data tables that once were common when traditional mathematical statistics was developed. Computerised statistical methods for re-sampling have already reduced the need for teaching complicated mathematical statistics to unwilling and terrified non-statisticians. Interactive computer graphics combined with computer-generated model summaries, provide both overview and detailed inspections.

The simplest, most central concepts in statistics, e.g. parameter estimation and uncertainty, are treated with reverence in this book. And the reader is advised to get help from a professional statistician in particularly difficult or critical cases.

There are a number of other mathematical modelling traditions in science, some of which will be briefly discussed here.

## A1.1. Causal Modelling

Traditional *physical and chemical* approaches to modelling of experimental data is based on causal description, often in terms of first principles relationships, of what is *known* to be true. This is good as long as the *known* theory is in fact true, but grossly misleading if our *knowledge* is wrong or incomplete, and such mistakes are difficult to detect inside the mathematical and mental *jail* of the *known* theory.

In contrast, the data analysis method explained in this book does not make much mathematical assumptions about the *"truth"*. The *soft* models are normally just regarded as *approximation* methods (Wold, 1975), which make the main information content of data cognitively accessible. The important modelling takes place in the mind of the scientists, not in the computer.

However, traditional causal modelling approaches may well be used in conjunction with the *soft* methods, either as a linearisation or compression tool during data pre-processing, or in terms of classical chemical and physical modelling, after the *soft* method has revealed that the data behave as expected from the theory. An interesting alternative is to fit the input data to the causal model, and then pass both the parameters and the lack-of-fit residuals from the causal model to *soft modelling*.

It should be mentioned that experiences in round-robin ring tests in chemistry indicate that scientists tend to overestimate their own analytical reliability, primarily because they forget to include some error sources in their calculations. Some researchers even mistake instrument repeatability for scientific reproducibility (ISO, 1995a).

## A1.2. Classical Statistics

Traditional *hypothesis testing* methods in mathematical statistics focus more on the *noise* than on the interesting *signal* in data. Therefore, such methods act primarily as scientific auditing to guard against Type I errors (being fooled by errors in the data). Such methods are less efficient as tools for the curious researcher who wants to discover the world. Type II errors (overlooking something important) have been down-played as far less important than Type I errors in science, presumably because it is worse to draw a false conclusion than to draw no conclusion at all. But both errors are indeed important from a practical point of view. Moreover, since the assumptions underlying classical hypothesis tests (e.g. additive, independent, identically and normally distributed random errors) are hardly ever fulfilled, the resulting significance probabilities are also just approximations and should be treated as such.

Classical statistical methods, based extensively on theoretical assumptions and mathematical deduction about error distributions, can be very useful in the hands of experts. In some cases it is the only approach possible, because of lack of available data.

With the advent of computer-intensive re-sampling methods (Monte Carlo simulation and bootstrapping), data analysis has liberated itself from some of the restrictions and difficulties of traditional mathematical statistics. The present book uses Monte Carlo simulation for simplified power assessment in experimental design (Chapter 11). The statistical bootstrapping methods may focus too strongly on the *noise* instead of the interesting *signal* in the data. If used uncritically, they may create false impressions about the distributional properties of data, e.g. in the presence of outliers. Therefore this book instead employs the *little brother* of bootstrapping, the pragmatic, visually-oriented cross-validation/jack-knifing (Chapter 10).

It is questionable whether contemporary multivariate analysis of real-world data is sufficiently well handled by mathematical statisticians, if they are primarily motivated by distribution theory and not by the discovery process. While being taught even to non-statisticians at every university around the world, the traditional statistical distribution theories and hypothesis test methods are often perceived as difficult and alien, and they are easily misused by non-statisticians who do not understand the concept of degrees of freedom and random errors.

However, model parameters have to be *estimated* on the basis of uncertain empirical observations. Therefore, some central statistical concepts and tools are important in data analysis.

The most important *defensive* concept from statistics is the awareness of *uncertainty* in input data as well as in estimated parameter output. For assessing how uncertainty affects the data analytical output from a given data set, the traditional concept of *statistical significance* at, e.g. 5, 1 or 0.1% level is here replaced by more pragmatic uncertainty measures, in terms of $\pm 2$ standard deviations, yielding roughly 95% confidence regions. These uncertainty measures are estimated by the visually oriented cross-validation/jack-knifing approach. More formal significance tests are still available if really needed and warranted. The book stresses the importance of explicitly stating which validity level is being used (repeatability, reproducibility, etc.). For assessing how uncertainty will affect a *future* experiment, the difficult traditional theory of statistical *power* of experimental designs is replaced by the simpler Monte Carlo method: simulating the future real data sets.

The most important *offensive* tool from statistics is the *least squares regression* modelling for estimating relationships. This is here used in the general, multivariate bi-linear framework of linear and bi-linear *soft modelling*.

In relation to the determination of quality, two particular branches of statistics are important: statistical quality control (SQC) and statistical process control (SPC). The former has, among other things, a strong focus on important problem of *representative sampling*. This was given some attention in Chapters 3 and 11. The latter concerns, among other things, how to make efficient and understandable early-warning systems for detecting quality problems in production processes. Chapter 15 shows an example of how the bi-linear regression may be implemented for multivariate SPC.

## A1.3. Dynamic Process Modelling

A third tradition is *time series analysis* and *dynamic systems identification based on state space modelling*. This is commonly used in, e.g. signal processing and in industrial process control. The use of large sets of *differential equations* is common in other fields, e.g. econometrics.

The bi-linear regression method from this book has also demonstrated its usefulness for more explicit time series analysis as well as for subspace identification (Di Ruscio, 1998). And the bi-linear models are easy to update dynamically over time (Helland et al., 1991; Dayal and MacGregor, 1997), in close analogy to, e.g. the so-called Kalman filter (Ergon, 1998; Di Ruscio, 1998). In many cases the systems of differential equations are of a linear nature, and may be studied by the present bi-linear method.

However, in the present introductory book, the only time-relevant modelling shown is the extraction of systematic process developments and plotting them graphically, leaving it to the human eye to see development trends etc. (cf. Chapters 15 and A15).

## A1.4. Modern Informatics Modelling

A fourth group of general modelling tools, *artificial neural nets* (ANN) and *genetic algorithms* (GA), are very powerful for elucidating relationships from heterogeneous, strongly non-linear data sets. But their results can be very difficult to interpret, and they are therefore often used as a *black box*. Also, they are difficult to optimise with respect to statistical reliability: they are so flexible that they tend to model the noise in the data too well. Unless the amount of data is very high or the number of input variables is very low, their models have too many independent parameters, and their algorithms visit and reject too many alternative solutions.

The bi-linear method to be taught here may be considered a simplified, linear version of ANN, with the *hidden nodes* called *latent variables*, and these are to be revealed, not hidden! Heterogeneous topologies are modelled by splitting the data sets in classes, and serious non-linearities are treated by linearisation or by polynomial expansion.

## A1.5. Econometrics Path Modelling

In economy and related disciplines there is a tradition to use multiblock path modelling to study complicated data sets (Lohmöller, 1989; Höskuldsson, 1999). The bi-linear regression method used in the present book, the two-block partial least squares regression (PLSR) was originally developed from one of these econometric traditions, PLS path-modelling (Jöreskog and Wold, 1982) in order to solve the multivariate calibration problem in chemistry (Wold et al., 1983a). The two PLS approaches are distinct, but philosophically very similar.

## A1.6. Psychometrics

With its focus on the pragmatic study of latent variables, the present *soft modelling* approach is related to data analysis traditions within experimental psychology and similar applied modelling traditions in social sciences. For simplicity, this introductory book largely ignores topics like optimal scaling (for improved linearisation, Gifi, 1990) and N-way modelling (Tucker, 1966; Kruskal 1989; Bro, 1996, 1998). However, these extensions can well be added to the present method, as discussed, e.g. by Martens and Næs (1989), pp. 159–163. Chapters 9 and 14 demonstrate how many types of non-linearities may be handled even within the present bi-linear model framework.

The data analytical method chosen in this book – the cross-validated bi-linear modelling, has emerged out of two fields – chemometrics and sensometrics:

## A1.7. Chemometrics

This field mainly concerns multivariate analysis applied to data from chemistry. New ways to apply multivariate data analysis in, e.g. analytical chemistry, organic chemistry and process chemistry, as well as data analytical method developments, are published, e.g. in *Journal of Chemometrics* and in *Chemometrics and Intelligent Laboratory Systems*.

The field of chemometrics tends to be characterised by a love/hate relationship to classical mathematical statistics, because the two fields share the basic problems concerning uncertainty, and also some basic tools (like weighted least squares regression). But the fields have traditionally differed w.r.t. competence about statistics and competence about the application fields, and consequently in the way to think about data.

## A1.8. Sensometrics

This relatively young field concerns the analysis of data from sensory science. It is distinctly interdisciplinary and combines people and modelling techniques from both classical mathematical statistics, psychometrics, psychophysics and chemometrics. The journals *Food Quality and Preference* and *Journal of Sensory Studies* are two of the main arenas for sensometrics.

# Appendix A2
# Sensory Science

Whenever quality is evaluated or interpreted (Chapter 2) the human senses and mind are in action. Sensory science is a multidisciplinary field comprising measurements, interpretation and understanding of human responses to e.g. product properties, as perceived by the senses, such as sight, smell, taste, touch and hearing. Sensory analysis is a way to measure human responses to chemical and physical properties (stimuli). When done systematically and in accordance with contemporary cognitive theory, it constitutes a powerful tool for obtaining data from the objects and surroundings that we want to analyse (Martens, 2000).

The most efficient and informative sensory method that may be analogous to chemical methods, is called *descriptive sensory analysis*. A set of objects is assessed quantitatively with respect to a set of descriptive variables (qualities, verbal descriptors) by a panel of trained sensory assessors, according to an experimental design with sufficient redundancy. Reliable descriptive analysis requires a professional sensory laboratory setting, and professionally selected and trained panelists (assessors).

It also requires a proper choice of sensory descriptor words that separates emotionally neutral descriptions from hedonic preferences and subjective expectations. Within a sensory analysis project, *each descriptor word* must be cognitively clear with respect to its meaning and its quantitative anchoring along some sensory scale, and it must discriminate between the objects. The set of words (*the sensory profile*) must have enough words to span the important dimensions of quality variation in the set of samples at hand, but not more words than the assessors can easily handle from a cognitive point of view (usually <15). Finally, reliable sensory analysis requires proper experimental design and data analysis.

In sensory science, factorial or mixture designs are often used in order to cancel out individual human differences and drift over time. With sufficient assessors (7–12) and sufficient replication (2–4 reps separated in time), the sensory results can be as reliable as those from many physical analytical instruments.

Professional sensory descriptive analysis can complement, and often replace, a wide range of traditional electronic laboratory instruments, ranging from spectrophotometers (colours) via rheological instruments (textures) to HPLC (taste) and GC (smell). Sensory analysis is most extensively used in food science, where it often forms the very basis for the measurement of quality, in terms of colour, texture and flavour. But it is also used in the perfume industry (odour and colour composition), in the electronic industry (optimisation of audio-visual equipment), car industry (design of product sound and appearance) etc.

Sensory science has a large potential in the measurement of many *difficult* qualities in traditional laboratories. For instance, simple, but systematic visual assessment of growth experiments in microbiology can be very informative. But this potential is until now largely untapped, possibly due to some unclear, unspoken and unscientific concepts concerning *subjective* and *objective* measurements. Sensory science is also highly relevant in visual assessment of data analytical plots (cf. Chapter 2.5.4).

The need for multivariate data analysis in sensory science arises from the fact that human responses are never univariate: There are always multicollinearities and *noisy* relationships that traditional statistical regression methods cannot handle so well (Martens and Martens, 1986a; Martens et al., 1987; Næs and Risvik, 1996; Dijksterhuis, 1997). Additional information about the objects (production facts, chemical/physical instrument measurements) or consumer demographics are usually available and should then be interpreted along with the sensory data (Martens et al., 1994).

Tables 3.1 and 3.2 in Chapter 3.1.1 show how sensory data are analysed in this book.

# Appendix A3.1

# Bi-linear Modelling Has Many Applications

The bi-linear multivariate *soft modelling* method outlined in Chapter 3 is based on PLS regression, with PCA as a special case when there are no Y-variables. Table 3.3 lists a number of data analytical purposes that may be fulfilled by a variety of specialised, *classical* data analytical methods, but also by one single method, the present bi-linear method.

**Table 3.3.** A unified tool-box for multivariate data analysis: overview of different data analyses, their purpose, how they are done traditionally, and how they may instead be performed by the method of cross-validated bi-linear modelling (BLM).

| Purpose | Classical methods | BLM method |
|---|---|---|
| Find underlying patterns of co-variation within a single data table **X** | Factor analysis, canonical variate analysis | PCA |
| Find underlying patterns of co-variation between two data tables **X**, **Y** | Canonical correlation | PLSR |
| Find predictive relations of variables **Y** from **X**, and predict **Y** in new objects | OLS regression | PLSR |
| | Ridge regression (RR) | PLSR |
| | Stepwise regression | PLSR |
| | Response surface modelling | PLSR (w/squares, interact.) |
| | Artificial neural nets (ANN) | SIMCA/PLSR |

Table 3.3 (*continued*)

| Purpose | Classical methods | BLM method |
|---|---|---|
| Find the differences between two or more classes of objects Symmetric classification | Linear discriminant analysis (LDA) | DPLSR, $Y = 0/1$ |
| Assymmetric classification Segmentation | Cluster analysis? PrefMap | SIMCA PLSR |
| Study effects of known factors on observed variables | ANOVA MANOVA Covariance analysis | APLSR, $X =$ e.g. $0/1$ APLSR, $X =$ e.g. $0/1$ APLSR, some X-variable measured, some from design |
| Resolve mixed signals into known constituent signals | LS unmixing/curve resolution | PLSR ($X$ = pure component spectra, $Y$ = identity matrix) |
| Remove effect of irrelevant variables | Blocking in ANOVA | PLSR, project input $Y$ on irrelev. variables $X$, replace $Y$ by residual $F$ |
| Statistical quality control or process control | Many univariate plots | PLSR, few plots of scores/ leverage and residuals/ RMSEP. Automatic outlier warnings |
| Analysis of N-way data Many data tables | N-way (e.g. Tucker) | PLSR (unfolded) [hierarchical or N-PLS] |
| Study patterns of change in time, and forecast new points in time: Time series analysis Update models in time | AR, MA, ARX, ARMA etc. State space model/Kalman filter | PLSR (with time shifts) PLSR [Updated, dynamical] |

# Appendix A3.2

# Common Problems and Pitfalls in Soft Modelling

In general, the versatility of the BLM method allows the researchers to trust their own intuition about where to start. But like any other methodology, it has to be learnt: the method principles must be understood, and the right software buttons must be remembered. Here are some problems that arise again and again. The proposed remedies refer to various techniques which are explained in detail in other chapters.

### A.3.2.1. Experimental Design

Even the most perfectly planned experiment may end up showing nothing of interest. At least this indicates that there is nothing of interest to be observed in the present system. However, if an experiment was *not* properly designed or the *wrong* variables were measured, the result is often frustrating:

*Under-dimensioned experiments seldom reveal anything*: The data from a spontaneous, small experiment show no clear effects, and you do not know why.

<u>Remedy</u>: Design your experiments better next time, as described in Chapter 11 and 16. In particular, assess the statistical power to guard against unacceptable risk.

*Over-dimensioned experiments cost too much*: Sometimes it is tempting to make large experiments and take a lot of replicates, *just to be sure*, especially if somebody else pays for the experiment. This is an ethical problem as well as an economical problem for the research institute. It is also a practical problem for the researcher if the experiment takes too long, and when the project is finally over, the researcher feels a little foolish if the results are obvious.

<u>Remedy</u>: Same as above.

## A.3.2.2. Scaling of Input Variables

*Dictatorship of the few*: Small, but valid structures in some variables may be swamped by large noise levels in other variables, if the variables have not been properly weighted *a priori*. This is relevant when e.g. combining variables from very different sources, and given in different units.

*Remedy*: Standardise the variables (Chapter 8), and/or plot correlation loadings instead of conventional loadings (Chapter 6).

*Dictatorship of the many*: A lot of variables that contain mostly noise hide valid structures in a few precise variables. This is relevant when e.g. many variables with virtually no variation have mistakenly been standardised, e.g. *baseline* regions in spectra from a multichannel analytical instrument.

*Remedy*: Reduce the number of *baseline* variables by averaging adjacent variables in the less interesting regions. If standardisation of the variables has been used, then modify or remove it (cf. Chapter 9).

*Useless variables adding confusion*: At the outset it is important to include a generous selection of variables in the multivariate modelling. But variables found to be irrelevant, useless or unreliable may decrease the quality of the model.

*Remedy*: In the end, such variables should be made passive or eliminated all together before the final modelling is made. The *jack-knifed* stability of the variables estimated during cross-validation (Chapter 10) may be used for this end, as illustrated in Chapter 14.

## A.3.2.3. Outlier Objects

*Bad actors dominating the scene*: Extreme, erroneous objects strongly affect the data analytical model, giving *strange* loadings and scores, and bad predictive performance of the cross-validation.

*Remedy*: Look for outliers by checking residuals in **X** and **Y** (preferably from cross-validation), score summaries (*leverage*, (eq. 9.3)) and cross-validated score perturbations (eq. (10.8a) in Appendix A.10). At least *try* remodelling without the outlier(s). Inspect the outliers' data, and correct or remove them if they are clearly erroneous. Repeat until all really bad actors have been removed. But do not throw out *all* your informative objects! Normally, do not remove more than 5–10% of your objects. And try to predict **Y** from **X** even in the eliminated outliers, using the obtained model.

*Shooting the lone hero*: A unique and particularly informative object (one that spans a certain type of valid variability all by itself) is eliminated as if it were an erroneous *bad actor*, and the resulting model therefore becomes less informative and less applicable for future unknown samples of the same general kind.

*Remedy*: Do not permanently eliminate a good outlier. But it may be informative to remove it *temporarily* just to see its influence on the model; that is very easy in modern software. Particularly informative, reliable *lone heroes* may be cloned and/or held fixed in the training set during the cross-validation.

### A.3.2.4. Heterogeneous Sample Sets

*Comparing apples and pears*: If objects which have little in common are pressed into one common model, this model will require many components, show bad predictive ability, or both.

*Remedy*: If score plot and/or prior knowledge indicates distinct and/or interpretable clusters or classes, try to model each of these classes separately (Chapters 9 and 14).

### A.3.2.5. Problems in the Internal Data Analytical Validation

*Overfitting*: If both **X** and **Y** have noise, and you accept too many PCs into your bi-linear model, then the model includes and amplifies small and noise-related effects, and this corrupts your predictive ability and can make interpretation difficult.

*Remedy*: Use cross-validation to find the optimal number of PCs, $A_{Opt}$ (cf. Chapter 10).

*Mistaking repeatability for reproducability etc.*: Cross-validating at the wrong validation level.

*Remedy*: Think before you define the cross-validation (Chapter 10.5)! What is the purpose of your validation? Which error source do you want to guard against?

*Having too few samples in a test set*: Setting aside some of the available samples as *independent test set* for model validation may give a good feeling of not having cheated. But cross-validation, performed as described in Chapter 10, gives more precise estimates of predictive ability and model uncertainty than the use of a small independent test set. The reasons is that you cannot have a good calibration model and a precise test-set assessment of this model at the same time, unless you have lots and lots of samples.

*Remedy*: Use cross-validation instead of pulling out a separate test set. In addition, try to validate the interpretation of the model by comparing it to previous knowledge. Finally, apply standard scientific norms for reproducing important results under fully independent conditions.

### A.3.2.6. Problems in the External, Interpretational Validation

*Over-interpretation of the loading and score plots*: Blowing up a small PC to see it better in a plot, and forgetting that it is still a *small* PC.

*Remedy*: Be conscious of the plot scaling: blow up the plots for visualisation of details. But scale plots according to their actual importance in order to maintain a realistic perspective!

*Loading or score plots are too crowded*: When the number of variables or samples is high, and/or their names long and unsystematic, the loading or score plots will be cluttered and difficult to read.

<u>Remedy</u>: Use shorter and more systematic naming convention. Show only characters in the names important to the problem at hand. Use colour codes and interactive activation/deactivation of names to avoid cluttering plots.

### A.3.2.7. Missing Values

*Getting funny results because of missing values*: Problem may arise if the input data have very many missing values (say, $>10\%$) and the missing values occur in some semi-systematic pattern that is neither balanced nor randomly distributed throughout the data matrices. Depending on how the missing values handling is implemented in the BLM software (cf. Appendix A5.6), the data analytical model may then become unnecessarily unstable and/or reveal some structures that are just artefacts.

<u>Remedy</u>:

1. Funny patterns of missing values can be seen easily in data table maps like Figure 3.6a).
2. Try to compare the results obtained *with* missing values, with results obtained *without* missing values (i.e. when the objects and variables with the most missing values have been temporarily deactivated).
3. Try to *obtain measurements* also for the values initially defined as missing.

# Appendix A4

# Mathematical Details

## A4.1. Some More Vector Algebra

❏ VECTOR MULTIPLICATION

*An outer vector product*: Assume that a data matrix $\mathbf{X}$ ($10 \times 48$) (cf. Chapter 4.1) is reconstructed by the outer vector product of two vectors: $\hat{\mathbf{X}} = \mathbf{t} \times \mathbf{p}'$. The multiplication of a column vector $\mathbf{t}$ ($10 \times 1$) by a row vector $\mathbf{p}'$ ($1 \times 48$) yields a matrix $\hat{\mathbf{X}}$ ($10 \times 48$).

*An inner vector product:* The multiplication of a column vector's transpose by, e.g. the column vector itself, yields a *scalar* called $c$

$$c = \mathbf{t}' \times \mathbf{t} = \Sigma t_i^2 \tag{4.3}$$

❏ A MATRIX PRODUCT

In $\hat{\mathbf{Y}} = \mathbf{X} \times \mathbf{B}$, the expression $\mathbf{X} \times \mathbf{B}$ is a matrix multiplication. The multiplication of a matrix $\mathbf{X}$ ($N \times K$) by a matrix $\mathbf{B}$ ($K \times J$) is the predicted Y-matrix, $\hat{\mathbf{Y}}$ ($N \times J$). Note that the *inner dimension*, the number of columns in the first matrix and the number of rows in the second matrix (here $K$), is the same. The multiplication means a summation of the individual product elements over this inner dimension

$$\hat{\mathbf{Y}} = \mathbf{XB}$$

because

$$\hat{y}_{ij} = x_{i1} \times b_{1j} + x_{i2} \times b_{2j} + \ldots + x_{ik} \times b_{kj} + \ldots + x_{iK} + b_{Kj} = \Sigma x_{ik} \times b_{kj}$$

❐ AN ILLEGAL MATRIX PRODUCT

The multiplication of two matrices is not defined if their inner dimensions do not fit. So, if $\mathbf{X}$ is $(N \times K)$ and $\mathbf{C}$ is $(M \times J)$, we cannot do $\mathbf{D} = \mathbf{X} \times \mathbf{C}$ if $K \neq M$.

❐ MATRIX INVERSION

Division by a matrix is called matrix *inversion*. The inverse of a matrix $\mathbf{C}$ is written $\mathbf{C}^{-1}$. Matrix inversion is analogous to the inversion of a scalar $c$, e.g.

$$c^{-1} \quad (= 1/c) \tag{4.4}$$

Just like $c^{-1}$ blows up if $c$ approaches zero, $\mathbf{C}^{-1}$ blows up if $\mathbf{C}$ has an *eigenvalue* that approaches zero (i.e. some columns or rows in $\mathbf{C}$ are too intercorrelated).

Implicitly or explicitly, matrix inversion is required in order to make multivariate regression models. Otherwise, it is impossible to ascribe $\mathbf{y}$'s regressand-information to different regressors in $\mathbf{X}$.

The BLM method is designed to avoid blow-up due to small eigenvalues when inverting the regressor covariance matrix; the inversion in made in a sequence of simple steps, each requiring only the inversion of a scalar, $1/c_a$, $a = 1, 2, ...., A_{Opt}$ (eq. (4.3)). Hence, matrix inversion is not used much in this book, and only in the technical appendices. For more detail, cf. Martens and Næs (1989).

## A4.2. Some Useful Statistical Expressions

❐ THE MEAN AND ITS USE FOR MEAN CENTRING OF VARIABLES

The mean of a single variable, e.g. $\mathbf{y} = [y_1, y_2, ..., y_N]' = [y_i, i = 1, 2, ..., N]'$, is defined, as usual, by

$$\bar{y} = (y_1 + y_2 + ... + y_N)/N \equiv \Sigma y_i/N \tag{4.5}$$

It measures the *average value*, and also the expected value for new samples of the same general kind.

When mean-centred, a variable $\mathbf{y}$ is here called $\mathbf{f}_0$. Correspondingly, the mean-centred X-variable, $\mathbf{x}$ is here called $\mathbf{e}_0$. They are defined by

$$\mathbf{e}_0 = \mathbf{x} - \bar{x}, \quad \mathbf{f}_0 = \mathbf{y} - \bar{y} \tag{4.6}$$

The mean-centred variables represent the between-samples variations that are of primary interest in data analysis. The means themselves are usually of less interest.

❐  STANDARD DEVIATION, S AND THE VARIANCE $s^2$

The standard deviation of a variable, e.g. **x**, is defined, as usual, by

$$s_x = \sqrt{\{[(x_1 - \bar{x})^2 + (x_2 - \bar{x})^2 + \ldots + (x_N - \bar{x})^2]/(N - 1)\}}$$

The standard deviation may equivalently be obtained from the mean-centred **x**:

$$s_x = \sqrt{\{\mathbf{e}_0', \mathbf{e}_0/(N - 1)\}} \tag{4.7}$$

The *variance* $s^2$ is simply the square of the standard deviation $s$ in eq. (4.7). This standard deviation $s$ is here usually called the *total initial standard deviation*, to distinguish it from the standard deviation of an error estimate, which is here called the *standard uncertainty*. In plots, etc., $s$ is sometimes abbreviated to SD, std. dev. or S.Dev. It measures the expected deviation from the mean in new samples of the same general kind. The standard deviation is primarily applied for standardization of variables (eqs. (8.1) and (8.2)).

The expression $N - 1$ reflects the fact that one parameter, $\bar{x}$, has already been estimated from the $N$ data. Hence, there are only $N - 1$ so-called *degrees of freedom* left. Due to the complexity of the concept of degrees of freedom, it is not used much in this book. For description of uncertainty and for assessment of model performance, the standard uncertainty $s_{err}$ and the root mean square error (RMSE) is used, as explained in Chapters 9 and 10.

❐  CORRELATION COEFFICIENT

The conventional (linear product-moment) *correlation coefficient* between two variables **x** and **y** is defined by

$$r_{xy} = \mathbf{f}_0'\mathbf{e}_0/\sqrt{(\mathbf{f}_0'\mathbf{f}_0 \times \mathbf{e}_0'\mathbf{e}_0)} \tag{4.8}$$

The correlation coefficient $r_{xy}$ is a unit-free measure of the goodness of fit between the two variables **x** and *y* within a certain set of $N$ samples. $r = 1$ means a perfect fit with positive slope, $r = -1$ means a perfect fit with negative slope, and $r = 0$ means no fit at all.

(In contrast to the correlation coefficient $r$, the regression coefficient, i.e. the slope $b$ of the regression line between two variables (see below), is *not* dimension free, and does *not* measure the goodness of the fit; it *defines* the model.)

❐  CROSS-PRODUCT AND COVARIANCE

For two variables **x** and **y**, the cross-product is simply defined as the inner products of their mean-centred column vectors, e.g.

$$w_{xy} = \mathbf{f}_0' \times \mathbf{e}_0 = \Sigma f_{0,i} \times e_{0,i} \tag{4.9a}$$

The covariance between two variables $\mathbf{x}$ and $\mathbf{y}$ is defined, via their cross-products, by

$$Cov_{xy} = \mathbf{f}_0' \times \mathbf{e}_0/(N - 1) \tag{4.9b}$$

Although the BLM method in this book relies on implicitly decomposing covariances between many variables, this explicit expression for covariance is not needed at the present technical level. So, cross products and covariances do not have to be considered explicitly by the reader.

### A4.3. Other Statistical Summaries: Median, Quartiles, Percentiles

For graphical inspection of raw data (see Figure 3.5), these distribution measures may be useful as alternatives to the mean and standard deviation. A *percentile* represents the value of a variable below which a certain percentage of the observations fall.

The 0, 25, 50, 75 and 100% percentiles are called the *quartiles*; they represent the values below which none, a quarter, half, three-quarters and all of the samples lie. The *median* is the 50% percentile where half of the samples have lower values and half of them have higher values. This is illustrated in Figure 3.5f).

### A4.4. More on Linear Least Squares Regression

❐ THE ORDINARY LEAST SQUARES SOLUTION

In matrix algebra, the regression model (eq. (4.2a,b), p. 90) could be written

$$\mathbf{y} = b_0 + \mathbf{x}b + \mathbf{f} = \mathbf{X}\mathbf{b} + \mathbf{f} \tag{4.10}$$

where, in this univariate case, $\mathbf{X} = [\mathbf{1}\,\mathbf{x}]$, with $\mathbf{1} =$ a column vector of $N$ ones; here: $\mathbf{1} = [11111]'$, and $\mathbf{b} = [b_0, b]'$ contains the model parameters to be estimated from the data $[\mathbf{x},\mathbf{y}]$.

The so-called ordinary LS (OLS) regression solution to eq. (4.10) is

$$\hat{\mathbf{b}} = (\mathbf{X}'\mathbf{X})^{-1}\mathbf{X}'\mathbf{y} \tag{4.11a}$$

For the univariate case (only one x-variable), the BLM regression and the linear LS solution in eq. (4.10) are identical. An alternative formulation of the LS solution is obtained if we use the mean-centred variables, defined in eq. (4.6).

The slope is then simply estimated by

$$\hat{b} = (\mathbf{e}_0'\mathbf{e}_0)^{-1}\mathbf{e}_0'\mathbf{f}_0 \equiv \mathbf{e}_0'\mathbf{f}_0/\mathbf{e}_0'\mathbf{e}_0 \tag{4.11b}$$

The offset may later be estimated from

$$\hat{b}_0 = \bar{y} - \bar{x}\hat{b} \tag{4.11c}$$

## A4.5. Effect Correction

The effect of $\mathbf{x}$ is effectively removed from $\mathbf{y}$ residuals $\mathbf{f}$. This makes the LS regression a nice pre-processing tool for removing the effects of one set of variables from another set of variables. That may simplify the interpretation of particularly complicated data sets, and is done by defining the former as X-variables and the latter as Y-variables. This also works with more than one X-variable and/or more than one Y-variable, when extending the LS regression to multivariate BLM regression (Figure 6.3, eq. (10.4d); used in Figures 16.5d and 16.11c,d).

## A4.6. Three Different Regression Methods

As mentioned, the *best* way to draw a straight line through a swarm of points depends on the purpose and the setting. Figure 4.3 compares three of them. Readers, who now panic, may be comforted by the facts that (1) they are in good company; statisticians have fought over it for decades, and (2) the choice does not matter very much in practice.

❒   REGRESSION OF $\mathbf{y}$ ON $\mathbf{x}$: *OD* ($\mathbf{y}$) ON *%LITMUS* ($\mathbf{x}$)

The ordinary LS regression from Figure 4.2c is shown again by the solid line in Figure 4.3a. It allows us to directly predict $y_i$ from $x_i$ in new samples, based on eq. (4.2c): $\hat{y}_i = 0.27 + x_i \times 0.014$

❒   SWAPPING THE DEFINITION OF $\mathbf{x}$ AND $\mathbf{y}$: *%LITMUS* ($\mathbf{y}$) ON *OD* ($\mathbf{x}$)

Figure 4.3b shows the result of reversing the definition of $\mathbf{x}$ and $\mathbf{y}$. It minimises the sum-of-squares of the residuals represented by the horizontal dotted lines. The resulting LS estimation yielded the dashed line in Figure 4.3b: $\hat{y}_i = 24 + x_i \times 28.8$. If so desired, the same line may be expressed the other way

$$\hat{x}_i = -0.83 + y_i \times 0.035$$

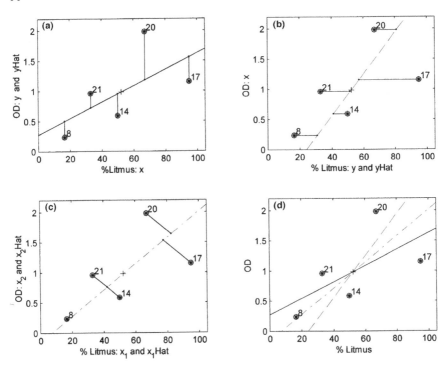

**Figure 4.3.** Three different ways to fit least squares models. (a) Conventional LS regression line (Figure 4.2c): regression of $\mathbf{y} = OD_{512}$ on $\mathbf{x} = \%Litmus$. (b) Alternative LS regression line: regression of $\mathbf{y} = \%Litmus$ on $\mathbf{x} = OD_{512}$. (c) Alternative total least squares regression (TLS): projection of $\mathbf{x}_1 = \%Litmus$ and $\mathbf{x}_2 = OD_{512}$ on their first weighted PC1, $\mathbf{t}_1$. (d) The three models compared.

◻ REGRESSING BOTH *%LITMUS* ($\mathbf{x}_1$) AND *OD* ($\mathbf{x}_2$) ON THEIR PRINCIPAL COMPONENT **t**

Figure 4.3c shows a third way to find a line through the same data points, namely via a latent variable **t**. This is really a simple, bivariate example of principal component analysis (PCA), (Chapter 5). A joint bi-linear model is then used

$$[\mathbf{x}_1, \mathbf{x}_2] = [\bar{x}_1, \bar{x}_2] + \mathbf{t} \times [p_1, p_2] + [\mathbf{e}_1, \mathbf{e}_2]$$

or, rewritten as matrices and vectors

$$\mathbf{X} = \bar{\mathbf{x}} + \mathbf{t}\mathbf{p}' + \mathbf{E} \tag{4.12}$$

where $\mathbf{X} = [\mathbf{x}_1, \mathbf{x}_2]$ is the ($5 \times 2$) table of input data [*%Litmus, OD(512)*], $\bar{\mathbf{x}} = [\bar{x}_1, \bar{x}_2]$ contains the two means, $\mathbf{t} = [t_1, t_2, ..., t_5]'$ is the *scores* of a latent variable (principal component, *PC*), to be estimated from the data $\mathbf{X}$, $\mathbf{p}' = [p_1, p_2]$

contains the so-called *loadings* that define how this *super-variable* **t** relates to **X**, and **E** = [**e**₁, **e**₂] are the residuals.

By PCA, the *perpendicular distances* from the data points to the line representing the first PC **t**, are minimised in the LS sense. But since the input data for $x_1$ and $x_2$ are given in totally different input units (% and OD units), they were first *standardised* to obtain equal total initial standard deviations, as defined by eq. (8.1).

The resulting model line is given by the dot-dashed line in Figure 4.3c. It was obtained as a line through the mean $[\bar{x}_1, \bar{x}_2]$ (marked as +) with the slope $p_2/p_1 = 0.022$. The resulting line is defined by

$$\hat{x}_{2,i} = -0.15 + x_{1,i} \times 0.022$$

The reason why the residual line segments in Figure 4.3c do not appear to be perpendicular to the principal component line is, that the so-called *aspect* ratio between ordinate and abscissa is not 1:1.

◻ WHICH MODEL LINE IS BEST?

Figure 4.3d compares the three lines. Obviously, they give different predictions of OD from *%Litmus* (or vice versa, if so desired). Which line should we use?

Theoretically, they have different statistical properties which depend on assumptions about the choice of samples analysed, on the distribution of samples for which the model will be applied later on, the error levels in **x** and **y**, etc. There is an extensive theoretical statistical literature on comparison of the two extreme ones (cf. Martens and Næs, 1989, pp. 77–85). The intermediate regression line reflects what in numerical analysis is called *total least squares* and in statistics *latent root regression*. By changing the relative weighting of the two X-variables used in eq. (4.12), one may in fact change this intermediate solution to anywhere between the two extreme solutions.

In practical data analysis, *it does not matter* which of the three lines is chosen: When the input data lie close to a straight line, as in Figure 4.1d, the three models almost coincide, so the results are very similar. When not, as in Figure 4.2b, the results may be quite different, but none of them are trustworthy or quantitative predictions, in any way. This pragmatic view on a traditionally statistical hot topic is supported by recent calibration literature (see Sundberg, 1999, 2000).

This book advocates (Chapter 1.4) that the model formulation should be defined by the intended use, with no extra operations like reversing the model to predict **x** from **y**. That is also the case when there are more than one X- and/or Y-variables. If some variables in the future are to be predicted from some other variables, the former are defined as **Y** and the latter as **X**. More detail on the choice of **X** and **Y** are given in Chapters 3.3 and 8.

For the rest of the LITMUS example (Chapters 4–7), we define *%Litmus* as a Y-variable and the OD spectra as X–variables.

# Appendix A5

# PCA Details

### A5.1. Two Ways to do Univariate Regression

The one-component bi-linear model in Chapter 5, $\mathbf{X}=\mathbf{tp}' + \mathbf{E}$ (eq. (5.2)) may be used as a linear regression model (Chapter 4.3) in two different ways, depending on what is known and what is unknown.

If the vector of amounts (the scores $\mathbf{t}$) of the component is considered *known*, the characteristic spectrum (the transposed loading vector $\mathbf{p}'$) may be estimated by projection of $\mathbf{X}$ on $\mathbf{t}$

$$\mathbf{p}' = (\mathbf{t}'\mathbf{t})^{-1}\mathbf{t}'\mathbf{X} \tag{5.7}$$

This corresponds to a regression on $\mathbf{t}$ over samples $i = 1, 2, ..., N$ for the X-data of each individual X-variable $k$, $\mathbf{x}_k$.

Conversely, if the characteristic spectrum (the loading $\mathbf{p}'$) is considered *known*, then the unknown amounts (score vector $\mathbf{t}$) may be estimated by projection of $\mathbf{X}$ on $\mathbf{p}$

$$\mathbf{t} = \mathbf{X}\mathbf{p}(\mathbf{p}'\mathbf{p})^{-1} \tag{5.8}$$

This corresponds to a regression on $\mathbf{p}$ over variables $k = 1, 2, ..., K$ for the X-data for each individual sample $i$, row-vector $\mathbf{x}_i$.

Of course, in PCA, the whole point is that we know neither $\mathbf{t}$ nor $\mathbf{p}$. But by successive approximations the scores and loadings may be estimated from the X-data, for one PC at a time:

### A5.2. The NIPALS Algorithm for PCA

In PCA, each PC $\mathbf{t}_a$ is defined as the linear combination of X-variables that account for as much as possible of the covariance remaining after the previous $a - 1$ PCs, i.e. as much as possible of sum-of-squares in $\mathbf{E}_{a-1}$. Hence, PCA is a least squares method. There are numerous equivalent algorithms to estimate PCs in PCA. The method is so fundamental that it has been re-discovered and re-named a number of times (Martens and Næs, 1989).

Herman Wold's *power method*-based NIPALS algorithm is one way to understand how to obtain scores $\mathbf{T}$ and loadings $\mathbf{P}$. For each PC, it is built on applying the univariate linear regression modelling in (eqs. (5.7) and (5.8)). In accordance with the notation in Figure 5.2, let the model for one given PC number $a$ be $\mathbf{E}_{a-1} = \mathbf{t}_a\mathbf{p}_a' + \mathbf{E}_a$ (eq. (5.4)). Guess some starting values for $\mathbf{t}_a$. Then iterate back and fourth between estimating preliminary loadings by

$$\mathbf{p}_a' = (\mathbf{t}_a'\mathbf{t}_a)^{-1}\mathbf{t}_a'\mathbf{E}_{a-1} \tag{5.9a}$$

and preliminary scores by

$$\mathbf{t}_a = \mathbf{E}_{a-1}\mathbf{p}_a(\mathbf{p}_a'\mathbf{p}_a)^{-1} \tag{5.9b}$$

After some iterations (typically, 20–30), this process converges to a sufficiently stable solution. Then the effect of this PC number $a$ is subtracted

$$\mathbf{E}_a = \mathbf{E}_{a-1} - \mathbf{t}_a\mathbf{p}_a' \tag{5.9c}$$

The PC number counter is incremented: $a = a + 1$, and the process is repeated for the next PC.

When enough (preferably, *more than enough*) PCs have been thus extracted from $\mathbf{X}$, the optimal number of PCs, $A_{\text{Opt}}$ is chosen, and PC numbers $a = 1, 2, ..., A_{\text{Opt}}$ are collected in matrices $\mathbf{T}$ and $\mathbf{P}$, and yield the bi-linear model $\mathbf{X} = \bar{\mathbf{x}} + \mathbf{TP}' + \mathbf{E}$ as described in eq. (5.5b).

### A5.3. Equivalent PCA Representations

Numerically, the fundamental operation in PCA is a so-called singular value decomposition of the mean-centred X-matrix, $\mathbf{E}_0$. This is based on the model: $\mathbf{E}_0 = \mathbf{U} \times \mathbf{S} \times \mathbf{V}'$, where $\mathbf{S}$ is a diagonal matrix containing the *singular values* (sizes) of PCs $a = 1, 2, ...$, while $\mathbf{U}$ are the basis vectors that span the samples and $\mathbf{V}$ the basis vectors that span the variables of these PCs, i.e. their directions in the $N$-dimensional sample space and the $K$-dimensional variable space, respectively. Equivalently, $\mathbf{U}$ may be seen as the eigenvectors of $\mathbf{E}_0\mathbf{E}_0'$ and $\mathbf{V}$ as the eigenvectors of $\mathbf{E}_0'\mathbf{E}_0$: The squared values of $\mathbf{S}$, $\mathbf{S}^2$, are the eigenvalues of $\mathbf{E}_0'\mathbf{E}_0$ (=eigenvalues of $\mathbf{E}_0\mathbf{E}_0'$).

Statistically, a key element is the choice of how many PCs to regard as valid, and the subsequent removal of all non-significant PCs $a = A_{\text{Opt}} + 1, A_{\text{Opt}} + 2, ...$ into residual $EA_{\text{Opt}}$, (Chapter 10).

## A5.4. Scaling and Rotation of Bi-linear Models

Graphically, the accepted PCs $a = 1, 2, ..., A = A_{Opt}$, may be scaled in different ways, depending on how the size factors (the singular values of elements $s_{a,a}$ in **S**) are used in the scaling of the scores and loadings. A common way to scale the PCA solution is to define the loadings

$$\mathbf{P} = \mathbf{V} \quad \text{and} \quad \mathbf{T} = \mathbf{U}'\mathbf{S} \tag{5.10}$$

For the loadings, the conventional scaling in eq. (5.10) means that since $\mathbf{V}'\mathbf{V} = \mathbf{I}, \mathbf{P}'\mathbf{P} = \mathbf{I}$, where $\mathbf{I}$ is the identity matrix. The orthogonal loading vectors are thus scaled to a length of 1: $\sqrt{\mathbf{p}_a'\mathbf{p}_a} = 1$. For the scores, this correspondingly means that $\mathbf{T}'\mathbf{T} = \mathbf{S}^2$. The orthogonal score vectors are thus scaled to a length equalling the singular value: $\sqrt{\mathbf{t}_a'\mathbf{t}_a} = s_{a,a}$.

❐ SCALING OF EACH INDIVIDUAL PC

Each PC is defined by its product $\mathbf{t}_a\mathbf{p}_a'$. This product may be scaled arbitrarily in many other ways, just like $12 = 3 \times 4 = 6 \times 2$

$$\mathbf{t}_a\mathbf{p}_a' = \mathbf{t}_a c_a \times c_a^{-1}\mathbf{p}_a' \tag{5.11}$$

where the coefficient $c_a$ may be chosen to have any value except 0. All choices give exactly identical contributions to the modelling of **X**. Once $c_a$ has been chosen, we may transform an old set of scores and loadings into new scores $\mathbf{t}_a = \mathbf{t}_{a,old}c_a$ and new loadings $\mathbf{p}_a = \mathbf{p}_{a,old}c_a^{-1}$.

There are several commonly used conventions for choosing $c_a$. One choice, used in Figures 5.1 and 5.2, is to choose a scaling so that the largest element in each t-vector has the value 1.0

$$c_a = 1/t_{a,Largest} \tag{5.12a}$$

Another is to let the length of each loading vector $\mathbf{p}_a$ be $\sqrt{\mathbf{p}_a'\mathbf{p}_a} = 1$ (as implied in eq. (5.10))

$$c_a = \sqrt{(\mathbf{p}_{a,old}'\mathbf{p}_{a,old})} \tag{5.12b}$$

A third way is to scale the scores to have length of $\sqrt{\mathbf{t}_a'\mathbf{t}_a} = 1$

$$c_a = \sqrt{(\mathbf{t}_{a,old}'\mathbf{t}_{a,old})^{-1}} \tag{5.12c}$$

❐ ROTATION OF A SET OF PCS

The bi-linear model with $A_{Opt}$ PCs (eq. (5.5b)) has yet another ambiguity that needs to be controlled, namely that of axis rotation. In the model

$$\mathbf{X} = \bar{\mathbf{x}} + \mathbf{TP}' + \mathbf{E}$$

the bi-linear contribution $\mathbf{TP}'$ may equally well be written as

$$\mathbf{T} \times \mathbf{P}' = \mathbf{T}(\mathbf{C} \times \mathbf{C}^{-1})\mathbf{P}' = (\mathbf{TC}) \times (\mathbf{C}^{-1}\mathbf{P}')' = \mathbf{T}_{\text{Rot}} \times \mathbf{P}'_{\text{Rot}} \qquad (5.13)$$

as long as the rotation matrix $\mathbf{C}$ has full rank and can thus be inverted.

In classical psychometrics literature, various methods for rotating the PCs from PCA (*varimax*, etc.) are well described. They were considered as important elements in the more ambitious statistical tradition of factor analysis (Harman, 1967), for extracting PCs that could be meaningfully interpreted individually, e.g. as different types of *intelligence*.

The bi-linear modelling in this book is not used at that high ambition level. Instead, the *plot a lot* rule is applied. The bi-linear subspaces are inspected visually for meaning, and also in the off-axis directions. Moreover, by extending the one-table PCA to the two-table PLS regression, external variables may be used to steer the extraction of PCs and to convey meaning to them in the plots.

❑  Visual Scaling in Plots

The visual appearance of bi-linear models in two-way plots may be affected by the choice of ordinate and abscissa scaling as well as the aspect ratio of the plot (the ordinate size/abscissa size). Often, the ordinate (e.g. PC2) is amplified compared to the abscissa (PC1) in order to reveal details. In addition, plots with a longer abscissa than ordinate are often found aesthetical (with two eyes, our visual field is wider than it is high).

When comparing scores $\mathbf{T}$ and loading $\mathbf{P}$ in different bivariate plots, these phenomena should be taken into account when interpreting directions other than purely horizontal and purely vertical. One way to simplify this is to use so-called *bi-plots*, where scores and loadings are represented in the same plot.

There are many different ways to make bi-plots. One is to replace eq. (5.10) with $\mathbf{T}_{\text{plot}} = \mathbf{U} \times \mathbf{S}$ and $\mathbf{P}_{\text{plot}} = \mathbf{V} \times \mathbf{S}$ and plot, e.g. $\mathbf{t}_{1,\text{plot}}$ vs. $\mathbf{t}_{2,\text{plot}}$ and $\mathbf{p}_{1,\text{plot}}$ vs. $\mathbf{p}_{2,\text{plot}}$ in the same figure. Another is to extend $\mathbf{X}$ with one object indicator variable for each sample, collected in the non-informative identity matrix: $\mathbf{X}_{\text{extended}} = [\mathbf{X}, \mathbf{I}]$ so that each object also has an indicator variable. Submitting $\mathbf{X}_{\text{extended}}$, instead of $\mathbf{X}$, to bi-linear modelling does not change the original bi-linear model ($\mathbf{T}$ is the same). But it yields $\mathbf{P}_{\text{extended}}$ with *dummy* loadings even for the samples, so plotting $\mathbf{p}_{1,\text{extended}}$ vs. $\mathbf{p}_{2,\text{extended}}$ yields a bi-plot.

### A5.5. Correlation Loadings for X

The correlation loading may either be computed like a conventional correlation coefficient (eq. (4.8) between $\mathbf{x}_k$ and $\mathbf{t}_a$), or by a transformation of the already obtained loading $p_{ak}$

$$r_{ak} = p_{ak} \times \sqrt{(t_a't_a/e_{0k}'e_{0k})} \qquad (5.14)$$

where $e_{0k} = x_k - 1\bar{x}_k$. When missing values are encountered in $e_{0k}$, the corresponding elements in $t_a$ must also be skipped. The correlation loadings between PC number $a$ and the $K$ X-variables are collected in vector

$$r_a = [r_{ak}, \ k = 1, 2, ..., K]$$

## A5.6. Missing Values

There are several ways to handle missing elements in the input data. A pragmatic method that does not require any additional statistical modelling, is to use the following modification of the NIPALS algorithm.

If, for a certain variable $k$, a missing value is encountered in $X$ for a certain object $i$, then the corresponding elements in $t_{ia}$ must also be skipped in eq. (5.9a), which for X-variable $k$ looks like

$$p_{ak} = (t_a't_a)^{-1}t_a'e_{k,a-1} \qquad (5.15a)$$

Likewise, if, for a certain sample $i$, a missing value is encountered in $X$ for a certain variable $k$, then the corresponding elements in $p_{ka}$ must also be skipped in eq. (5.9b), which for sample $i$ is

$$t_{ia} = e_{i,a-1}p_a(p_a'p_a)^{-1} \qquad (5.15b)$$

This method may give some convergence problems if the number of missing values in $X$ is high. But its advantage is its simplicity; other methods based on imputation tend to require the user to decide on the optimal number of PCs *before* the actual PCA.

*But missing data is missing information.* Therefore, the user who has a data set with many missing data is recommended to compare the outcome of modelling of all the data, with the outcome of modelling after the variables and/or samples with many missing data have been eliminated. If the outcome is so different that the conclusions are affected, then the user is recommended to consult a professional statistician.

# Appendix A6

# PLS Regression Details

### A6.1. BLM by PLSR

The bi-linear modelling by PLSR (Chapter 6) is, formally speaking very similar to that of PCR, (Figure 6.1). PCR consists of two independent steps: (1) PCA of $\mathbf{X}$, extracting PCs as functions of $\mathbf{X}$, $\mathbf{T} = w(\mathbf{X})$; followed by (2) regression of $\mathbf{Y}$ and $\mathbf{X}$ on $\mathbf{T}$. PLSR has the same major steps, but they are connected, in order to ensure that the first few PCs from $\mathbf{X}$ are relevant for $\mathbf{Y}$. Mathematically, $w(\mathbf{X})$ in PLSR is based on maximising the explained covariance between $\mathbf{X}$ and $\mathbf{Y}$, while in PCA/PCR it is based on maximising the explained covariance within $\mathbf{X}$.

Bi-linear regression modelling will now be defined in more detail. As shown in Figure 6.3, the bi-linear regression consists of extracting a small series of latent variables $\mathbf{t}_a$, $a = 1,2,...$ (the PCs) from $\mathbf{X}$ via functions $\mathbf{t}_a = w_a(\mathbf{X})$, and use these PCs to model both $\mathbf{X}$ and $\mathbf{Y}$. The definition of PC number $a$ is most easily understood in terms of the X- and Y-residuals after having subtracted the previous $a - 1$ PCs, $\mathbf{E}_{a-1}$ and $\mathbf{F}_{a-1}$.

Initially ($a = 0$), the residual matrices $\mathbf{E}_0$ and $\mathbf{F}_0$ just contain the mean-centred X- and Y-variables.

### A6.2. Loading Weights: $w_a$

In PLSR, the function $\mathbf{t}_a = w_a(\mathbf{X})$ for each PC is defined by the loading weight vector $\mathbf{w}_a$, which is extracted as the eigenvector of the remaining X–Y covariance, $\mathbf{E}_{a-1}'\mathbf{F}_{a-1}$ that has the largest eigenvalue

$$\mathbf{w}_a = \text{first eigenvector of } (\mathbf{E}_{a-1}'\mathbf{F}_{a-1}\mathbf{F}_{a-1}'\mathbf{E}_{a-1}) \tag{6.5a}$$

Thereby, $\mathbf{w}_a$ will strike a balance between modelling $\mathbf{X}$ and modelling $\mathbf{Y}$.

See Martens & Næs (1989, p. 120) for equivalent PLSR algorithms.

## A6.3. Scores: $t_a$

The scores of PC number $a$ are defined as linear functions of $\mathbf{X}$:

$$\mathbf{t}_a = \mathbf{E}_{a-1}\mathbf{w}_a \tag{6.5b}$$

## A6.4. Y-loadings: $q_a$

The Y-loadings of PC number $a$ are defined from the model (6.1c)

$$\mathbf{F}_{a-1} = \mathbf{t}_a\mathbf{q}_a' + \mathbf{F}_a$$

The least squares solution of this is

$$\mathbf{q}_a' = (\mathbf{t}_a'\mathbf{t}_a)^{-1}\mathbf{t}_a'\mathbf{F}_{a-1} \tag{6.5c}$$

## A6.5. X-loadings: $p_a$

The X-loadings of PC number $a$ are defined from the model (eq. 6.1b)

$$\mathbf{E}_{a-1} = \mathbf{t}_a\mathbf{p}_a' + \mathbf{E}_a$$

The least squares solution of this is

$$\mathbf{p}_a' = (\mathbf{t}_a'\mathbf{t}_a)^{-1}\mathbf{t}_a'\mathbf{E}_{a-1} \tag{6.5d}$$

## A6.6. Y-residuals: $F_a$

The effect of PC number $a$ is subtracted from the table of Y-data

$$\mathbf{F}_a = \mathbf{F}_{a-1} - \mathbf{t}_a\mathbf{q}_a' \tag{6.5e}$$

## A6.7. X-residuals: $E_a$

The effect of PC number $a$ is likewise subtracted from the table of X-data

$$\mathbf{E}_a = \mathbf{E}_{a-1} - \mathbf{t}_a\mathbf{p}_a' \tag{6.5f}$$

Increment PC number $a$ until more than enough PCs have been extracted

$$a = a + 1 \tag{6.5g}$$

The optimal number of PCs, $A_{Opt}$, is then determined by some validation method (Chapter 10).

## A6.8. Comments on the PLSR Algorithm

❏ COMPARISON WITH PCA

In contrast to the PLSR computation of $w_a$ in eq. (6.5a), in PCA this is done from the remaining X–X covariance, $E_{a-1}'E_{a-1}$. Hence, $t_a$ in PCA may or may not model Y. Therefore, PLSR is more versatile than PCA/PCR. For instance, in a designed experiment with many responses (Chapter 14), design variables X may be orthogonal and thus have no correlation structure, so PCA is difficult to use. But the PLSR will instead be controlled by the structure between the responses in Y.

❏ STABILISATION OF PLSR

*Stabilised loading weights*: In order to ensure maximum versatility and safety in bi-linear modelling, with minimum danger of overfitting and of numerical problems, a few minor adjustments may be implemented in the original PLS regression algorithm (Wold et al., 1983).

If there is little X–Y covariance remaining after $a - 1$ PCs, the obtained eigenvector $w_a$ from eq. (6.5a) will be noise-sensitive and numerically unstable, and may lead to needless overfitting problems if used by mistake. In such cases it is advantageous to switch towards the PCA definition, instead using the first eigenvector of the remaining X-structure, $E_{a-1}'E_{a-1}$.

However, if the first eigenvalue of $E_{a-1}'E_{a-1}$ is also very small (because $a > \text{rank}(X)$), $w_a$ will still be numerically unstable. Then it is necessary to ensure explicitly that $w_a$ is orthogonal to the previous $a - 1$ loading weights.

*Stabilised Y-loading*: If, for some PC number $a$, the cross product $t_a't_a$ is very small, eq. (6.5c) may give large and erratic values for $q_a'$. This may lead to needless variance inflation in the final regression model.

That may be overcome by so-called *ridging*. A small offset is added to $t_a't_a$ before inverting it. Eq. (6.5c) may then be rewritten

$$q_a' = (t_a't_a + d_a)^{-1}t_a'F_{a-1} \tag{6.5h}$$

## A6.9. Estimates of the Regression Coefficients $B_A$ and offset $b_{0,A}$

For a model with $A$ PCs, the regression coefficient matrix (Figure 6.3) in the linear model (eq. (6.1d), Figure 6.3) is estimated from the bi-linear PLSR parameters

$$\hat{B}_A = W_A(P'_A W_A)^{-1}Q'_A \tag{6.6a}$$

When X has full column rank and the number of PCs equals the number of X-variables ($A = K$), then eq. (6.6a) gives identical results to the OLS regression solution (cf. eq. (4.11a)):

$$\hat{\mathbf{B}}_{OLS} = (\mathbf{E}_0'\mathbf{E}_0)^{-1}\mathbf{E}_0\mathbf{Y} \tag{6.6b}$$

But the PLS regression is easier to use than the OLS regression when $\mathbf{X}$, and hence the mean centred $\mathbf{X}$, $\mathbf{E}_0$ do not have full column rank because then we let $A < K$.

The corresponding offset vector is determined from the means

$$\hat{\mathbf{b}}_{0,A} = \bar{\mathbf{y}} - \bar{\mathbf{x}}\hat{\mathbf{B}}_A \tag{6.6c}$$

The *Hat* convention is used for $\hat{\mathbf{b}}_{0,A}$ and $\hat{\mathbf{B}}_A$ but not for the rest of the model parameters. The reason is that the parameters in the linear model, $\hat{\mathbf{b}}_{0,A}$ and $\hat{\mathbf{B}}_A$, sometimes are intended to represent statistical estimates of some well-specified, causal model parameters, e.g. in analysis of effects in designed experiments (Chapter 14).

In contrast, the parameters of the bi-linear model ($\mathbf{W}$, $\mathbf{T}$, $\mathbf{P}$ and $\mathbf{Q}$) are just approximation tools that together may describe important physical phenomena. But the individual parameter elements (e.g. $t_{ia}$, $q_{ja}$) are not intended to represent individual physical phenomena.

### A6.10. PLSR Draws on Two Sources of Structure

If $\mathbf{X}$ has clear structure, as it often has when $\mathbf{X}$ represents multivariate empirical observations, the PLSR will use this to stabilise the regression model against noise in $\mathbf{Y}$. This is a reason why PLSR gives regression models with good statistical properties, e.g. for multivariate calibration of multichannel instruments, (Burnham et al., 1996; Helland, 1988).

On the other hand, if $\mathbf{X}$ has no correlations between its variables, but $\mathbf{Y}$ has, then the PLSR model will reveal this *factor structure* in $\mathbf{Y}$, and just map the X-variables into it. That is the reason why PLSR is useful for analysis of effects in designed experiments with many responses $\mathbf{Y}$.

### A6.11. Correlation Loadings for Y

The elements $q_{aj}$ in the Y-loadings $\mathbf{q}_a$ may be rescaled to correlation loadings $r_{aj}$, in analogy to the X-loadings (eq. (5.10)), and with the same precautions for missing values

$$r_{aj} = q_{aj} \times \sqrt{(\mathbf{t}_a'\mathbf{t}_a/\mathbf{f}_{0j}'\mathbf{f}_{0j})} \tag{6.7}$$

### A6.12. Loading Weights vs. Loadings

There are several equivalent ways to express PLSR models (see e.g. Martens and Næs, 1989). The present way (originally developed by Svante Wold; Wold et al.,

1983a based on NIPALS iterations, and reformulated by Höskuldsson (1988, 1996) to an eigenvector solution) ensures that the parameters from one PC $a$, $\mathbf{w}_a$, $\mathbf{t}_a$, $\mathbf{p}_a$ and $\mathbf{q}_a$, are not affected by how many more PCs $a + 1$, $a + 2,\ldots$ are being used. This is, in practice, an advantage.

The price to be paid for this is the double set of parameters for the X-variables, loading weights $\mathbf{w}_a$ and loadings $\mathbf{p}_a$. The former defines the impact each X-variable has on the PC $\mathbf{t}_a$. The latter defines the impact PC $\mathbf{t}_a$ has on each X-variable. In PCA, the two are identical. In practical PLSR work, $\mathbf{w}_a$ and loadings $\mathbf{p}_a$ are usually quite similar; the same conclusions are drawn by looking at the loading weights or the loadings.

In order to keep things simple in this book, we here only look at the loadings $\mathbf{p}_a$, $a = 1,2,\ldots, A$. These are compatible with the Y-loadings $\mathbf{q}_a$, $a = 1, 2,\ldots, A$ and may therefore be plotted together.

If the loading weights $\mathbf{W}_A = [\mathbf{w}_a, a = 1, 2, \ldots, A]$ are to be studied graphically (Esbensen, 2000) in bivariate plots together with the Y-loadings, the original Y-loadings for $A$ PCs, $\mathbf{Q}_A = [\mathbf{q}_a, a = 1, 2, \ldots, A]$ should first be rotated slightly to $\mathbf{Q}'_{A,W} = (\mathbf{P}'_A \mathbf{W}_A)^{-1} \mathbf{Q}'_A$, in order to be compatible with $\mathbf{W}_A$.

## A6.13 Disagreement Between X and Y

In rare cases, the X- and Y-data each have clear correlation structures, but these correlation structures are not intercorrelated. As a result the loading weights and the loadings may differ a lot. This may then be seen in the estimated regression coefficients $\hat{\mathbf{B}}_A$, which in eq. (6.6a) is dominated by $\mathbf{W}_A$ and not by $\mathbf{P}_A$ and therefore will differ from the pattern expected from the loadings $\mathbf{P}_A$.

Disagreement between $\mathbf{X}$ and $\mathbf{Y}$ may also be studied by reanalysing the data after having swapped what is $\mathbf{X}$ and what is $\mathbf{Y}$. Since $\mathbf{T}$ comes from $\mathbf{X}$ and not from $\mathbf{Y}$, the two solutions will then be very different. That is an indication that the data in $\mathbf{X}$ and $\mathbf{Y}$ *disagree* about what is the important patterns among the samples, which of course is a problem that requires closer scrutiny of the data.

When the two swapped $\mathbf{X/Y}$ solutions are found to be similar, the user may rest assured that the modelling is sufficiently stable and complete; this is demonstrated in Chapter 8.

## A6.14. Special Versions of PLSR

Figure 3.4 pointed out two special cases of PLSR, when one block of variables consists of binary indicator (0/1) variables. When $\mathbf{Y}$ contains class indicator (0/1) variables, the PLSR represents a discriminant analysis, and is here called *DPLSR*. In some literature it is called *PLS-discriminant analysis* (Wold et al., 1983b). When instead $\mathbf{X}$ contains design indicator (e.g. 0/1) variables, the PLSR represents analysis of effects, and is here called *APLSR*. But the basic algorithm is the same as in conventional PLSR.

The PLSR may be modified in a number of different ways (Chapter 3.3, cf. also Martens and Næs, 1989, p. 158), but more detail is beyond the scope of the present book.

# Appendix A7

# Modelling the Unknown

With the example in Chapter 7 in mind, let us think about the 4-PC solution in Figure 4.1i, p. 86, for a moment. In this data-driven modelling we did not *a priori* need to *know* the reason why 4 PCs were needed. The bi-linear modelling *found* the PCs for us, from the empirical data **X** and **y**. Thus, bi-linear modelling allows mathematical modelling of *unknown* phenomena.

But this multivariate analysis then *indicated* that 4 PCs were needed because the OD data seem to span the following:

1. variation in total *litmus concentration* (**y**, known, major);
2. variation in *litmus colour* (unknown, major);
3. variation in sample *turbidity* (unknown, major);
4. different non-linear responses at different X-variables (unknown, minor).

The real reasons why 4 PCs were needed, of course, was the presence of three independently varying physical phenomena, namely concentration variations in *blue* (unprotonated) *litmus*, *red* (protonated) *litmus* and *zinc oxide* powder. In addition, stray light in the instrument probably caused more response curvature at high OD than at low OD wavelengths.

If all of this had been *known*, we could possibly have made a "*real scientific model*", like it is usually done in, e.g. physics, based on causal first-principles. But that model may still have been wrong (cf. the comparison of Figure 5.1a, b).

Soft modelling is neither a *black box* operation, nor *black magic*. In contrast, believing too strongly in *a priori* theory and scientific *laws* can give the researcher a *black eye*.

The strength of *data-driven* bi-linear modelling is that it allows us to detect and model even unexpected phenomena, for which we inductively can then search for an explanation, based on graphical inspection of the results. (Liang et al., 1993).

# Appendix A8

# Non-linearity and Weighting

## A8.1. Non-Linear Relationships Handled by Polynomial BLM

Although the bi-linear PLSR technically belongs to the class of statistical models referred to as *linear*, it can still handle many types of *non-linear* **X**–**Y** relationships. This just reflects a confusion of terms: a *linear model* is a statistical model in which the effects of the various model *parameters* are mathematically *added* together (as opposed to multiplied or any other operation). A *linear relationship* between *variables* is a *straight-lined relationship*. The confusion arises because a linear model is often able to describe a non-linear relationship as well.

The term *non-linear* is even more ambiguous, both in its use for statistical models and in its use for relationships between variables. In this book we stick to linear models. But when needed (Chapter 8.2.8), we extend the linear models with *polynomial* terms.

A polynomial model is still a linear model, but it explicitly tries to account for non-linear **X**–**Y** relationships, by the use of square terms and interaction terms in **X**. Say that the effect of X-variable $\mathbf{x}_k$ on **Y** depends on the value of another X-variable, $\mathbf{x}_m$. A new variable $\mathbf{x}_c$ may then be defined by the product of $\mathbf{x}_k$ and $\mathbf{x}_m$. The data analysis of the new variable is most easy if they are based on mean-centred versions of the original variables

$$\mathbf{x}_c = \left(\mathbf{x}_k - \bar{\mathbf{x}}_k\right)\left(\mathbf{x}_m - \bar{\mathbf{x}}_m\right) \tag{8.4}$$

When $k = m$, this generates a so-called *square term*. Otherwise, it is called a *two-factor interaction term*.

Such polynomial terms are often used in the analysis of data from designed factorial experiments, in complex systems such as biology. **X** may then describe the actual design factors, as well as their interactions and square terms, and **Y** be the actual response variables from the experiment. This is discussed in Chapters 8.2.8, 11.3 and illustrated in Chapter 16.

The square- and interaction terms $\mathbf{x}_c$, $c = 1, 2,....$ may be generated for any combination of X-variables, included in the X-matrix and submitted to PLSR. The number of X-variables then increases, which again increases the risk of modelling chance correlations between noise in $\mathbf{Y}$ and these transforms of $\mathbf{X}$. So, in order to keep the models simple and stable, it is wise to eliminate again those extra terms in $\mathbf{X}$ found to be unnecessary, e.g. by cross-validation/jack-knifing (Westad and Martens, 2000). This is explained in Chapter 10.

Even third-order interaction terms, e.g. involving the product of three or more X-variables, may thus be generated and used. However, third-order and higher-order interactions are usually difficult to interpret, and therefore not pursued here.

### A8.2. A *Priori* Weighting: the Use of Prior Knowledge to Scale the Input Variables for BLM

Being based on the principle of least squares, the PCA and PLSR give sub-optimal results if the input data cover widely different ranges. This may be improved by *re-scaling* the variables. The *a priori* weighting of a variable $\mathbf{x}_{k,\ input}$ was expressed as $\mathbf{x}_k = (\mathbf{x}_{k,\ input} - \bar{\mathbf{x}}_{k,\ input}) \times v_k$ in eq. (8.1). Different weighting strategies may be used, by choosing different definitions of weights $v_k$, for the X-variables $k = 1, 2,...,K$. Y-variables may be weighted likewise.

❐ STANDARDISATION: SCALING ALL VARIABLES TO SAME TOTAL INITIAL VARIATION

In conventional standardisation, these scaling weights were simply defined (eq. (8.2)) as the inverse of the total initial standard deviation $s_k$, $v_k = 1/s_k$, with $s_k$ defined by eq. (4.7), p. 372.

❐ MODIFIED STANDARDISATION: AVOID AMPLIFYING VARIABLES WITH LITTLE OR NO VALID INFORMATION

In order to avoid amplifying *baseline variables* (Figure 4.1), whose total initial standard deviation mainly reflects measurement noise, eq. (8.2) may be modified, e.g. into

$$v_k = 1/(s_k + c) \tag{8.5}$$

where scalar $c$ is a small number. Variables with large total initial variations $s_k$ will be more or less unaffected, while we avoid amplifying variables with small total initial variations $s_k$ by division by almost zero (cf. Appendix A3.2.2).

❐ Uncertainty Weighting: Scaling All Variables to Approximately the Same Uncertainty

An alternative method of weighting, which is statistically more optimal than the standardisation, is to weight with respect to the *error* standard deviations, the standard uncertainties, $s_{err, k}$, $k = 1, 2,...,K$

$$v_k = 1/s_{err, k} \tag{8.6}$$

It is usually difficult *a priori* to know the precise uncertainty levels $s_{err, k}$, $k = 1, 2,...,K$. But that is not a problem, because this is a pre-processing that does not have to be precise; the important thing is to avoid that large errors in some variables overshadow real variations in other variables.

## A8.3. Graphical Consequences of the Weighting

The scaling weights affect the X- and Y-loadings **P** and **Q**, as well as the estimates of the regression coefficients **B** and $\mathbf{b}_0$. This may, in turn affect how the X- and Y-variables appear in some model parameter plots.

❐ Re-Scaling the Loadings

The correlation loadings (eqs. (5.14) and (6.7)) remain unaffected by the scaling. The estimated loadings for each scaled X-variable, $\mathbf{p}_k$, may be returned to the scale of the input variables by the inverse of eq. (8.1)

$$\mathbf{p}_{k, \text{ input}} = \mathbf{p}_k/v_k \tag{8.7a}$$

Likewise, the Y-loadings $\mathbf{q}_j$ may be re-scaled for each Y-variable $j$, by

$$\mathbf{q}_{j, \text{ input}} = \mathbf{q}_j/v_j \tag{8.7b}$$

❐ Re-Scaling the Regression Coefficients

The regression coefficients $\hat{\mathbf{B}} = [\hat{b}_{kj}]$ and $\hat{\mathbf{b}}_0 = [\hat{b}_{j0}]$ (eq. (6.6a)) are also affected by the weights $v_k$ and $v_j$. For visual inspection, two re-scaled versions of the regression coefficients are useful
  1. De-scaled coefficients

$$\hat{b}_{kj, \text{ input}} = \hat{b}_{kj} \times v_k/v_j \tag{8.8a}$$

$$\hat{b}_{0j, \text{ input}} = \hat{b}_{0j}/v_j \tag{8.8b}$$

These parameters can be applied directly to new raw data **X** to predict **Y**

$$\hat{\mathbf{y}}_{i, \text{ input}} = \mathbf{x}_{i, \text{ input}} \times \hat{\mathbf{B}}_{\text{input}} + \hat{\mathbf{b}}_{0, \text{ input}} \tag{8.8c}$$

(see eq. 8.9c)

2. Standardised coefficients: $\hat{\mathbf{B}}_{standardised}$ and $\hat{\mathbf{b}}_{0, standardised}$, corresponding to the *standardised* (eq. (8.2)) X- and Y-variables

$$\hat{b}_{kj, standardised} = \hat{b}_{kj} \times v_k/v_j \times \left(s_k/s_j\right) \tag{8.9a}$$

$$\hat{b}_{0j, standardised} = \hat{b}_{0j}/\left(v_j \times s_j\right) \tag{8.9b}$$

These parameters can be easier to *interpret*, especially for interdisciplinary data. They represent the relative contributions from the different X-variables to the different Y-variables, when each of these variables has a standard deviation of 1 independently of the *a priori* weights $v_k$, $v_j$ used. They correspond to the prediction model

$$\hat{y}_{i, standardised} = \mathbf{x}_{i, standardised} \times \hat{\mathbf{B}}_{standardised} + \hat{\mathbf{b}}_{0, standardised} \tag{8.9c}$$

## A8.4. Bi-Linear Modelling by Ordinary Least Squares (OLS), Weighted Least Squares (WLS) and Generalised Least Squares (GLS)

❒  PCA AND PLSR AS OLS METHODS

Model parameter estimation by PCA and PLSR is based on the principle of least squares (LS). When no *a priori* weighting is applied to the input variables, as in Chapters 4–7, the estimation may be regarded as being based on *ordinary least squares* (OLS). OLS is here used in the general meaning of the word, not in the special statistic abbreviation (Sundberg, 2000) for ordinary least squares regression (eq. (6.6b)).

❒  FROM OLS TO WLS

An *a priori* weighting (e.g. in eqs. (8.2)–(8.6)) converts the PCA and PLSR into a so-called *weighted least squares* (WLS) methods. But contrary to conventional WLS modelling, the weighting is done *outside* the LS estimation algorithm, not inside.

For a matrix of variables $\mathbf{X}_{input} = [\mathbf{x}_{k, input}, k=1, 2,...,K]$, eq. (8.1) may equivalently be written

$$\mathbf{X} = \left(\mathbf{X}_{input} - \bar{\mathbf{x}}_{input}\right) \times \mathbf{V} \tag{8.10}$$

where the weight matrix $\mathbf{V}$ is *diagonal*, i.e. its only non-zero elements are the diagonal elements, called *diag*(**V**). These are the elements in vector $\mathbf{v}=[v_k, k=1, 2,...,K]'$.

Let us here define **S** as the diagonal *uncertainty variance* matrix with diagonal elements $diag(\mathbf{S})=[s_{err,k}^2, k=1, 2,...,K]'$. Eq. (8.10) may then be used, based on eq. (8.5), with weight matrix

$$\mathbf{V} = \mathbf{S}^{-1/2} \tag{8.11}$$

Thereby the K-dimensional X-space is shrunk along each of the $K$ axes in X-space, in proportion to the uncertainty of that direction.

## ❑ FROM WLS TO GLS

In WLS methods, each input variable is scaled *separately* (eq. (8.6)), even though we can write the scaling for all of them in the same expression (eqs. (8.10) and (8.11)). However, if we know *a priori* that the uncertainty error in the different input variables are inter-correlated, even more optimal weighting may be attained, by extending the WLS into *generalised least squares* (GLS) estimation.

The error variance/covariance matrix **S**, describing the *a priori* known error structure, is no longer diagonal. Still, eq. (8.10) may be applied, as before. In brief, the only difference is that instead of just defining the inverse square root of the variance matrix as $\mathbf{V}=\mathbf{S}^{-1/2}=diag(1/s_{err,k})$ (eq. (8.11)), we must write the inverse square root of the whole variance/covariance matrix as

$$\mathbf{V} = \mathbf{S}^{-1/2} = \mathbf{G} \tag{8.12}$$

where $\mathbf{G}'\mathbf{G}=\mathbf{S}^{-1}$.

Thereby the $K$-dimensional X-space is shrunk also in off-axis directions in X-space where the uncertainty covariance is particularly high. This allows us to simplify and stabilise the BLM, by reducing the impact of error sources that we already know and want to get rid of.

Weight matrix **G** may be obtained e.g. by singular value decomposition of $\mathbf{S}^{-1}$.

# Appendix A9

# Classification and Outlier Detection

### A9.1. SIMCA Classification

Appendix A9.1–3 gives more details on the methodology from Chapter 9.

1. *Define each class model separately*: For different data sets with the same number of X-variables, but different samples, develop different bi-linear models $c = 1, 2,...$ by PCA or PLSR. Determine class boundaries (classification limits), e.g. $2 \times RMSE(X)_c$, where $RMSE(X)_c$ (eq. (9.6a) below), in each model summarises the sample-to-model distances $RMSE(X)_{i,c}$ (eq. (9.4)) for the normal calibration samples. Use e.g. cross-validation in this estimation, eq. (10.5f), avoid over-fitting. Look out for useless variables and outliers in each calibration set.

2. *Cross-classify and evaluate*: Fit each sample $i$ in each of the sets to each of the *other* models $c = 1, 2,....$ Determine each sample-to-model distance $RMSE(X)_{i,c}$. For pairs of models $c = [1,2]$ etc., draw a Cooman's plot (Figure 9.4a), study how well the classes are separated, and look for unexpected patterns among the samples. Plot the discriminative ability for each X-variable (Figure 9.4b). (Optionally, remove useless variables and samples and repeat steps 1 and 2).

3. *Classify new samples*: Fit the X-data of each new sample $i$ to each of the models $c = 1, 2,...$ and classify it by checking if it falls inside the boundaries of any of the classes.

### A9.2. Discriminant PLS Regression (DPLSR)

1. *Define one common multiclass discrimination model*: For different data sets

with the same number of X-variables, but different samples $N_1$, $N_2$,..., develop one unified bi-linear model by PLSR of the $N = N_1+N_2+...$ samples, with **Y** having one binary class indicator variable for each class. Define model boundaries (outlier detection limits, below). Look out for useless variables and outliers in the calibration set.

2. *Interpret and evaluate*: Validate and study the model like any other PLSR model.
3. *Classify new samples*: Fit the X-data of each new sample $i$ to the DPLSR model: (a) Predict the classification Y-variables. (b) Check if the sample is an outlier, i.e. falls outside the model boundaries.

## A9.3. Outlier Analysis

1. *Build a population model*: For a given calibration data set, develop a bi-linear model by PCA or PLSR. Look out for useless variables and outliers in the calibration set.
2. *Define what is normal*: Validate and study the obtained model. Define model boundaries for what is considered *normal* (outlier detection limits, below).
3. *Check new samples*: Fit the X-data of each new sample $i$ to the bi-linear model, and check if the sample is an outlier, i.e. falls outside the model boundaries, with respect to any of the X- and Y-measures.

There are several ways to detect a sample $i$ as an outlier relative to a bi-linear model, say, for class $c$. Here are some suggested limits for automatic outlier detection, based on the subsequent statistics estimates in A9.4–A9.6:

### ❒ OUTLIER DETECTION IN **X**

*Leverage in X, $h_i$ eq. (9.3) for 'extremists':*   Suggested outlier detection limit: A sample is reported automatically as an outlier if leverage is greater than two times the average leverage among the calibration samples, $(A+1)/N$. This may be written: $h_i > 2(A+1)/N$.

*Residual in X, RMSE(X)$_i$ eq. (9.4) for 'aliens':*   Suggested outlier detection limit: Two times the *average* distance among the calibration samples, RMSE(X) (eq. (9.6a)): RMSE(X)$_i > 2$ RMSE(X).

### ❒ OUTLIER DETECTION IN **Y**

*Outlier-sensitive standard prediction uncertainty, YDev$_i$ eq. (10.6):*   Suggested outlier detection limit YDev$_i > 3$ RMSE(Y) (eq. (9.6b)).

*Prediction errors RMSE(Y)ᵢ eq. (9.5b):* Only for samples with known Y-data. Suggested limit: $\text{RMSE(Y)}_i > 2\,\text{RMSE(Y)}$.

In this outlier analysis, the *predictive* version of the RMSE(X) and RMSE(Y) after $A$ PCs, called $\text{RMSEP(X)}_A$ and $\text{RMSEP(Y)}_A$ are most reliable. They are given in eq. (10.5a–g).

## A9.4. Statistics for One Sample

Appendix A9.4–6 gives detailed formulae for A9.1–3.

For sample $i$ and class $c$, the following statistics may be estimated.

*Scores, residuals and Y-predictions:* Vector $\mathbf{t}_{i,c}$, X-residual vector $\mathbf{e}_{i,c}$ and prediction $\hat{\mathbf{y}}_{i,c}$ for a given number of PCs, $A$, may be estimated in different ways; here is one algorithm (cf. eqs. (6.2) and (6.3), pp. 118–119):

Start-values: $\mathbf{e}_{i,c,0} = \mathbf{x}_i - \mathbf{x}_c$; $\hat{\mathbf{y}}_{i,c,0} = \mathbf{y}_c$

for PC $a = 1, 2,..., A$

$$t_{i,c,a} = \mathbf{e}_{i,c,a-1}\mathbf{w}_{a,c} \tag{9.2a}$$

$$\mathbf{e}_{i,c,a} = \mathbf{e}_{i,c,a-1} - t_{i,c,a}\mathbf{p}'_{a,c} \tag{9.2b}$$

$$\hat{\mathbf{y}}_{i,c,a} = \hat{\mathbf{y}}_{i,c,a-1} + t_{i,c,a}\mathbf{q}'_{a,c} \tag{9.2c}$$

◻ LEVERAGE OF SAMPLES IN X

$$h_{i,c} = \sum\left(t_{i,a}^2/\mathbf{t}'_a\mathbf{t}_a\right) + g \tag{9.3}$$

where $g = 1/N$ for calibration samples and zero for other samples and $\mathbf{t}_a$ is the score vector of the $N$ calibration samples. For calibration samples, leverage $h_i$ lies between $1/N$ (an average sample) and 1 (a dominant unique calibration sample). For new samples, $h_i$ may be larger than 1 if they are outliers.

◻ DISTANCE TO MODEL IN X

Only defined as long as the number of PCs, $A$, is lower that the number of X-variables, $K$.

$$\text{RMSE(X)}_{i,c} = \sqrt{\left\{\mathbf{e}_{i,c,A}\mathbf{e}'_{i,c,A}/(K-A)\right\}} \tag{9.4}$$

❐ DISTANCE TO MODEL IN **Y**: (ONLY IF **y**ₗ IS KNOWN)

Y-residual after $A$ PCs : $\mathbf{f}_{i,c,A} = \mathbf{y}_i - \hat{\mathbf{y}}_{i,c,A}$ (9.5a)

Summary of errors in the $J$ Y-variables

$$RMSE(Y)_{i,c} = \sqrt{\{\mathbf{f}_{i,c,A}\mathbf{f}'_{i,c,A}/J\}}$$ (9.5b)

❐ VALIDATION

For calibration samples, $RMSE(X)_{i,c}$ and $RMSE(Y)_{i,c}$ are preferably based on residuals estimated by cross-validation, and are then called $RMSEP(X)_{i,c}$ and $RMSEP(Y)_{i,c}$ (cf. eq. (10.5a)).

### A9.5. Statistics for One Model (Class)

For a class $c$ of $N$ samples, the X- and Y-errors are summarised as the root-mean-square averages

$$RMSE(X)_c = \sqrt{\left\{\sum RMSE(X)_{i,c}^2/N\right\}}$$ (9.6a)

$$RMSE(Y)_c = \sqrt{\left\{\sum RMSE(Y)_{i,c}^2/N\right\}}$$ (9.6b)

### A9.6. Statistics for One Variable

❐ DISCRIMINATIVE ABILITY

Write, for simplicity, $d_{k,c,s} = \sqrt{\{\mathbf{e}_{k,c,A}'\mathbf{e}_{k,c,A}/N\}}$ as the X-residual distance of a certain set $s$ of $N$ samples to class model $c$ at a certain X-variable $k$, using the optimal, $A$, number of PCs.

For two-class SIMCA modelling, involving sample sets $s=1$ and $s=2$, and their class models $c = 1$ and $c = 2$, respectively, the *discriminative ability* for each X-variable is defined as its mean ratio of the squared between- and within-class distances:

Discriminative ability$_k = \left\{d_{k,\,c=1,\,s=2}^2 + d_{k,\,c=2,\,s=1}^2\right\} / \left\{d_{k,\,c=1,\,s=1}^2 + d_{k,\,c=2,\,s=2}^2\right\}$ (9.7)

The within-class distances in the denominator are most realistically estimated on the basis of cross-validated residuals $\mathbf{e}_{k,c,A}$; Chapter 10.

□  LEVERAGE OF X-VARIABLES

$$h_{k,c} = \sum \left( w_{k,a}^2 \right) = \mathbf{w}_a' \mathbf{w}_a \qquad\qquad (9.8)$$

where, in PCA loading weight $\mathbf{w}_a$ equals loading $\mathbf{p}_a$. Leverage $h_k$ lies between 0 (a variable not affecting the model) and 1 (a dominant X-variable).

Appendix A.10.3 gives statistical details on the MSE estimation.

# Appendix A10

# Cross-validation Details

## A10.1. Different Ways to Define Cross-validation Segments

☐ LEAVE-ONE-TREATMENT-OUT: THE DEFAULT

The preferred data analytical approach advocated here is to remove input replicates by averaging them, during the initial data pre-processing and quality control stage (Figure 3.2) and then cross-validate (Chapter 10) between *independent sets of samples* (e.g. treatments) used as cross-validation segments $m = 1, 2, \ldots, M$.

In Figures 10.4–10.5 the individual treatments, generated from an experimental mixture design, were used as *independent samples*, with $M = N$ segments: *leave-one-treatment-recipe-out*. That validation was intended to show how well the obtained model seems to describe the underlying $\mathbf{X}$–$\mathbf{Y}$ relationship, assuming that the experiment was performed correctly.

The same leave-one-treatment-out segmentation is used in Chapter 16, although for other types of designs (factorial designs). Similarly, the first half of Chapter 13 uses this $M = N$ segmentation (*leave-one-bank-department-out*) and in Chapter 14 (*leave-one-molecule-out*). But there are several other ways to define the segments:

☐ RANDOM SPLIT-HALF CROSS-VALIDATION: TWO INDEPENDENT SUB-SETS

If the number of samples $N$ is high, we may choose to use $M = 2$ segments with *N/2 randomly selected samples in each*, that is called a *random split-half* cross-validation. The two local sub-models $m = 1$ and $m = 2$ are independent of each other, since their local calibration sets are non-overlapping. But this validation is usually quite conservative.

❐  LEAVE-SOME-RANDOMLY-SELECTED-SAMPLES-OUT: SAVE COMPUTER TIME

In Chapter 13.3 the number of samples is rather high ($N = 649$ people). To save computation time, the $N$ samples were allocated at random to $M = 100$ cross-validation segments.

If there is sufficient redundancy in the sample set, random segmentation with anywhere between $M = 10$ and $M = 100$ segments usually give about the same result as full leave-one-out cross-validation ($M = N$, $N \geq 100$). If there have been sampling problems when the samples were obtained, the random segmentation is usually slightly more conservative, the fewer segments $M$ we use.

With the advent of faster computers, the computation time will no longer be an issue. Then it will be advisable to run, repeatedly, a number of *different* randomly segmented cross-validations, with $M = 2,3,5,10,25,100,...,N$. By plotting how RMSEP(Y)$^2$, $\hat{s}(\hat{b})$ and their uncertainties increase with decreasing $M$, an analogue to what the geologists call the *variogramme* is obtained, which shows more about nature of the sampling problems at hand.


❐  LEAVE-SOME-OUT, AFTER SORTING: SAVE COMPUTATION TIME WITH LESS RISK

In Chapter 12, a different approach is used, with $M < N$ in order to save computation time, but with the segments chosen in order to minimise risk of accidentally losing too much information in any segment.

A set of $N = 151$ samples are sorted for increasing value of a variable **y**, and then allocated into $M = 16$ cross-validation segments with nine samples in each. Since 151 is not a multiple of 16, the last segment has only seven samples. These segments are organised so that samples $i = 1,17,33...$ are collected in segment $m = 1$, samples $i = 2,18,34,...$ in segment $m = 2$, and so on until samples $i = 16,32,48,...$ are collected in segment $m = 16$. This segmentation scheme may also be denoted as $m = 1,2,3,...,15,16, 1,2,3,...,15,16, 1,2,3,...$ for the $N = 151$ samples.

This *leave-some-sorted-samples-out* cross-validation is more slightly conservative than the full leave-one-out cross-validation, but is a lot quicker to compute.


❐  REPRESENTATIVE SPLIT-HALF CROSS-VALIDATION: INDEPENDENT SUBSETS WITH LESS RISK

If we had split the y-sorted sample set into $M = 2$ segments according to the scheme $m = 1,2,1,2,1,2,...$ in the $N$ samples, this *representative split-half* segmentation may do almost as well as using more than two segments, with even less computation. It may also be easier to present to others who are not familiar with cross-validation. But since each sub-model only has $N/2$ samples, this require many samples, $N$.

In Chapter 14.2 this approach is compared to leave-one-out cross-validation, and also to a split-half cross-validation between two non-representative sample sets, somewhat oddly chosen as *calibration* and *test* sets in a previous publication.

❐ NON-REPRESENTATIVE SPLIT-HALF CROSS-VALIDATION: CHECKING NON-LINEARITIES

The opposite split-half approach may be used for checking the **X–Y** linearity by cross-validation: after sorting with respect to **y**, the samples are allocated into e.g. $M = 2$ cross-validation segments, this time according to the scheme $m = 1,1,1,...,1; 2,2,2,...,2$ in the $N$ samples.

This *non-representative split-half* cross-validation between the high-**y** and low-**y** samples is even more conservative than the previous one, because it checks the *extrapolation* ability between the two sub-models $m = 1$ and $m = 2$ (high and low **y**). In case of a curved **X–Y** relationship in the input data, it will be revealed by a much higher RMSEP(Y) from the *non-representative split-half* than from the *representative split-half* cross-validation (cf. also Figure 14.7a).

❐ LEAVE-ONE-PERIOD-OUT: DECOUPLE THE SEGMENTS IN TIME

In Chapter 15 the cross-validation is made conservative in another way: the first $N = 60$ sequential sampling times in an industrial process are split into $M = 20$ consecutive segments, with three samples in each: $m = 1,1,1,2,2,2,3,3,3,...,$ 20,20,20. The reason for this is to have cross-validation segments that are more independent of each other, given the presumed time constant of the auto-correlated process noise.

❐ LEAVE-ONE-REPLICATE-OUT AND LEAVE-ONE-REPLICATE-SAMPLE-OUT: CHECKING THE REPEATABILITY

If replicates are included in the final data analysis (instead of being averaged out in the pre-processing stage, (Figure 3.2), then they may be used in the cross validation as was illustrated in Figures 10.6 and 10.7.

## A10.2. Passive Variables

There are two ways to treat passive variables (cf. Figures 10.5 and 10.7). One way is to *ignore* these variables during the bi-linear model estimation, and *afterwards* estimate their loadings by passively regressing them on the scores **T**.

The other way is to *scale* these variables by a small factor, e.g. 1/1000 during the *a priori* weighting of the input variables. Thereby, their total initial standard deviation becomes very small. They have little or no impact on the modelling of the first couple of PCs, because their contributions in the least-squares estimation of scores $\mathbf{T}$ is virtually nil. Their loadings in $\mathbf{P}$ or $\mathbf{Q}$ will then be very small as well, so they will be plotted near the origin in conventional loading plots. But e.g. in the correlation loading plots they will be amplified back to the scale-free unit of correlation coefficients, and can be compared to the other variables.

## A10.3. Cross-Validation with Jack-knifing in BLM

❐ Full Model

Using the input data $\mathbf{X}$ and $\mathbf{Y}$ in all $N$ available calibration samples, the full bilinear model is developed (Chapter 6) according to eq. (6.2). For $A = 0,1,2,...,$ $A_{\mathrm{Max}}$ PCs, this model has the parameters $\bar{\mathbf{x}}, \bar{\mathbf{y}}, \mathbf{T}_A, \mathbf{W}_A, \mathbf{P}_A$ and $\mathbf{Q}_A$, where e.g. $\mathbf{T}_A = [\mathbf{t}_1,\mathbf{t}_2,...\mathbf{t}_A]$. It may be written

$$\mathbf{T}_A = W_A(\mathbf{X}) \tag{10.2a}$$

where, in PCA, $\mathbf{T}_A = (\mathbf{X} - \bar{\mathbf{x}})\mathbf{P}_A$; in PLSR, $\mathbf{T}_A = (\mathbf{X} - \bar{\mathbf{x}})\ \mathbf{W}_A(\mathbf{P}_A{}'\mathbf{W}_A)^{-1}$.

$$\hat{\mathbf{X}}_A = \bar{\mathbf{x}} + \mathbf{T}_A\mathbf{P}_A', \qquad \hat{\mathbf{Y}} = \bar{\mathbf{y}} + \mathbf{T}_A\mathbf{Q}_A' \tag{10.2b}$$

$$\mathbf{E}_A = \mathbf{X} - \hat{\mathbf{X}}_A, \qquad \mathbf{F}_A = \mathbf{Y} - \hat{\mathbf{Y}}_A \tag{10.2c}$$

The linear model summary is

$$\hat{\mathbf{Y}}_A = \hat{\mathbf{b}}_0 + \mathbf{X}\hat{\mathbf{B}}_A$$

where (eq. (6.6a)) $\hat{\mathbf{B}}_A = \mathbf{W}_A\left(\mathbf{P}_A'\mathbf{W}_A\right)^{-1}\mathbf{Q}_A', \qquad \hat{\mathbf{b}}_0 = \bar{\mathbf{y}} - \bar{\mathbf{x}}\hat{\mathbf{B}} \tag{10.2d}$

❐ Calibration Fit RMSEC for Y and X, Based on the Full Model

In accordance with Figure 10.1, the residual matrices $\mathbf{E}_A$ and $\mathbf{F}_A$ are summarised for each sample (row) $i = 1,2,...,N$, in analogy to eqs. (9.4) and (9.5), based on its X-residual row vector $\mathbf{e}_{i,A}$ and Y-residual row vector $\mathbf{f}_{i,A}$ (eqs. (9.4) and (9.5a)

$$\mathrm{RMSEC(X)}_{i,A} = \sqrt{\left\{\mathbf{e}_{i,A}\mathbf{e}_{i,A}'/(K - A)\right\}}, \qquad \mathrm{RMSEC(Y)}_{i,A} = \sqrt{\left\{\mathbf{f}_{i,A}\mathbf{f}_{i,A}'/J\right\}} \tag{10.3a}$$

The calibration fit for each X-variable $k = 1,2,....,K$ is estimated from its column vector $\mathbf{e}_k$

$$\mathrm{RMSEC(X)}_{k,A} = \sqrt{\left\{\left[\mathbf{e}_{k,A}'\mathbf{e}_{k,A}/N\right] \times [K/(K - A)]\right\}} \tag{10.3b}$$

and for each Y-variable $j = 1,2,\ldots J$ by from its column vector $\mathbf{f}_j$

$$\text{RMSEC(Y)}_{j,A} = \sqrt{\left\{\mathbf{f}'_{j,A}\mathbf{f}_{j,A}/N\right\}} \qquad (10.3c)$$

The calibration fit for the whole X-matrix is summarised by

$$\text{RMSEC(X)}_A = \sqrt{\left\{\sum \text{RMSEC(X)}^2_{i,A}/N\right\}} \equiv \sqrt{\left\{\sum \text{RMSEC(X)}^2_{k,A}/K\right\}} \qquad (10.3d)$$

and for the whole Y-matrix by

$$\text{RMSEC(Y)}_A = \sqrt{\left\{\sum \text{RMSEC(Y)}^2_{i,A}/N\right\}} \equiv \sqrt{\left\{\sum \text{RMSEC(Y)}^2_{j,A}/J\right\}} \qquad (10.3e)$$

For the relationship between RMSEC(Y) and the error $s_{err}$, in full-rank models, see eq. (16.1), p. 334.

*Correcting the X-residuals:* The individual X-residuals after the subtraction of $A$ PCs, may be corrected for the loss of $A$ degrees of freedom in each sample, in analogy to eqs. (9.4), (10.3b) and (10.5d). Otherwise these X-residuals decrease towards zero when the number of PCs increases towards its maximum possible, which normally is the number of X-variables $K$. The reason is that the scores of the $A$ PCs have to be estimated, stealing $A$ of the $K$ *degrees of freedom* in each sample's X-vector.

❑ CROSS-VALIDATION TO ESTIMATE PREDICTION ABILITY

*Local calibration:* Keeping out the samples in segment $m$, all the model parameters in eq. (10.2a–d), using $A = 0,1,2,\ldots,A_{Max}$ PCs are re-estimated from the local calibration sample subset remaining, symbolised by subscript $-m : \bar{\mathbf{x}}_{-m}, \bar{\mathbf{y}}_{-m}, \mathbf{W}_{-m,A}, \mathbf{P}_{-m,A}$ and $\mathbf{Q}_{-m,A}$. The linear model summaries are $\hat{\mathbf{B}}_{-m,A} = \mathbf{W}_{-m,A}(\mathbf{P}_{-m,A}'\mathbf{W}_{-m,A})^{-1}\mathbf{Q}_{-m,A}'$, $\mathbf{b}_{0,-m,A} = \bar{\mathbf{y}}_{-m} - \bar{\mathbf{x}}_{-m}\hat{\mathbf{B}}_{-m,A}$.
The scores for this local calibration subset remaining are

$$\mathbf{T}_{-m,A} = (\mathbf{X}_{-m} - \bar{\mathbf{x}}_{-m})\mathbf{W}_{-m,A}\left(\mathbf{P}'_{-m,A}\mathbf{W}_{-m,A}\right)^{-1} \qquad (10.4a)$$

*Local test set:* The scores, X- and Y-reconstructions and residuals for the segment of samples left out, symbolised by subscript $m$, are estimated from the model where they had been left out, $-m$ (cf. eq. (6.3)).

$$\mathbf{T}_{m,-m,A} = (\mathbf{X}_m - \bar{\mathbf{x}}_{-m})\mathbf{W}_{-m,A}\left(\mathbf{P}'_{-m,A}\mathbf{W}_{-m,A}\right)^{-1} \qquad (10.4b)$$

$$\hat{\mathbf{X}}_{m,-m,A} = \bar{\mathbf{x}}_{-m} + \mathbf{T}_{m,-m,A}\mathbf{P}'_{-m,A}, \ \hat{\mathbf{Y}}_{m,-m,A} = \bar{\mathbf{y}}_{-m} + \mathbf{T}_{m,-m,A}\mathbf{Q}'_{-m,A} \qquad (10.4c)$$

$$\mathbf{E}_{m,-m,A} = \mathbf{X}_m - \hat{\mathbf{X}}_{m,-m,A}, \qquad \mathbf{F}_{m,-m,A} = \mathbf{Y}_m - \hat{\mathbf{Y}}_{m,-m,A} \qquad (10.4d)$$

Sometimes, $\mathbf{E}_{m,-m,A}$ and $\mathbf{F}_{m,-m,A}$ are simply called the cross-validated residuals $\mathbf{E}_A$ and $\mathbf{F}_A$, for simplicity.

## ☐ RMSEP FOR **Y** AND **X**, BASED ON THE *M* SUB-MODELS

When all *M* sub-models have been estimated, that means that each of the *N* samples in the full calibration set have been predicted in some test set *m* from a local model estimated with it left out, $-m$.

In accordance with Figure 10.1, and in analogy to eq. (10.3a), the cross-validated residual matrices $\mathbf{E}_{-m,A}$ and $\mathbf{F}_{-m,A}$ from eq. (10.4d) are then summarised for each sample (row) $i = 1,2,\ldots,N$, based on its X-residual, row vector $\mathbf{e}_{i,-m,A}$ and Y-residual, row vector $\mathbf{f}_{i,-m,A}$

$$\mathrm{RMSEP(X)}_{i,A} = \sqrt{\left\{ \mathbf{e}_{i,-m,A}\mathbf{e}'_{i,-m,A}/(K-A) \right\}},$$

$$\mathrm{RMSEP(Y)}_{i,A} = \sqrt{\mathbf{f}_{i,-m,A}\mathbf{f}'_{i,-m,A}/J} \tag{10.5a}$$

The cross-validated prediction error for each cross-validation segment *m*, consisting of $N_m$ samples, is summarised in **X** by

$$\mathrm{RMSEP(X)}_{m,A} = \sqrt{\left\{ \sum \mathrm{RMSEP(X)}^2_{i,A}/N_m \right\}} \tag{10.5b}$$

and in **Y** by

$$\mathrm{RMSEP(Y)}_{m,A} = \sqrt{\left\{ \sum \mathrm{RMSEP(Y)}^2_{i,A}/N_m \right\}} \tag{10.5c}$$

The cross-validated prediction error for each X-variable $k = 1,2,\ldots,K$ is estimated from its column vector $\mathbf{e}_{k,-m,A}$:

$$\mathrm{RMSEP(X)}_{k,A} = \sqrt{\left\{ \left[ \mathbf{e}'_{k,-m,A}\mathbf{e}_{k,-m,A}/N \right] \times [K/(K-A)] \right\}} \tag{10.5d}$$

and for each Y-variable $j = 1,2,\ldots,J$ by from its column vector $\mathbf{f}_{j,-m,A}$:

$$\mathrm{RMSEP(Y)}_{j,A} = \sqrt{\left\{ \mathbf{f}'_{j,-m,A}\mathbf{f}_{j,-m,A}/N \right\}} \tag{10.5e}$$

The cross-validated prediction error for the whole X-matrix is equivalently obtained by averaging over the samples $i = 1,2,\ldots, N$, the cross-validation segments $m = 1,2,\ldots, m$ or the variables $K = 1,2,\ldots, K$:

$$\text{RMSEP(X)}_A = \sqrt{\left\{\sum \text{RMSEP(X)}^2_{i,A}/N\right\}} \equiv \sqrt{\left\{\sum \text{RMSEP(X)}^2_{k,A}/K\right\}}$$

$$\equiv \sqrt{\left\{\sum \text{RMSEP(X)}^2_{m,A}/M\right\}} \tag{10.5f}$$

and for the whole Y-matrix by

$$\text{RMSEP(Y)}_A = \sqrt{\left\{\sum \text{RMSEP(Y)}^2_{i,A}/N\right\}} \equiv \sqrt{\left\{\sum \text{RMSEP(Y)}^2_{j,A}/J\right\}}$$

$$\equiv \sqrt{\left\{\sum \text{RMSEP(Y)}^2_{m,A}/M\right\}} \tag{10.5g}$$

❒ $A_{\text{Max}}$: THE MAXIMUM SENSIBLE NUMBER OF PCs TO COMPUTE

The number of PCs computed should be more than necessary, in order to make sure that the optimal number of PCs, $A_{\text{Opt}}$, later can be found. But when RMSEC(X) has reached 0, no more PCs can be meaningfully extracted. Usually, $A_{\text{RMSEC(X)}=0} = K$ (minus dimensions lost by closure between X-variables). However, in a *short, fat* calibration set with fewer samples than X-variables ($N < K$), the minimum number of non-zero PCs, $A_{\text{RMSEC(X)}=0}$ is not $K$, but $N - 1$.

❒ $A_{\text{Opt}}$: THE OPTIMAL NUMBER OF PCs TO USE

The optimal number of PCs is decided after an inspection of the cross-validated results. The prediction error of **Y** is usually of primary importance. RMSEP(Y)$_A$ is plotted against # of PCs, $A = 0,1,2,...,A_{\text{Max}}$, as in Figure 10.4b (solid curve).

The optimal number of PCs, $A_{\text{Opt}}$, is usually taken at minimum RMSEP(Y)$_A$, or as the lowest number of PCs that give adequately low RMSEP(Y)$_A$. But other criteria may also be used: interpretability, smoothness of loading and scores, etc. Two situations require special consideration.

*Choosing a robust model when the X-measurements are very precise:*  In some types of empirical high-precision data, RMSEP(Y)$_A$ falls abruptly and then only slowly decreases with $A$. In such cases one should stop before the error has attained its very minimum, in order to get a model that is robust in practical applications; this is demonstrated in Figure 12.2b.

*Choosing an exhaustive model for designed experiments:*  When the X-matrix consists of design variables from a factorial experiment, a PCA of **X** would show that all, or at least many, directions in X-space are equally large. In that case, **X** provides no covariance for stabilising the modelling of **Y**, and APLSR has modelled **Y** already with a few PCs, $A_{\text{Opt}}$. After that, the model usually remains constant, and it does not matter if more than $A_{Opt}$ PCs are used in the final model, cf. Figure 16.3.

Only the $A_{Opt}$ PCs need to be inspected graphically. However, for APLSR of designed experiments, the number of PCs used for computation of the linear regression coefficients **B** and for the jack-knifing may be taken as the minimum number of PCs that makes $RMSEC(X) = 0$ (Chapter 16). This ensures compatibility with traditional ANOVA. On the other hand, if **Y** contains many intercorrelated, but noisy response variables, the stability and the *significance* of the APLSR model may actually deteriorate again if that many PCs are used. Then the $A_{Opt}$ solution should be used instead.

## ❐ ESTIMATED STANDARD Y-UNCERTAINTY FOR PREDICTION, RMSEP(Y)

The estimate of the apparent standard uncertainty in Y-predictions, RMSEP(Y) is the *average* level of apparent prediction error, $|y_i - \hat{y}_i|$ to be expected in new samples of the same general kind. This is simply defined as the value of the curve $RMSEP(Y)_A$ at $A = A_{Opt}$.

## ❐ ESTIMATED STANDARD X-UNCERTAINTY FOR PREDICTION, RMSEP(X)

The estimate of the apparent error in the fit of the X-data to the model, RMSEP(X) is the expected distance of the X-vectors of new samples, $x_i$, to the model. This is similarly defined as the value of the curve $RMSEP(X)_A$ at $A = A_{Opt}$, and is used in SIMCA classification and detection of outliers (Appendix A9, eq. (9.6)).

## ❐ UNCERTAINTY OF THE ESTIMATED STANDARD UNCERTAINTY, RMSEP(Y)

Since RMSEP(Y) is an estimate, it has its own estimation uncertainty. This may be pragmatically assessed graphically from the variability between the individual prediction errors in the $M$ segments, $RMSEP(Y)_m, m = 1,2,...,M$ from eq. (10.5c) at $A = A_{Opt}$. A crude statistical summary is to estimate the standard error of the mean squared error of prediction, $MSEP(Y) = RMSEP(Y)^2$, based on the $M$ perturbations

$$s(MSEP(Y)) = \sqrt{\left\{\sum (MSEP(Y) - RMSEP(Y)^2{}_m)^2/[(M-1)M]\right\}} \qquad (10.5h)$$

Provided that $s(MSEP(Y)) < MSEP(Y)/2$, an approximate reliability range for RMSEP(Y) may then be defined (cf. Chapter 12.2.8) by

$$\left[\sqrt{[MSEP(Y) - 2s(MSEP(Y))]}, \qquad \sqrt{[MSEP(Y) + 2s(MSEP(Y))]}\right] \qquad (10.5i)$$

## ❑ THE UNCERTAINTY OF EACH INDIVIDUAL PREDICTION $\hat{y}_i$

*The general expected error level:* For simplicity, assume that we only have one Y-variable, $\mathbf{y}$ ($J = 1$). When a new sample number $i$ has its Y-value(s) predicted from its X-data, $\hat{y}_i = \hat{b}_0 + \mathbf{x}_i \hat{\mathbf{b}}$, we have no direct way of knowing the error in this prediction, $\hat{y}_i$. Of course, if we already have obtained its input data $y_i$, like for the calibration samples, we can compare the two and check the prediction precision (apart from the error in $y_i$ itself). But that is usually not the case for future *unknown* samples.

Instead, we have to rely on our assessment of prior experience: if input $\mathbf{x}_i$ seems to be satisfactory, then we may expect prediction $\hat{y}_i$ to be satisfactory as well: if we had measured $y_i$ by the old reference method, $y_i$ would probably lie somewhere within $\hat{y}_i \pm 2\text{RMSEP(Y)}$ (eq. (10.5g)).

*The individual expected error level:* Theoretically, different individual samples have somewhat *different* uncertainty in their predictions $\hat{y}_i$, even if they are *normal*. The reason is (Martens and Næs, 1989, pp. 79–85) that for the type regression modelling used in this book, the uncertainty in $\hat{y}_i$ depends somewhat on how the X-vector for each sample $i$, $\mathbf{x}_i$, is situated relative to the *cloud* of the $N$ calibration samples in the X-space.

If we know a lot about the various noise types and levels (sampling noise, measurement noise, modelling error, etc.), this prediction uncertainty may be estimated explicitly (Faber and Kowalski, 1997). Unfortunately, in most practical cases we do not know enough about the various noise sources to be able to estimate the individual prediction errors theoretically.

Instead, we may combine the general RMSEP(Y) from the $N$ calibration samples and the individual outlier test based on $\text{RMSEP(X)}_i$ and leverage $h_i$ into a pragmatic, outlier-sensitive estimate of the standard uncertainty in $\hat{y}_i$, which in this book called $\text{YDev}_i$ (Høy and Martens, 1998). This name alludes to its similarity to an estimated uncertainty standard deviation of $\hat{y}_i$, but with some hesitation, because it is also outlier-sensitive.

$\text{YDev}_i$ gives a rough indication of what level of *uncertainty* to expect in each prediction $\hat{y}_i$, and at the same time alerts the user for possible *outliers*. Examples of the use of $\hat{y}_i \pm \text{YDev}_i$ are seen in Figures 3.6g, 7.3, 9.5c, 12.6b and 15.7. The expression for $\text{YDev}_i$ used in this book is:

$$\text{YDev}_i =$$

$$\sqrt{[\text{RMSEP(Y)}^2 \times \{(1 - (A + 1)/N)\} \times \{h_i + \text{RMSEP(X)}_i^2/\text{RMSEP(X)}^2 + 1/N\}]}$$

$$(10.6)$$

❏ Jack-Knifing to Estimate Model Parameter Stability

*Summing, not averaging the squared errors:*   The idea behind the present jack-knifing is to summarise the deviations between the $M$ sub-models and the full model. But that has to be done differently from how the X- and Y-prediction residuals were treated in order to estimate RMSEP(X) and RMSEP(Y).

The cross-validated prediction *residuals* in **X** and **Y** represent $N$ *independent* observations, because they were based on local models where the $N$ samples in turn had been left out. But because the local models themselves are not independent of the full model, their parameters' deviations from the full model cannot be *averaged*. They represent *partial* perturbations that instead have to be *summed*.

The version of jack-knifing used for the bi-linear modelling was developed specially for this book (Martens and Martens, 2000). It is similar to the jack-knifing developed for traditional full-rank regression modelling (cf. Efron, 1982; Efron and Tibshirani, 1993; Shao and Tu, 1995), but deviates at two points:

1. While traditional jack-knifing compares the individual sub-models to the *average of the sub-models*, this book compares them to the *full model*. The choice has been made in order to (a) reduce the number of concepts and parameters involved, and (b) to tie the estimated uncertainties closer to the actual parameters being assessed (the full model). The difference appears to very small, and for some parameters (means and offsets), there is no difference at all. However, research is in progress to study the theoretical consequences.

2. Bi-linear models have an ambiguity (affine transformation, cf. eq. (5.13)) that makes it useless to compare the sub-models directly. There are two sources of ambiguity: (a) The sign of a PC is arbitrary, and may change from one model to the next. (b) The relative importance of various directions in X-space may change between models, which causes rotation problems. The affine transformation problem is here solved by rotating each sub-model orthogonally towards the full model before being plotted and having their partial perturbations summarised.

*Assessment of the bi-linear model parameters:*   Let **W**, **T**, **P** and **Q** represent the bilinear parameters of the full model at its optimal number of PCs, $A_{Opt}$, and let $\mathbf{W}_{-m}$, $\mathbf{T}_{-m}$, $\mathbf{P}_{-m}$ and $\mathbf{Q}_{-m}$ represent the parameters of sub-model number $m$ at the same number of PCs, $A_{Opt}$.

The parameters of each model are rotated (eq. (5.13)) towards those of the full model, by the following: First, an orthogonal rotation matrix $\mathbf{C}_m$ is estimated. That is here done in the score space

$$\mathbf{T} = \mathbf{T}_m \times \mathbf{C}_m + \mathbf{D}_m \qquad (10.7)$$

where $\mathbf{D}_m$ represents residuals, and $\mathbf{T}_m$ represents all $N$ samples, both the local calibration samples for the sub-model, $\mathbf{T}_{-m}$ and $\mathbf{T}_{m,-m}$, the local prediction samples in segment number $m$. The orthogonal rotation matrix $\mathbf{C}_m$ is estimated from eq. (10.7), as described by Martens and Martens (2000). The estimate $\hat{\mathbf{C}}_m$ is used for rotating the scores, and its inverse $\hat{\mathbf{C}}_m^{-1}$ are used for rotating the loadings, since $\hat{\mathbf{C}}_m \times \hat{\mathbf{C}}_m^{-1} = \mathbf{I}$ (the identity matrix)

$$\mathbf{T}_{(m)} = \mathbf{T}_m \times \hat{\mathbf{C}}_m \tag{10.8a}$$

$$\mathbf{P}_{(m)} = \mathbf{P}_{-m} \times \hat{\mathbf{C}}'^{-1}_m, \qquad \mathbf{Q}_{(m)} = \mathbf{Q}_{-m} \times \hat{\mathbf{C}}'^{-1}_m \tag{10.8b}$$

The rotated partially perturbed scores $\mathbf{T}_{(m)}$ are plotted together with the full-model scores $\mathbf{T}$, and the rotated partially perturbed loadings $\mathbf{P}_{(m)}$ and $\mathbf{Q}_{(m)}$ are plotted together with the full-model loadings $\mathbf{P}$ and $\mathbf{Q}$. Alternatively, these may be expressed as correlation loadings, as shown in Figures 10.5 and 10.7. The standard uncertainties of the parameters are then estimated from the summed partial perturbations

$$\hat{s}(t_{ia}) = \sqrt{\left[\sum (t_{ia} - t_{(m),ia})^2 \times g\right]} \tag{10.9a}$$

$$\hat{s}(p_{ka}) = \sqrt{\left[\sum (p_{ka} - p_{(m),ka})^2 \times g\right]} \tag{10.9b}$$

$$\hat{s}(q_{ka}) = \sqrt{\left[\sum (q_{ja} - q_{(m),ja})^2 \times g\right]} \tag{10.9c}$$

where the correction factor $g$ is close to 1; in this book, $g$ (Efron and Tibshirani, 1993) is defined as

$$g = (N - 1)/N \tag{10.9d}$$

*Linear model:*   The stability of the elements in the linear model parameters $\hat{\mathbf{B}}$ are assessed by their partial perturbations to the sub-models $\hat{\mathbf{B}}_{-m}$

$$\hat{s}(\hat{b}_{kj}) = \sqrt{\left[\sum (\hat{b}_{kj} - \hat{b}_{-m,kj})^2 \times g\right]} \tag{10.10}$$

*Method extensions:*   Optionally, the model centres, $\bar{\mathbf{x}}, \bar{\mathbf{y}}, \bar{\mathbf{x}}_{-m}, \bar{\mathbf{y}}_{-m}$, may also be included as extra loadings in the rotation, by extending the model eq. (10.7) to

$$[1\mathbf{T}] = [1\mathbf{T}_m] \times \mathbf{C}_m + \mathbf{D}_m$$

in order to reduce the jack-knife perturbations of the high-leverage samples. Moreover, irrelevant rotations of $\hat{\mathbf{B}}_{-m}$ in the null-space of $\mathbf{X}$ may be removed. But for simplicity, these details have not been implemented here.

# Appendix A11

# Power Estimation Details

## A11.1. Monte Carlo Assessment of the Power of an Experimental Design

❑ THE POWER ESTIMATION METHOD

The main steps in the Monte Carlo-based power estimation of effects in experimental designs (Chapter 11) are outlined below. The present version of the method (Martens H. et al., 2000) was implemented by the authors in Matlab™ to run under The Unscrambler™.

Monte Carlo based power estimation of regression model parameters in a design

1. Choose the design and generate design variables in $\mathbf{X}$, based on the design factors (and, optionally, also their squares and interactions)
2. Choose the design variable $\mathbf{x}_k$ whose effect is most critical to the success/failure of the experiment
3. Choose the response variable $\mathbf{y}$ that is most critical to the success/failure of the experiment
4. Assume as an experimental hypothesis H1 a value for $b_{k,true}$, e.g. reflecting a current theory to be put to falsification test, or reflecting the smallest *true* effect that $\mathbf{k}$ must have on response $\mathbf{y}$ for the next experiment to be worth doing. The b-value for the other X-variables may be set to zero for simplicity, for the additive models used in this book. Offset terms are for simplicity ignored in the following.
5. Assume a standard uncertainty for the noise in response $\mathbf{y}$, $s_{err}$
6. Generate "*true*" y-responses according to design matrix $\mathbf{X}$

$\mathbf{y}_{true} = \mathbf{X}\mathbf{b}_{true}$

7. For each virtual experiment $m = 1,2,..., M$ (e.g. 10 000), generate artificial data $\mathbf{y}$ and regress $\mathbf{y}$ on $\mathbf{X}$:
(a) Add representative artificial measurement noise to the *true* response variable

$\mathbf{y}_m = \mathbf{y}_{true} + \mathbf{f}_{m,true}$

where residuals $f_{m,true}$ is, e.g. independent normally distributed random numbers with an expected mean of 0 and expected standard uncertainty of $s_{err}$.

(b) Submit $y_m$ and $X$ to full-rank linear or bi-linear regression modelling, just as if they were normal data, based on the regression model

$$y_m = Xb_m + f_m$$

(c) Pick out and store the $M$ elements $\hat{b}_{k,m}$ from the estimated regression coefficients $\hat{b}_m$

8. Assume as null hypothesis H0 a value for $b_{k,true} = 0$, and repeat steps 4–6

9. Study how the estimated elements $\hat{b}_{k,m}$, $m = 1,2,...,M$ are distributed around the "*true*", expected value $b_{k,true}$, both for the experimental hypothesis H1 and for the null hypothesis H0; (a) Make relative frequency distribution histograms of each; (b) Summarise these into cumulative frequency distributions $\alpha$ and $\beta$; (c) Plot risk of false positive (Type I error) $\alpha$ vs. risk of false-negatives (Type II error) $\beta$. Estimate the equal-risk probability $p_{\alpha=\beta}$. Estimate the risk of false-positives at a chosen risk of false negatives, e.g. $\beta|_{\alpha=0.05}$; (d) Estimate the approximate power of the design under these assumptions as $1 - \beta$.

## ❐ VISUAL COMPARISON OF DISTRIBUTION HISTOGRAMS

Figure 11.6 illustrates how three designs can be compared for statistical power by Monte Carlo simulation. The figure concerns the reduced mixture design (Figure 11.3) with seven treatments. The three alternative designs in (a), (b) and (c) have one, three and five replicates of this mixture design, respectively. Thus they have different cost, because they generate a need for seven, 21 and 35 samples to be produced and analysed.

For each of the three designs, two probability-distribution histograms are shown, one for the experimental hypothesis H1 and one for the corresponding null-hypothesis, H0.

The experimental hypothesis H1 (grey histogram) is that the unknown, but true Y-effect of the design variable of $x_k$ to be assessed, $b_{k,true}$, equals a certain chosen minimum value

H1 : $b_{k,true} = b_{k,Min}$

In the present example, the effect to be assessed is that of design variable $x_2$ (*%SUGAR−%MILK*), and the minimum interesting value is set to $b_{k,Min}=0.05$.

The null-hypothesis H0 (black histogram) is simply that $x_k$ has no effect

H0 : $b_{k,true} = 0$

An assumption behind the simulations is that the standard uncertainty of $y$ is from previous experience expected to be $s_{err}=1$ sensory units (s.u.). Moreover, the error in $y$ is presently assumed to be random, equally and independently distributed: $f$ is $N(0,s_{err})$ (i.e. *normal* distribution).

Figure 11.6a shows that with only one replicate, the two distributions strongly overlap: it is hardly possible to distinguish the real effect $b_{2,true}=0.05$ from effects

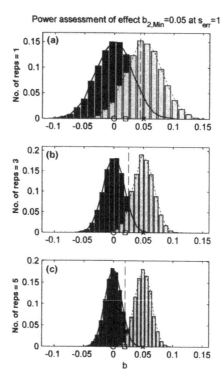

**Figure 11.6.** Effect of replicates on the probability distributions of $\hat{b}_2$, for the alternatives: grey histograms, H1, when $x_2$ changes from $-20\%$ to $+20\%$, then $y$ changes at least 2 s.u. (i.e. $b_{2,\text{true}} = b_{2,\text{Min}} = 2/40 = 0.05$); black histograms, H0, changes in $x_2$ have no effect on $y$ (i.e. $b_{2,\text{true}} = 0$). (a) one replicate; (b) three replicates; (c) five replicates. Assumptions: independent, normally distributed errors with standard uncertainty of $s_{\text{err}} = 1$ s.u.

purely caused by random noise. If we want to have only 5% risk of *false positives* ($\alpha = 0.05$ in the black histogram), any estimate values of $\hat{b}_2$ below the critical value $b_{2,\text{Lim}}=0.044$ (dashed vertical line) would have to be rejected as unreliable. This corresponds to overlooking the real effect $b_{2,\text{true}}=0.05$ in almost 50% of the cases ($\beta \approx 0.49$)! This was also seen in Figure 11.4a (p. 222).

With three replicates the two distributions are much more distinct. Here, the critical value $b_{2,\text{Lim}}=0.024$, which leads to an estimated risk of only about 12% ($\beta \approx 0.12$) at $\alpha = 0.05$. Hence the power of this design is 0.88.

With five replicates the critical value is $b_{2,\text{Lim}}=0.019$, with only about 2% risk of overlooking the effect $b_{2,\text{true}}=0.05$ ($\beta \approx 0.02$). The power of this design is as high as 0.98.

◻ QUANTITATIVE ASSESSMENT OF THE DISTRIBUTION HISTOGRAMS FOR A
DESIGN

Figure 11.7 repeats the histograms for the one-replicate design from Figure
11.6a, but for a repeated MC simulation. The histograms for H1 (grey) and
H0 (black) are shown in Figure 11.7a. Each is based on 10 000 MC simulations.
Figure 11.7b shows their cumulative sums (integrated in opposite directions).
The solid curve marked with circles represents the risk $\alpha$ of *false positives*, i.e.
accepting an estimate $\hat{b}_2$, when in fact $b_{2,\text{true}} = 0$, at various choices of the critical

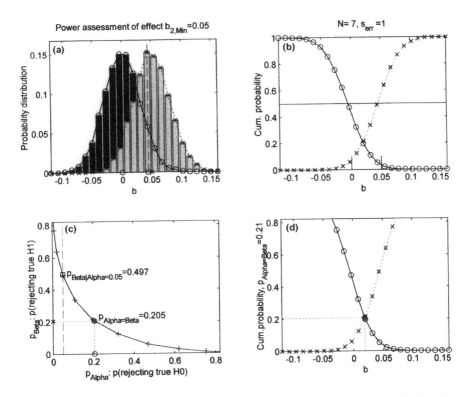

**Figure 11.7.** Power assessment details. How to estimate $\beta$ (the risk of overlooking the
minimum interesting effect of $x_2 = \%SUGAR - \%MILK$, $b_{2,\text{Min}} = 0.05$) vs. $\alpha$, under the
assumptions from Figure 11.4. (a) Probability distributions of H1: $b_{2,\text{true}} = 0.05$ (grey) and H0:
$b_{2,\text{true}} = 0$ (black). (b) Cumulative frequency distributions of H0 (solid, O) and H1 (dotted, ×).
(c) Risk of Type I error, $\alpha$ (abscissa) vs. risk of Type II error, $\beta$ (ordinate). Dotted lines,
equal-risk probability $p_{\alpha=\beta}$. Dashed line, lack of power at 95% significance level, $\beta|\alpha =$
0.05. (d) Amplification of (b) around the cross-over point and $p_{\alpha=\beta}$ (dotted lines). Assumptions: **y**=sensory *colour* darkness in the reduced mixture design (Figure 11.3) without any
replication ($N = 7$ in input data). Error in **y** assumed to be random, identically and normally
distributed with zero mean and standard uncertainty $s_{\text{err}}=1$ s.u.

value $b_{2,\text{lim}}$. The dotted curve marked with crosses represents the corresponding risk $\beta$ of *false negatives*, ignoring an estimate $\hat{b}_2$ as a possible noise effect, when in fact $b_{2,\text{true}} = 0.05$.

Figure 11.7d is just an amplification of Figure 11.7b in the region where $\alpha$ and $\beta$ intersect. The dotted line shows how the equal-risk probability $p_{\alpha = \beta}$ is estimated. This risk is illustrative if equal importance is ascribed to errors of Type I and Type II, and may indicate the risks involved in purely explorative data analysis, if the validity of the results are assessed by interpretation rather than by statistical means.

Figure 11.7c shows the trade-off between $\alpha$ and $\beta$. The equal-risk probability $p_{\alpha = \beta}$ is again shown; it is here seen to be about 21%. In addition, the *false positive* risk is shown at a chosen *false negative* risk of 5%, $\beta_{|\alpha = 0.05}$. This risk of overlooking the effect $b_{2,\text{true}} = 0.05$ is seen to be about 49% in the present case.

This illustrates how the MC method works. In the present illustration, homoscedastic, normally distributed noise was assumed. But heteroscedastic noise could equally well have been used, if so desired. This distinguishes the MC power estimation from classical power estimation. Further details on the MC power estimation method may be found in Martens H. et al., (2000).

# Appendix A12

# What Makes NIR Data So Information-rich?

Karl Norris at the US Department of Agriculture and others in the 1970s showed that diffuse multi-wavelength NIR spectroscopy (Chapter 12), properly calibrated, could replace old-fashioned reference methods such as Kjeldahl-N. Since then, the simple and rapid NIR method has found numerous applications for determination of chemical and physical qualities in a number of product types. Starting in quality assessment in food and agriculture, it spread to the pharmaceutical industry, and to quality monitoring and process control in the paper, petrochemical and polymer industry, and is now being increasingly used in medicine, e.g. for brain monitoring. NIR may even be used for high-speed prediction of human sensory perception of food quality (Martens and Martens, 1986b).

Today, a number of different commercial NIR instrument makers have delivered thousands of instrument units, operating in various wavelength-ranges between 700 and 2500 nm, to a wide variety of applications around the world. NIR analysis dominates quality control in, e.g. the grain trade, and is also extensively used in slaughterhouses and dairies. The development has been fuelled by the development of a new industry delivering NIR instruments and software.

In each of these applications, the more or less non-destructive NIR measurements ($X$) replace one or more traditional reference methods ($Y$) which are slower, more expensive, noxious or more imprecise. Moreover, multivariate outlier analysis of the NIR spectra is often used for quality assurance of raw materials, intermediates and final products, particularly in the pharmaceutical industry; this particular aspect may still have unused potentials in, e.g. medical intensive-care units.

However, being application driven rather than theory driven, the field of NIR spectroscopy has been slow to gain academic acceptance. When Norris started his pioneering work, NIR was called 'God's own garbage bin', because of the selectivity problems in NIR measurements. No single wavelength channel shows any selective signal for any single chemical constituent. Only through the empirical combination of several wavelength channels did it become possible to attain selectivity (Williams and Norris, 1987).

With his NIR background knowledge, Norris himself preferred to use a specialised, highly interactive multichannel correlation analysis as a calibration method (Norris, 1983). The present bi-linear regression approach was developed (Martens and Jensen, 1983) to make the calibration technique more generally applicable. This was facilitated by the advent of PLS regression (Wold et al., 1983a).

Fundamentally, NIR spectroscopy is based on differences in molecular vibrations of dominant constituents like water, starch, protein, etc. These vibrations give rise to strong light absorbance patterns at certain wavelengths of light in the infra-red (IR) range. The nature and intensity of these absorbance patterns reflect the chemical composition of the samples, as well as its physical state (e.g. the water temperature and the balance between free and bound water). Unfortunately, these IR light absorbances are too strong to be used for simple reflectance measurements of untreated samples: there is virtually no light reflected back from samples at these wavelengths!

However, the fundamental IR vibrations give various overtones and combination bands in the NIR range. The lower the wavelength, the broader and weaker are these secondary bands. Hence, e.g. in the 1000–2500 nm range used here, it is possible to shine light through several millimetres of material (be it grain, muscle or some industrial polymer). This makes it suitable for so-called *diffuse reflectance*, whereby light passing into a sample is scattered inside the sample and reflected back out and detected, carrying information about *where it has been and what it has seen*, in terms of different relative loss of light at different wavelengths.

The *global fingerprint* reflected from samples with different chemical compositions will therefore differ, and this is the reason why NIR reflectance spectra can be used for determining chemical quality criteria like *protein* content in wheat flour. But the *global fingerprints* are also sensitive to other major changes in chemical constituents (e.g. water content) as well as to sample temperature. Moreover, these signals are scrambled together strongly by variations in the light scattering properties of the powders (i.e. particle size), which in turn are related to other important qualities, such as milling *hardness*.

What makes the NIR spectra so information-rich is that the many *different* phenomena (e.g. in wheat: protein, starch, cellulose, fat and water content, water binding, sample temperature and particle size/hardness, etc.) have *different*, individual contributions to the *global fingerprint*. By multivariate *soft modelling* of an informative *set of samples*, it is possible to *unscramble* the contributions from each other. Even *unexpected, unidentified* contributions can thus be extracted and studied, as long as they are systematic enough to stick out from random background noise. This makes the present *soft unscrambling* applicable even for incompletely understood systems (see Appendix A7), as opposed to *hard unmixing* of mixtures of known constituent contributions (cf. Chapter 6.5).

For instance, in the present example, we are primarily interested in determining the main quality criterion, the *protein* percentage. But other, more or less unknown variation phenomena in the samples will also be picked up in the NIR instrument, and extracted from the data by the PCs. These PCs are combined automatically in the bi-linear regression, to ensure selectivity for the *protein* calibration. These PCs may also be interpreted graphically, for unexpected, but meaningful structures.

The multivariate data analysis actually changed researchers' view of NIR measurements. The many effects in the spectra were no longer seen primarily as *selectivity problems*, but as potentially useful sources of information. This approach to multichannel measurement has later been applied to numerous other types of instrument data with success: image analyzers, sound analyzers, chromatograms, fluorescence (Chapter 15), UV, IR and NMR spectra, rheological instruments, sensory analysis (Chapters 8–11), etc. (Martens and Næs, 1989; Martens and Geladi, 1999).

# Appendix A13

# Consequences of the Working Environment Survey

In practice, the analysis of the data in Chapter 13 actually had consequences. The data set came from a real survey conducted in the beginning of the 1990s. A report to the management, the unions, etc. was based on this PLSR analysis and provided clear enough results that the central management in the company got the message and took action.

- The problem areas revealed by the data analysis in the individual departments were followed up. Subsequent surveys were conducted in order to monitor points of improvement in the quality of the working environment, particularly at the middle management level.
- Focus was set on how important the quality of management is for the employees' *job satisfaction*, and thereby for job performance. A program for educating management was initiated.
- The survey improved the awareness of the quality of the working environment. Managers were stimulated to improve this, and top management got a quality criterion for measuring the *soft* values in the company.
- Yearly plans of action were made. In addition to the usual topics like cost efficiency, sick leave and productivity, clear goals were also formulated for improving the quality criterion concerning *job satisfaction*.
- The employees felt that their perception of the working environment quality was taken seriously.

By subsequently monitoring departments later on, it was found that after 5 years the frequency of sick-leave in the problematic departments had clearly reduced. It is reasonable to believe that the organisational changes taken were at least in part responsible for this.

# Appendix A14

# Details of the Molecule Class Models

In Chapter 14, it was found that the toxicity/quantum chemistry modelling improved when the three molecular classes were treated separately. In brief, here is a summary of each of the class models to describe their toxicity ($\mathbf{y}$):

❐ CLASS 1: NO2A

$N = 22$ samples. Final $\mathbf{X} = 8$ linear and four square terms. The standardised b-vector was dominated ($+$) by *HOMO–LUMO, LUMO\*\*2, Dipole\*\*2, HOMO, CoreCoreE, Pol., Ref.* and *HeatOfForm*, and ($-$) by *LUMO* and *Binding E* (cf. Figure 14.7b).

❐ CLASS 2: PAH

$N = 6$ samples. Final $\mathbf{X} = 13$ linear terms. The b-vector was here dominated ($+$) by *Dipole* and *HeatOfFormation*, and ($-$) by *LUMO* (cf. Figure 14.7c).

❐ CLASS 3: AA

$N = 19$ and four outliers. Final $\mathbf{X} = 7$ linear terms and one square term. The b-vector was even here dominated ($+$) by *Dipole, Dipole\*\*2* and *HeatOfFormation*, and ($-$) by *LUMO*, but even *Pol.* and *Mass* gave positive contributions to the prediction of $\mathbf{y} = $ *toxicity* (cf. Figure 14.7d).

# Appendix A15

# Forecasting the Future

The process modelling in Chapter 15 was used for determining $y = Ash$ from $X =$ auto-fluorescence in each sugar sample, without having to wait for the sugar sample to be brought up to the laboratory for traditional analysis. This means that the Y- and X-values are sampled at the *same time*.

In principle it is possible to be even more ambitious: calibrating for *future* Y-values from *past* and *present* X- and Y-values. Simple auto-regressive models are obtained by time-shifting the Y-values to become *future* relative to the *present* and *past* X-values (e.g. fluorescence spectra, plus *present* and *past* Y-values).

To explain this in words, let us assume that our process had only changed very slowly, so that it was sufficient to sample it *once a day*. Then, at one point in time ($i =$ today), the value of $y_i$ could be the *Ash* content of today, $Ash_i$. The vector $x_i$ could for instance consist of measurements from *yesterday* ($i - 1$) and the *day before yesterday* ($i - 2$). Let the *auto-fluorescence* spectrum here be abbreviated to $u$. We then have

$$y_i = Ash_i, \quad x_i = [u_{i-1}, u_{i-2}, Ash_{i-1}, Ash_{i-2}]$$

One such row of Y- and X-data may be defined for each point in time, $i$. The rows from a series of points in time, $i = 3, 4, \ldots, N$ then generate a vector $y$ and a matrix $X$, suitable for bi-linear regression to estimate regression coefficients $b$.

If the modelling is successfully, it can later be used to provide *forecasts* of *future* unknown y-values from *present* and *past* X-data

$$\hat{y}_{i+1} = x_{i+1}\hat{b} + \hat{b}_0$$

i.e. the forecast

$$Ash_{i+1} = [u_i, u_{i-1}, Ash_i, Ash_{i-1}]\,\hat{b} + \hat{b}_0$$

Unfortunately, in the present case the process data were too noisy to give reliable forecasting ability, so this will not be reported here. Wold et al. (1984) and MacGregor (1997) give more detail on the use of PLS regression for time series analysis.

# Appendix A16

# Significance Testing with Cross-validation vs. ANOVA

The leave-one-sample-out cross-validation/jack-knifing of PLS regression will here be compared to traditional analysis of variance (ANOVA), in this case based on the examples in Chapter 16, using full-rank OLS regression (eqs. (4.11a) and (6.6b)).

## ☐ RESPONSE SURFACE MODEL

Figure 16.9 showed the estimates of the effects $\hat{\mathbf{b}}$, (the standardised regression coefficients, i.e. for standardised X- and Y-variables, cf. eq. (8.9c), p. 392, using 6 PCs), and the rough significance test based on whether or not zero was inside the reliability range $\hat{\mathbf{b}} \pm 2\hat{s}(\hat{\mathbf{b}})$. It showed that for the mould species, *G. can*, the following effects appeared to be clearly different from zero

- Main effects: $aW$, pH
- Interaction: $O2*CO2$
- Square effects: $aW**2$, $pH**2$

Table 16.4 shows these results from a more formal hypothesis testing, with the between-samples jack-knifing ($M = N = 43$) based on PLSR, compared to classical ANOVA based on full-rank OLS regression (eq. 6.6b), here called multiple linear regression (MLR). Each column represents one of the 14 X-variables.

Rows 1–3 show the estimated effects $\hat{\mathbf{b}}$, for PLSR with 6 and 14 PCs, and for the MLR. The former two are seen to be almost identical. The latter two are seen to be completely identical, as the theory requires.

Rows 4–5 show the standard uncertainty estimate using 6 and 14 PCs in the jack-knifed PLSR model, and in row 6 using traditional MLR/ANOVA formalism

$$\hat{s}(\hat{\mathbf{b}}) = \sqrt{\text{diag } (\mathbf{E}_0'\mathbf{E}_0)^{-1} s_{\text{err}}^2} = \tag{16.2}$$

$$\sqrt{\text{diag } (\mathbf{E}_0'\mathbf{E}_0)^{-1}} \times \sqrt{(N/(N-1-K))} \times \text{RMSEC}(Y)_{A=14}$$

These results show that the uncertainty estimates are also quite similar, but the jack-knifed estimates are a little more conservative than the classical MLR estimates. Also, the 6 PC estimates in row 5 are even more conservative than the full-rank 14 PC estimates in row 6.

Rows 7–9 show formal significance testing of these results. This was done by a $t$-test of the expression $\hat{\mathbf{b}}/\hat{s}(\hat{\mathbf{b}})$, with $N = 43$ degrees of freedom for the jack-knifed PLSR results and $N - 1 - K = 28$ degrees of freedom for the RMSEC-based MLR/ANOVA results. It shows the probability $p$ that the estimated effects in $\hat{\mathbf{b}}$ could have been caused by random errors only. The effects of design variables $aW$, $pH$, $aW**2$ and $pH**2$ are clearly significant in all methods. The only effects that show some method differences are two interactions, $pH*CO2$ ($p = 0.26$ in PLSR, $p = 0.064$ in the MLR) and $O2*CO2$ ($p = 0.064$ in PLSR and $p = 0.007$ in MLR). In particular the $pH*CO2$ effect is very small, and its status as almost significant in the MLR could be incidental.

Under idealised assumptions of random, independent and identically, normally distributed residuals, a more detailed, yet compact representation of the cross-validation results for the 14-PC PLSR solution is as follows: "the following effects were found to be significant: $aW$ $(+++)$, $pH$ $(---)$, $O_2*CO_2$ $(-?)$, $aW*2$ $(--)$, $pH**2$ $(--)$; the rest of the effects were not significantly different from zero".

❐ Screening Experiment for $\mathbf{y} = A.$ *fla* Diameter

Figure 16.4 showed the reliability ranges of the effects in the initial, smaller screening design ($N = 20$), for the initial mould species, *A. fla*. These results, based on PLSR/CV with 11 PCs, indicated that the following effects appeared to be clearly different from zero

* Main effects: $aW$, pH and $CO2$
* Square effect: *Centre*
* Interactions: aW*$O2$(?), $O2*CO2$

A more formal hypothesis testing of these PLSR/CV results was performed. The significance probabilities in Table 16.5 were obtained. For comparison, the same input data were also submitted to traditional MLR/ANOVA as described above.

The table shows again that the PLSR/cross-validation is a little more conservative than the MLR/ANOVA. But both approaches give the same conclusions, and these conclusions had already been drawn from the visual inspection of the reliability ranges in Figure 16.4.

**Table 16.4.** Significance testing: final response surface design for $\mathbf{y}_1$ = colony diameter of the mould species *G. can* for X-variables $k = 1,2,...,14$. [a]

| | | aW 1' | pH 2' | O2 3' | CO2 4' | aW*pH 5' | aW*O2 6' | aW*CO2 7' | pH*O2 8' | pH*CO2 9' | O2*CO2 10' | aW**2 11' | pH**2 12' | O2**2 13' | CO2**2 14' |
|---|---|---|---|---|---|---|---|---|---|---|---|---|---|---|---|
| $\hat{b}_{k,1}$ 6 PCs | 1 | 0.825 | −0.264 | −0.064 | 0.017 | −0.012 | 0.068 | 0.004 | 0.005 | 0.095 | −0.143 | −0.298 | −0.251 | −0.082 | −0.079 |
| $b$ 14 PCs | 2 | 0.823 | −0.262 | −0.074 | 0.021 | −0.014 | 0.062 | −0.001 | 0.009 | 0.094 | −0.143 | −0.319 | −0.255 | −0.064 | −0.062 |
| $b$ MLR | 3 | 0.823 | −0.262 | −0.074 | 0.021 | −0.014 | 0.062 | −0.001 | 0.009 | 0.094 | −0.143 | −0.319 | −0.255 | −0.064 | −0.062 |
| $\hat{s}(\hat{b}_{k,1})$: | | | | | | | | | | | | | | | |
| s 6 PCs | 4 | 0.103 | 0.095 | 0.097 | 0.089 | 0.093 | 0.106 | 0.103 | 0.114 | 0.076 | 0.086 | 0.116 | 0.105 | 0.107 | 0.091 |
| s14 PCs | 5 | 0.113 | 0.096 | 0.070 | 0.084 | 0.067 | 0.075 | 0.086 | 0.075 | 0.082 | 0.075 | 0.100 | 0.082 | 0.079 | 0.060 |
| s MLR | 6 | 0.049 | 0.049 | 0.049 | 0.050 | 0.049 | 0.049 | 0.049 | 0.049 | 0.049 | 0.049 | 0.052 | 0.052 | 0.052 | 0.055 |
| $\alpha$: | | | | | | | | | | | | | | | |
| p 6 PCs | 7 | 0.000 | 0.008 | 0.511 | 0.849 | 0.900 | 0.525 | 0.966 | 0.965 | 0.219 | 0.104 | 0.014 | 0.021 | 0.447 | 0.389 |
| p 14 PCs | 8 | 0.000 | 0.009 | 0.299 | 0.809 | 0.840 | 0.414 | 0.993 | 0.900 | 0.259 | 0.064 | 0.003 | 0.003 | 0.425 | 0.308 |
| p MLR | 9 | 0.000 | 0.000 | 0.144 | 0.682 | 0.783 | 0.215 | 0.988 | 0.847 | 0.064 | 0.007 | 0.000 | 0.000 | 0.232 | 0.268 |

[a] Rows 1–3, effect estimates $\hat{\mathbf{b}}_1$ from APLSR with 6 and 14 PCs, and from MLR. Rows 4–6, standard uncertainty $\hat{s}(\hat{\mathbf{b}}_1)$ estimated by jack-knifing (4,5) and ANOVA (6). Rows 7–9, probability $\alpha$ of being fooled by noise: $p(\hat{b}_{k,1}|b_{k,1,\text{true}} = 0)$.

**Table 16.5.** Significance testing: initial screening design for $y$ = colony diameter of the mould species *A. fla* for X-variables $k = 1, 2, \ldots, 11$.[a]

| | | aW 1' | pH 2' | O2 3' | CO2 4' | Centre 5' | aW*pH 6' | aW*O2 7' | aW*CO2 8' | pH*O2 9' | pH*CO2 10' | O2*CO2 11' |
|---|---|---|---|---|---|---|---|---|---|---|---|---|
| $p$ 11 PCs | 1 | 0.001 | 0.012 | 0.290 | 0.000 | 0.057 | 0.694 | 0.052 | 0.513 | 0.435 | 0.8941 | 0.032 |
| $p$ MLR | 2 | 0.000 | 0.004 | 0.084 | 0.000 | 0.030 | 0.530 | 0.008 | 0.306 | 0.226 | 0.832 | 0.004 |

[a] Probability of being fooled by noise: $p(\hat{b}_k | b_{k,\text{true}} = 0)$. Row 1, jack-knifed APLSR ($A = 11$ PCs). Row 2, traditional MLR/ANOVA.

Under the idealised assumptions concerning the residuals being random etc., the PLSR model (Figure 16.4b) may be summarised as: "the following effects were found to be significant: $aW(++)$, pH $(-)$, $CO_2(---)$, *Centre* $(+?)$, $aW*O_2$ $(+?)$, $O_2*CO_2$ $(-)$. The rest of the terms were not significantly different from zero".

In summary, the cross-validated/jack-knifed PLSR, with rough graphical assessment of reliability ranges, may be used for many different purposes (cf. Table 3.3, p. 364), even for assessing controlled experiments according to factorial designs. This eliminates the need for the users to maintain *two very different thought-models* in their data analysis: a soft modelling approach for exploratory analysis, with visually accessible cross-validation, and a classical ANOVA approach for confirmatory analysis, with its abstract tabulations of sums-of-squares, mean-squares, degrees-of-freedom and $F$-tests.

For researchers who already know how to use ANOVA correctly, it does not hurt to run both methods – they should give the same conclusions with respect to reliability.

However, with fewer thought models and more graphically oriented methodology, the researcher should be able to do better data analysis: maintaining overview, discovering unexpected patterns and ensuring statistical reliability. At least, that is the authors' intention with this book.

# References

Aastveit A.H. and Martens H. ANOVA interactions interpreted by partial least squares regression, *Biometrics*, **42**, 829–844 (1986).

Andersen H.C. *The Emperor's New Clothes*. Skandinavisk Bogforlag, Odense (1977).

Benigni R. and Richard A.M. QSARS of mutagens and carcinogens: two case studies illustrating problems in the construction of models for non-congeneric chemicals, *Mutation Research*, **37**, 29–46 (1996).

Berglund A. and Wold S. INLR; implicit non-linear latent variable regression, *Journal of Chemometrics*, **11**, 141–156 (1996).

Box G.E.P. Scientific statistics, *RSS News*, **21**(4), (1993).

Box G.E.P., Hunter W.G. and Hunter J.S. Statistics for experimenters. *An introduction to Design, Data Analysis and Model Building*. John Wiley & Sons, New York (1978).

Bro R. Multiway calibration, multi-linear PLS, *Journal of Chemometrics*, **10**, 47–61 (1996).

Bro R. *Multi-way Analysis in the Food Industry: Models, Algorithms and Applications*. Doctoral thesis. University of Amsterdam (1998).

Bro R. Exploratory study of the sugar production using fluorescence spectroscopy and multi-way analysis, *Chemometrics and Intelligent Laboratory Systems*, **46**, 133–147 (1999).

Bugelski B.R. and Alampay D.A. The role of frequency in developing perceptual sets, *Canadian Journal of Psychology*, **15**, 205–211 (1961).

Burnham A.J., Viveros R. and MacGregor J.F. Frameworks for latent variable multivariate regression, *Journal of Chemometrics*, **10**, 31–45 (1996).

Caplan P. Engineering knowledge: the politics of etnography, *Antropology Today*, **4**(5), 8–12 (1988).

Chalmers D.J. *The Conscious Mind: in Search of a Fundamental Theory*. Oxford University Press, New York (1996).

Cleveland W.S. *The Elements of Graphing Data*. Wadsworth Advanced Books and Software, Monterey, CA (1985).

Cohen J. The Earth is round (p $<$ .05), *American Psychologists*, **49**(12), 997–1003 (1994).

Cornell J.A. *Experiments with Mixtures: Designs, Models and the Analysis of Mixtures.* John Wiley & Sons, New York (1981).

Coomb C.H. *A Theory of Data.* John Wiley & Sons, Chichester (1964).

Dayal B.S. and MacGregor J.F. Recursive exponentially weighted PLS and its applications to adaptive control and prediction, *Journal of Process Control,* **7**(3), 169–179 (1997).

Deming W.C. *On the Statisticians Contribution to Quality.* Proceeding 46th Session of the International Statistical Institute, Tokyo (1987).

Dennett D.C. *Consciousness Explained.* Little, Brown & Co., Boston, MA (1991).

De Vries S. and Ter Braak C.J. Prediction error in partial least squares regression: a critique on the deviation used in the Unscrambler, *Chemometrics and Intelligent Laboratory Systems,* **30**(2), 239–245 (1995).

Dewar M.J.S., Zoebisch E.G., Healy E.F. and Stewart J.J.P. AM1: A new general purpose quantum mechanical molecular model, *JACS* **107**, 3902–3909 (1985).

Di Ruscio D. The partial least squares algorithm: a truncated Cayley–Hamilton series approximation used to solve the regression problem, *Modeling, Identification and Control,* **19**(3), 117–140 (1998).

Dijksterhuis G.B. *Multivariate Data Analysis in Sensory and Consumer Science.* Food Nutrition Press Inc., Trumbull, USA (1997).

Early R. *Guide to Quality Management Systems for the Food Industry.* Blackie, Glasgow (1995).

Efron B. *The Jack-knife, the Bootstrap and Other Resampling Plans.* Society for Industrial and Applied Mathematics, USA. ISBN 0-89871-179-7. (1982).

Efron B. and Tibshirani R.J. *An Introduction to the Bootstrap.* Chapman and Hall, London (1993).

Ellis H.C. and Hunt R.R. *Fundamentals of Cognitive Psychology.* W.M.C. Brown Publisher, Dubuque (1993).

EOQC, European Organization for Quality. *Glossary of the Terms Used in Management of Quality.* 6th ed. ZH ISBN 3-85669-006-9 (1989).

Ergon R. Dynamic system multivariate calibration, *Chemometrics and Intelligent Laboratory System.* **44**, 135–146 (1998).

Esbensen K.H. and Martens H. Predicting oil-well permeability and porosity from wire-line petrophysical logs – a feasibility study using partial least squares regression, *Chemometrics and Intelligent Laboratory Systems,* **2**, 221–232 (1987).

Esbensen K.H. *Multivariate Data Analysis – In Practice,* 4th ed. CAMO ASA, Oslo (2000).

Faber K. and Kowalski B.R. Propagation of measurement errors for the validation of predictions obtained by principal component regression and partial least squares, *Journal of Chemometrics,* **11**(3), 181–238 (1997).

Folkenberg D., Bredie W.L.P. and Martens M. What is mouthfeel? Sensory–rheological relationships in instant hot cocoa drinks, *Journal of Sensory Studies,* **14**, 181–195 (1999).

Geladi P. and Martens H. A calibration tutorial for spectral data. Part 1: data pretreatment and principal component regression using Matlab, *Journal of Near Infrared Spectroscopy*, **4**, 225–242 (1996a).

Geladi P., Martens H., Hadjiiski L. and Hopke P. A calibration tutorial for spectral data. Part 2: partial least squares regression using Matlab and some neural network results, *Journal of Near Infrared Spectroscopy*, **4**, 243–255 (1996b).

Gifi A. *Non-linear Multivariate Analysis*. John Wiley & Sons, Chichester (1990).

Gigerenzer G. and Murray D.J. *Cognition as Intuitive Statistics*. Lawrence Erlbaum, Hillsdale, NJ (1987).

Glichev A.V. 15 years of qualimetry. Problems and prospects, *Proceedings of the EOQC Conference on Quality and Development*, **2**, 259–269 (1985).

Gower J.C. Generalised Procrustes Analysis, *Psychometrika*, **40**, 33–51 (1975).

Gy, P. *Sampling for Analytical Purposes*. John Wiley & Sons, Chichester (1998).

Haasum I. and Nielsen P.V. Ecophysiological characterization of some food-born fungi in relation to pH and water activity under atmospheric compositions, *J. Applied Microbiology*, **44**, 451–461 (1998).

Harman H.H. *Modern Factor Analysis*. University of Chicago Press, Chicago, IL (1967).

Helgesen H., Solheim R. and Næs T. Consumer preference mapping of dry fermented lamb sausages, *Food Quality and Preference*, **18**(2), 97–109 (1997).

Helland I.S. On the structure of partial least squares regression, *Communications in Statistics, Simulation and Computation*, **17**(2), 581–607 (1988).

Helland K., Berntsen H.E., Borgen O.S. and Martens H. Recursive algorithm for partial least squares regression, *Chemometrics and Intelligent Laboratory Systems* **14**, 129–137 (1991).

Hoffman D.L. and Young F.W. Quantitative analysis of qualitative data: applications in food research, in *Food Research and Data Analysis*, Martens H. and Russwurm Jr. H., eds. Applied Science, Barking, pp. 69–93 (1983).

Höskuldsson A. PLS regression methods, *Journal of Chemometrics*, **2**, 221–228 (1988).

Höskuldsson A. *Prediction Methods in Science and Technology*. Thor Publishing, Denmark (1996).

Höskuldsson A. Causal and path modelling, in *Les Methodes PLS, Symposium International PLS'99*, Tenenhaus M. and Morineau A., eds. CISIA–CERESTA, France, pp. 93–122 (1999).

Høy M. and Martens H. Review of partial least squares regression prediction error in Unscrambler, *Chemometrics and Intelligent Laboratory Systems*, **44**, 123–133 (1998).

ISO. *Guide to the Expression of Uncertainty in Measurement (GUM)*. International Organization for Standardization ZH (1995a).

ISO. *Quality Management and Quality Assurance – Vocabulary (8402)*, 2nd ed., International Organization for Standardization ZH (1995b).

ISO. *Quality Management Systems-Fundamentals and Vocabulary* ZH (1999).

Juran J.M. and Gryna F.M. *Juran's Quality Control Handbook*, 4th ed., McGraw-Hill, New York (1988).

Jöreskog K.G. and Wold H. *Systems under Indirect Observation – Causality\* Structure\* Prediction Parts I and II*, North-Holland, Amsterdam (1982).

Kettaneh-Wold N. Analysis of mixture data with partial least squares, *Chemometrics and Intelligent Laboratory Systems*, **14**, 57–69 (1992).

Kettaneh-Wold N., MacGregor J.F., Dayal B.S. and Wold S. Multivariate design of process experiments (M-DOPE), *Chemometrics and Intelligent Laboratory Systems*, **23**, 39–50 (1994).

Kramer R. *Chemometric Techniques for Quantitative Analysis*, Marcel Dekker, New York (1998).

Kruskal J.B. *Rank, Decomposition and Uniqueness for 3-way and N-way Arrays in Multiway Data Analysis*. Coppi R. and Bolasco S., eds. Elsevier Science, Amsterdam, p. 7 (1989).

Liang Y.Z., Kvalheim O.M. and Manne R. White, grey and black multicomponent systems a classification of mixture problems and methods for their quantitative analysis, *Chemometrics and Intelligent Laboratory Systems*, **18**, 235–250 (1993).

Lohmöller J.B. *Latent Variables Path Modelling with Partial Least Squares*, Physica-Verlag, Heidelberg (1989).

Lundahl D.S. Influence of special consumer groups as a subset of respondents on the outcomes of consumer acceptance tests, in *Product Development and Research Guidance Testing with Special Consumer Groups*, 2nd Volume, Wu L. and Gelinas A., eds, pp. 37–54 ASTM STP 1155 (1992).

MacGregor J.F. Using on-line process data to improve quality: Challenges for statisticians, *International Statistical Review*, **65**, 309–323 (1997).

Manne R. Analysis of two partial-least-squares algorithms for multivariate calibration, *Chemometrics and Intelligent Laboratory Systems*, **2**, 187–197 (1987).

Martens H. and Jensen S.Å. *Partial Least Squares Regression: A New Two-stage NIR Calibration Method, Proceedings of the 7th World Cereal and Bread Congress*. Holas and Kratochvil, eds. Elsevier, Amsterdam, 607–647 (1983).

Martens H. and Næs T. *Multivariate Calibration*, John Wiley & Sons, Chichester (1989).

Martens H. and Dardenne P. Validation and verification of regression in small data sets, *Chemometrics and Intelligent Laboratory Systems*, **44**, 99–121 (1998).

Martens H. and Geladi G. Multivariate calibration, in *Encyclopedia of Statistical Sciences, Update Vol. 3*, Kotz S., ed., pp. 483–495 (1999).

Martens H. and Martens M. Validation of PLS Regression models in sensory science by extended cross-validation, in *Les Methodes PLS*, Tenenhaus M. and Morineau A., eds. CISIA-CERESTA, Montreuil (1999).

Martens H. and Martens M. Modified jack-knife estimation of parameter uncertainty in bi-linear modelling (PLSR), *Food Quality and Preference*, **11**(1–2), 5–16 (2000).

Martens H., Wold S. and Martens M. A layman's guide to multivariate data analysis, in *Food Research and Data Analysis*, Martens, H. and Russwurm Jr. H., eds. Applied Science, Barking, pp. 473–492 (1983).

Martens H., Izquierdo L., Thomassen M. and Martens M. Partial least squares regression on design variables as an alternative to analysis of variance, *Analytica Chimica Acta*, **191**, 133–148 (1987).

Martens H., Dijksterhuis G.B. and Byrne D.V. Estimation of statistical power by Monte Carlo simulations, *Journal of Chemometrics*, **14**, 441–462 (2000).

Martens M. Chemometricians and quality (in Norwegian), in *Anvendelse av Kemometri innen Forskning og Industri*, Nortvedt R., ed. Norsk Kjemisk Selskap, Tidsskriftforlaget Kjemi, Oslo, pp. 523–533 (1996).

Martens M. A philosophy for sensory science, *Food Quality and Preferences*, **10**, 233–244 (2000).

Martens M. and Martens H. Partial least squares regression, in *Statistical Procedures in Food Research*, Piggott, J.R., ed. Elsevier, UK, pp. 293–360 (1986a).

Martens M. and Martens H. Near-infrared reflectance determination of sensory quality of peas, *Applied Spectroscopy*, **40**(3), 303–310 (1986b).

Martens M., Risvik E. and Martens H. Matching sensory and instrumental analyses, in *Understanding Natural Flavours*, Piggott J.R. and Paterson A., eds. Blackie, Glasgow, pp. 61–76 (1994).

Martens M., Bredie W.L.P. and Martens H. Sensory profiling data studied by partial least squares regression, *Food Quality and Preference*, **11**(1–2), 147–149 (2000).

Massart D.L., Vandeginste B.G.M., Buydens L.M.C., DeJong S., Lewi P.J. and Smeyers-Verbeke J. *Handbook of Chemometrics and Qualimetrics*, Parts A and B. Elsevier, Amsterdam (1997).

Molnár P. The design and practical use of an overall quality index for food products, in *Food Research and Data Analysis*, Martens H. and Russwurm Jr. H., eds. Applied Science, Barking, pp. 115–146 (1983).

Munck L. Man as selector – a Darwin boomerang striking through natural selection, in *Environmental Concerns: An Interdisciplinary Exercise*, Hansen J.A., ed. Elsevier, London, pp. 211–227 (1991).

Munck L, Nørgaard L., Engelsen S.B., Bro R. and Andersson C.A. Chemometrics in food science – a demonstration of the feasibility of a highly exploratory, inductive evaluation strategy of fundamental scientific significance, *Chemometrics and Intelligent Laboratory Systems*, **44**, 31–60 (1998).

Norris K.H. Extracting information from spectrophotometric curves. Predicting chemical composition from visible and near-infrared spectra, in *Food Research and Data Analysis*, Martens H. and Russwurm Jr. H., eds. Applied Science, Barking, pp. 95–113 (1983).

Núñez R. and Freeman W.J., eds. Reclaiming cognition, *Journal of Consciousness Studies*, **6**, 11–12 (1999).

Næs T. and Risvik E., eds. *Multivariate Analysis of Data in Sensory Science*. Elsevier Science, UK (1996).

Nørretranders T. *Mærk Verden. En beretning om bevidsthed* (in Danish), Gyldendal, Denmark (1991).

Pirsig R.M. *Zen and the Art of Motorcycle Maintenance*, Transworld, London (1974).

Polanyi M. *Tacit Knowledge*, Routledge, London (1978).

Robson C. *Real World Research: a Resource for Social Scientists and Practitioner-Reseachers*. Blackwell, Oxford (1993).

Sagan C. *The Demon-Haunted World – Science as a Candle in the Dark*. Random House, New York (1995).

Schutz H.G. Appropriateness as a measure of the cognitive–contextual aspects of food acceptance, in *Measurement of Food Preferences*, MacFie H.J.H. and Thomson D.M.H., eds. Blackie, Glasgow, pp. 25–50 (1994).

Shao J. and Tu D. *The Jackknife and Bootstrap*. Springer-Verlag, New York (1995).

Sundberg R. Multivariate calibration – direct and indirect methodology (with disussion), *Scandinavian Journal of Statistics*, **26**, 161–207 (1999).

Sundberg R. Aspects of statistical regression in sensometrics, *Food Quality and Preference*, **11**, 17–26 (2000).

Taguchi G. *Introduction to Quality Engineering: Design Quality into Products and Processes*. Asian Productivity Organisation, Tokyo (1988).

Tenenhaus M. and Morineau A. *Les Methodes PLS, Symposium International PLS'99*, CISIA-CERESTA, France (1999).

Thompson W.L. 326 Articles/Books Questioning the Indiscriminate Use of Statistical Hypothesis Tests in Observational Studies, www.cnr.colostate.edu/~anderson/thompson1.html (2000).

Tucker L.R. Some mathematical notes on three-mode factor analysis, *Psychometrika* **31**, 279 (1966).

Tukey J.W. The future of data analysis, *Annals of Mathematical Statistics*, **33**, 1–67 (1962).

Tukey J.W. *Exploratory Data Analysis*. Addison-Wesley, Reading, MA (1977).

Von Wright G.H. *The Varieties of Goodness*, Routledge and Kegan Paul, London (1964).

Vrignaud P. Using PLS to study individual differences in the recall of knowlegde from semantic memory, in *Les Methodes PLS, Symposium International PLS'99*, Tenenhaus M. and Morineau A., eds. CISIA-CERESTA, France, pp. 251–273 (1999).

Westad F. PLS regresjon anvendt på arbeidsmiljodata, Nørgaard L. and Höskuldsson A., eds. *Proceedings, Symposium i Anvendt Kemometri, Denmark Technical University January 27–29, 1997*, Thor Publishing, Denmark, pp. 127–134 (1997).

Westad F. and Martens H. Variable selection in NIR based on significance testing in partial least squares regression (PLSR), *Journal of Near Infrared Spectroscopy* **8**, 117–124 (2000).

Williams P. and Norris K., eds. *Near-Infrared Technology in the Agricultural and Food Industries*. American Association of Cereal Chemists, Inc. St. Paul, MN (1987).

Wise B. and Gallagher N.B. The process chemometrics approach to process monitoring and fault detection, *Journal of Process Control*, **6**, 329–348 (1996).

Wold H. Soft modeling by latent variables: the non-linear iterative partial least squares approach, in *Perspectives in Probability and Statistics*, Gani, ed. Academic Press, London (1975).

Wold H. Partial least squares, in *Encyclopedia of Statistical Sciences*, Kotz S. and Johnson N.L., eds. John Wiley & Sons, New York, pp. 581–591 (1985).

Wold S. Cross-validatory estimation of the number of components in factor analysis, *Technometrics,* **20**, 397–406 (1978).

Wold S., Martens H. and Wold H. The multivariate calibration problem in chemistry solved by the PLS method, in *Proceedings of the Conference on Matrix Pencils, March 1982, Lecture Notes in Mathematics*, Ruhe A. and Kågstrom B., eds. Springer Verlag, Heidelberg, pp. 286–293 (1983a).

Wold S., Albano C., Dunn III W.J., Esbensen K., Hellberg S., Johansson E. and Sjostrom M. Pattern recognition: finding and using regularities in multivariate data, in *Food Research and Data Analysis*, Martens H. and Russwurm Jr. H. eds. Applied Science, Barking, pp. 147–188 (1983b).

Wold S., Albano C., Dunn W., Edlund U., Esbensen K., Geladi P., Hellberg S., Johansson E., Lindberg W. and Sjöström, M. Multivariate data analysis in chemistry, in *Chemometrics: Mathematics and Statistics in Chemistry*, Kowalski B.R., ed. Reidel Publishing Co., Dordrecht, The Netherlands, pp. 17–95 (1984).

Wold S., Sjostrom M., Carlson R., Lundstedt T., Hellberg S., Skagerberg B., Wikstrom C. and Öhman J. Multivariate design, *Analytica Chimica Acta*, **191**, 17–32 (1986).

# Index

Printed and bound by CPI Group (UK) Ltd, Croydon, CR0 4YY

27/10/2024

14580206-0004